ACS SYMPOSIUM SERIES **488**

Supercritical Fluid Technology
Theoretical and Applied Approaches to Analytical Chemistry

Frank V. Bright, EDITOR
State University of New York at Buffalo

Mary Ellen P. McNally, EDITOR
E. I. du Pont de Nemours and Company

Developed from a symposium sponsored
by the Division of Analytical Chemistry
at the 201st National Meeting
of the American Chemical Society,
Atlanta, Georgia,
April 14–19, 1991

American Chemical Society, Washington, DC 1992

Library of Congress Cataloging-in-Publication Data

Supercritical fluid technology: theoretical and applied approaches to analytical chemistry / Frank V. Bright, editor, Mary Ellen P. McNally, editor.

p.   cm.—(ACS symposium series, 0097–6156; 488)

"Developed from a symposium sponsored by the Division of Analytical Chemistry at the 201st National Meeting of the American Chemical Society, Atlanta, Georgia, April 14–19, 1991."

Includes bibliographical references and indexes.

ISBN 0–8412–2220–7

1. Supercritical fluid chromatography—Congresses.   2. Supercritical fluid extraction—Congresses.

I. Bright, Frank V., 1960–   .   II.   McNally, Mary Ellen P.
III. American Chemical Society. Meeting (201st: 1991: Atlanta, Ga.)
IV. Series.

QD79.C45S89   1992
543'.0894—dc20   92–11434
                                                        CIP

The paper used in this publication meets the minimum requirements of American National Standard for Information Sciences—Permanence of Paper for Printed Library Materials, ANSI Z39.48–1984. ∞

Foreword

THE ACS SYMPOSIUM SERIES was founded in 1974 to provide a medium for publishing symposia quickly in book form. The format of the Series parallels that of the continuing ADVANCES IN CHEMISTRY SERIES except that, in order to save time, the papers are not typeset, but are reproduced as they are submitted by the authors in camera-ready form. Papers are reviewed under the supervision of the editors with the assistance of the Advisory Board and are selected to maintain the integrity of the symposia. Both reviews and reports of research are acceptable, because symposia may embrace both types of presentation. However, verbatim reproductions of previously published papers are not accepted.

Contents

SEPARATION SCIENCE

Preface

THE CHEMISTRY OF SUPERCRITICAL FLUIDS has been studied extensively in the past decade. Consequently, our understanding of this field has expanded significantly. Simultaneously, the number of applications in associated analytical technologies (for example, supercritical fluid chromatography and supercritical fluid extraction) has increased. Although the areas of fundamentals and applications are clearly interrelated, they are often discussed separately.

This volume spans these diverse areas and bridges the fields of modeling, spectroscopy, chromatography, and extraction. The contributors range from the academic, industrial, and governmental sectors. With this diversity, we have compiled an extensive volume that presents the current status of supercritical fluid technology.

The breadth of this volume results from a series of tandem symposia. The first, "Spectroscopic Investigations in Supercritical Fluids", was aimed at the discussion of fundamental aspects of solvation in supercritical fluid media. This symposium brought together individuals whose expertise ranged from the kinetics of solvation to chemical reaction processes to simulations of model solvation processes in supercritical fluids. The second symposium, "Supercritical Fluids in Analytical Chemistry", provided a forum on state-of-the-art technology as well as the current status and expectations of regulatory agencies toward the implementation of supercritical fluid technology. The goal of these symposia was to improve the basic understanding of the extraction and chromatographic processes. Specific topics dealt with the fundamental mechanisms of solute–fluid–matrix interactions, advances in instrumentation for reproducible and accurate chemical analyses, and novel applications of supercritical fluid technology. The broad-based versatility offered by supercritical fluids continues to be demonstrated and improved with the wide range of applications illustrated throughout this book.

This volume serves as a link between researchers studying the more fundamental aspects of supercritical fluids and researchers involved in the application of supercritical fluid technology to solve difficult chemical problems.

We thank the authors for their patience and perseverance. We also thank the reviewers, who took the time to painstakingly evaluate and critique each chapter of this book.

FRANK V. BRIGHT
Department of Chemistry, Acheson Hall
State University of New York at Buffalo
Buffalo, NY 14212

MARY ELLEN P. MCNALLY
Agricultural Products, Experimental Station, E402/3328B
E. I. du Pont de Nemours and Company
Wilmington, DE 19880–0402

November 25, 1991

Chapter 1

Fundamental Studies and Applications of Supercritical Fluids

A Review

Mary Ellen P. McNally[1] and Frank V. Bright[2]

[1]Agricultural Products, Experimental Station, E402/3328B, E. I. du Pont de Nemours and Company, Wilmington, DE 19880–0402
[2]Department of Chemistry, Acheson Hall, State University of New York at Buffalo, Buffalo, NY 14214

Over the course of the last several years, essential studies have been conducted in a wide variety of areas using supercritical fluids. The continued constructive development of the science employing these fluids is dependent upon a unique bridging of scientists from many disciplines involved in this work. This text, a compilation of several of the papers presented at the ACS meeting in Atlanta in the spring of 1991, attempts to illustrate the numerous scientific endeavors that have been and continue to be pursued. Individually, these works are able to stand alone in their viability. Collectively, they demonstrate the vast interest, the overwhelming potential and the extensive growth that is available.

This text is a snap-shot in time since the scientific developments are being driven rapidly by interest, funding and high expectations for the potentials that have been suggested. This introductory chapter presents a brief background synopsis of supercritical fluids in spectroscopic investigations, the loss of impetus in supercritical fluid chromatographic (SFC) developments but its vast potential, and the flourishing future status of supercritical fluid extraction (SFE).

Many physicochemical properties describe a chemical substance or mixture. For example, the boiling point, density, and dielectric constant can all be used to characterize a particular species or system as a solid, liquid, or gas. However, if a substance is heated and maintained above its critical temperature it becomes impossible to liquify it with pressure (1). When pressure is applied to this system a single phase forms that exhibits unique physicochemical properties (1-14). This single phase is termed a supercritical fluid and is characterized by a critical temperature and pressure (T_c and P_c).

Supercritical fluids offer a convenient means to achieve solvating properties which have gas- and liquid-like characteristics without actually changing chemical structure. By proper control of pressure and temperature one can access a significant

0097–6156/92/0488–0001$06.00/0

range of physicochemical properties (density, diffusivity, dielectric constants, etc.) without ever passing through a phase boundary, e.g., changing from gas to liquid form. That is, a supercritical fluid can be considered a continuously adjustable solvent. To illustrate how supercritical fluids compare to gases and liquids, Table 1 compiles some of the more important physicochemical properties of each.

TABLE 1. Properties of Supercritical Fluids vs. Gases and Liquids(13)

	Gas	SF	Liquid
Density(g/cm^3)	10^{-3}	0.1-1	1
Diff. Coeff.(cm^2/s)	10^{-1}	10^{-3}-10^{-4}	$< 10^{-5}$
Viscosity(g/cm·s)	10^{-4}	10^{-3}-10^{-4}	10^{-2}

Because of their unique characteristics, supercritical fluids have received a great deal of attention in a number of important scientific fields (1-14). Several reasons are given for choosing a supercritical fluid over another solvating system, but choice is governed generally by: 1) the unique solvation and favorable mass transport properties (5) and 2) the ease with which the chemical potential can be varied simply by adjustment of the system pressure and/or temperature (13).

Over the past decade, much progress in supercritical fluid technology has occurred. For example, supercritical fluids have found widespread use in extractions (2-5), chromatography (6-9), chemical reaction processes (10,11), and oil recovery (12). Most recently, they have even been used as a solvent for carrying out enzyme-based reactions (14). Unfortunately, although supercritical fluids are used effectively in a myriad of areas, there is still a lack of a detailed understanding of fundamental processes that govern these peculiar solvents.

In an effort to overcome this weakness, significant effort has been devoted to determining the fundamental aspects of solute-solute, solute-fluid, and solute-cosolvent interactions in supercritical fluids and supercritical fluid processes (15-45). Some of these efforts have used optical spectroscopy (17-34,43-45) and others have used chromatographic methods as a tool to probe the aforementioned interactions. Many of these experimental studies have been advanced by theoretical calculations and modelling (35-42). The general conclusion from these studies is that there is local density augmentation (i.e., solvent clustering) about the solute near the critical point. In addition, it has been suggested that there is enhanced solute-solute interactions near the critical point (5,31,43-45).

This introductory chapter is divided into two segments; its goal is to provide a framework for the remaining chapters of this volume. It has been structured to provide background information as a prelude to the technical chapters of this book. Given the above information, this chapter reviews:

1) spectroscopy-based investigations as a tool to study fundamental processes in supercritical fluids and

2) supercritical fluid chromatography (SFC) and supercritical fluid extraction (SFE) as methods to effect difficult separations and/or isolate specific analytes from complex milieu.

Spectroscopic Investigations of Supercritical Fluids

When electromagnetic radiation interacts with matter, changes in frequency, intensity, and/or phase provide insight into the material through which it has passed (46,47). Optical spectroscopy consists of a powerful battery of methods for probing many aspects of a chemical system. Infra-red (IR) and Raman spectroscopy, for example, yield information about the vibration modes within a molecule (46,47). Insight into chemical bonding and the effects of fluid conditions (e.g., density) on bonding is gained. Ultraviolet and visible (UV-Vis) absorption spectroscopy yield information on the electronic structure within a molecule (46,47). If the electronic structure is altered by the interaction of the solute with the fluid, UV-Vis absorbance spectroscopy can follow and quantify this. Fluorescence spectroscopy allows one to probe the ground- and excited-states of the solute molecule and learn how each is influenced by local solvent composition (48,49). Also, because of the time scale of the fluorescence process (10^{-9} - 10^{-12} s) kinetic aspects of solvation in supercritical fluids can be directly followed.

Steady-State Solvatochromism. The majority of the reports on supercritical fluid solvation have used steady-state solvatochromic absorbance measurements (21-28). The original aim of these experiments was to determine the solvating power of supercritical fluids for chromatography and extraction (SFC and SFE) (26,28). To quantify solvent strength, researchers (21-28) adopted the Kamlet-Taft π^* solvent polarity scale (50-55). This scale best correlates solvatochromic effects on $\sigma \rightarrow \pi^*$ and $\pi \rightarrow \pi^*$ electronic absorption transitions.

Initial supercritical fluid work was on CO_2 (21-23) and indicated weak interactions between the fluid and solute. Additional work has appeared on Xe, SF_6, C_2H_6, and NH_3 (24,25). For all fluids, spectral shifts were observed with fluid density. Yonker, Smith and co-workers (24-26,28) compared their results to the McRae continuum model for dipolar solvation (56,57), which is based on Onsanger reaction field theory (58). Over a limited density range, there was agreement between the experimental data and the model (24-26,28), but conditions existed where the predicted linear relationship was not followed (28). At low fluid densities, this deviation was attributed (qualitatively) to fluid clustering around the solute (28).

In more recent work, Johnston and co-workers (17,18,20,27,32) showed quantitatively that the local fluid density about the solute is greater than the bulk density. In these papers, results were presented for CO_2, C_2H_4, CF_3H, and CF_3Cl. Local densities were recovered by comparison of the observed spectral shift (or position) to that expected for a homogeneous polarizable dielectric medium. Clustering manifests itself in deviation from the expected linear McRae continuum model (17,18,20,27,32,56,57). These data were subsequently interpreted using an expression derived from Kirkwood-Buff solution theory (20). Detailed theoretical calculations and computer simulations from the Debenedetti (35-37,42) and Cochran (38) groups have been consistent with solvent-clustering and the local density augmentation concept.

Vibrational spectroscopy, too, has been used to study supercritical fluid systems. Buback reviewed (59) this area; however, much of his discussions are on fluid systems that are well removed from ambient conditions or difficult to handle easily (e.g., H_2O, HCl). In an early report, Hyatt (21) used IR absorbance spectroscopy to determine the influence of several solvent systems, including CO_2, on the vibrational frequencies (ν) of solute molecules. Specifically, he studied the $\nu_{C=O}$ of acetone and cyclohexanone and ν_{N-H} of pyrrole. The goal of this work was to determine the suitability of supercritical fluids as reaction solvent. Hyatt concluded that the ketones experienced an environment similar to nonpolar hydrocarbons in CO_2 and that there were no differences between liquid and supercritical CO_2. In contrast, the pyrrole studies indicated that the solvent strength of CO_2 was between ether and ethyl acetate. This apparent anomalous result was a manifestation of the, albeit weak, degree of pyrrole hydrogen bonding to CO_2.

Yonker and co-workers (60) used near-and mid-IR spectroscopy to study supercritical CO_2 and binary supercritical fluid systems composed of CO_2/H_2O, Kr/H_2O, and Xe/H_2O. The CO_2 results are consistent with increased intermolecular interaction between CO_2 molecules with increasing density. This parallels previous results using UV-Vis solvatochromism (21-28,32). For an ideal gas/water system an Onsanger electrostatic model (dipole-induced-dipole) sufficed to describe the spectral shifts. In contrast, the CO_2/H_2O system exhibited density-dependent changes in specific intermolecular interactions.

Smith and co-workers (61-64) have employed dynamic light scattering (DLS), small angle neutron scattering (SANS), and Fourier-Transform Infrared (FT-IR) spectroscopy to study AOT (sodium bis(2-ethylhexyl) sulfosuccinate; Aerosol-OT) reverse micelles in ethane, propane, and xenon. From their DLS experiments, they reported that the hydrodynamic volume increased with pressure, reached a maximum (depending on the water-AOT ratio and the actual fluid) at intermediate continuous phase pressures and asymptotically approached a limiting value at still higher pressures. It was proposed that the average micelle size varied with fluid density. However, detailed FT-IR experiments showed (61-64) the water core region remaining insensitive to fluid density. This was confirmed by Johnston and co-workers using other spectroscopic methods (65-67). By combining the DLS, SANS, and FT-IR information, Smith and co-workers came to two conclusions. First, a significant variation in the micelle-micelle interaction potential with fluid density exists that manifests in individual micelles forming aggregates. Second, the interior of the micelle, which makes up an aggregate, is essentially identical to individual micelles. The micelle interior is apparently unaffected by fluid density; however, it is strongly dependent on the amount of water in the system.

Jonas and co-workers (68) used Raman spectroscopy to study the naphthalene-CO_2 system under high pressure. The experimental results (a blue shift of the C-H stretching and C-H out-of-plane bending modes) and model calculations presented indicate that quadrupole-quadrupole coupling is significant in this system. The preferred orientation of the $O=C=O$ and naphthalene were determined to be face-to-face.

Steady-state fluorescence spectroscopy has also been used to study solvation processes in supercritical fluids. For example, Okada et al. (29) and Kajimoto and co-workers (30) studied intramolecular excited-state complexation (exciplex) and charge-transfer formation, respectively, in supercritical CHF_3. In the latter studies, the observed spectral shift was more than expected based on the McRae theory (56,57), this was attributed to cluster formation. In other studies, Brennecke and Eckert (5,31,44,45) examined the fluorescence of pyrene in supercritical CO_2, C_2H_6, and CHF_3. Steady-state emission spectra were used to show density augmentation near the critical point. Additional studies investigated the formation of the pyrene excimer (i.e., the reaction of excited- and ground-state pyrene monomers to form the excited-state **dimer**). These authors concluded that the observance of the pyrene excimer in the supercritical fluid medium was a consequence of increased solute-solute interactions.

Fox, Johnston, and co-workers (32) studied twisted-intramolecular-charge-transfer complexes in supercritical CHF_3. Again, the charge-transfer process was governed by the proximity to the critical point.

Most recently, Fox, Johnston, and co-workers (65) showed how an environmentally-sensitive fluorescent species could be used to probe the water pool of AOT reverse micelles in supercritical alkanes. The interesting observation was that the emission spectra were not shifting with fluid density. These results were consistent with the postulate of the micelle interior not changing with the supercritical fluid density; paralleling conclusions reached using other spectroscopic techniques (61-64).

Time-Resolved Spectroscopy. Steady-state solvatochromic techniques provide a reasonable means to study solvation processes in supercritical media (5,17-32,43-45,59-68). But, unless the interaction rates between the solute species and the supercritical fluid are slow, these "static" methods cannot be used to study solvation kinetics. Investigation of the kinetics requires an approach that offers inherent temporal resolution. Fortunately, time-resolved fluorescence spectroscopy is ideally suited for this task.

To date, there have been only a handful of time-resolved studies in dense fluid media (33,34,69-72). Of these, the bulk have focused on understanding a particular chemical reaction by adjusting the solvent environment (69-71). Only over the past two years have there been experiments directed toward studying the peculiar effects of supercritical fluids on these solvation processes (33,34,72). The initial work (33,34) showed that: 1) time-resolved fluorescence can be used to improve our understanding of solvation in supercritical fluids and 2) the local solvent composition, about a solute molecule, could change significantly on a subnanosecond time scale.

Most recently, Robinson and co-workers (72) used time-resolved decay of anisotropy experiments to probe the AOT reverse micelle system in ethane and propane. These authors conclude there was no local solvent density augmentation about the reverse micelle. In addition, the rotational dynamics of their probe (perylene tetracarboxylate) was independent of fluid density. This observation was consistent with results from Johnston and co-workers and Smith and co-workers (61-65,72).

Applications of Supercritical Fluids in Chromatography

The field of supercritical fluid chromatography (SFC) has dramatically regressed in its development over the last few years. From its inception, in the early 60's with Klesper's paper which illustrated the use of supercritical fluids for capillary chromatography, interest has peaked and waned successively (73). Specifically, in 1978 Conaway, Graham and Rogers demonstrated the effects of pressure, temperature, adsorbent surfaces, and mobile phase composition on the supercritical fluid retention characteristics of the oligomers of monodisperse polystyrene samples. Atypical to more current chromatographic separations using supercritical fluids, this report used n-pentane as the fluid instead of the more commonly used carbon dioxide. The n-pentane was modified with isopropanol or cyclohexane (74).

In the early 1980's Hewlett Packard introduced the first commercially available supercritical fluid chromatograph. This instrument was simply a Model 1084 B high performance liquid chromatograph modified for supercritical fluid mobile phases and SFC (75). However, the design was efficient and the instrument has withstood the test of time, since many of these original instruments are still actively used in research laboratories (76). Because of the commercial availability of this instrument, the direction that was initially pursued in SFC was packed column technology with ultraviolet detection (UV). This lead the pursuit of supercritical fluid chromatography to imitate the directional map that liquid chromatography had previously followed. Gradient elution with modified and pure supercritical fluids under isothermal temperatures was available. No adjustment of density or pressure during the chromatographic run was conducted easily. The pressure was elevated by the use of a Tescom® back pressure regulator. Pressure drops across the chromatographic column were obtained but measured with inlet and outlet pressure gauges. Supercritical fluid chromatography was present, but dramatic strides were not realized, less than 50 of these instruments were placed in analytical laboratories.

It was not until the introduction of the first capillary supercritical fluid chromatograph in 1986, that pressure and density programming became readily available. This instrumentation, available from three small companies, generally was commercialized as an alternative to gas chromatography, i.e. GC (77). The choice of SFC to replace GC was touted because of the unique advantages of a supercritical fluid, particularly, carbon dioxide. The solubilizing properties of the fluid at relatively low temperatures were a distinct advantage over the lack of solubilization obtained with the typical mobile phase gases used in GC. With GC, the solute is vaporized to achieve chromatographic separation. This was no longer necessary with the use of supercritical fluids, non-volatile and thermally labile compounds could be readily injected. The viscosity, diffusivity and density of supercritical fluids, as illustrated in Table 1, were intermediate between the properties of a liquid and a gas. Therefore, some loss in resolution obtained in a capillary gas separation would occur in a supercritical fluid separation. However, the gains, when molecules not separated by gas chromatography, were resolved and detected with these unique intermediate properties of the supercritical fluid, showed the technique to have a future in separation technology. Gas chromatographic detectors were readily interfaced. Separations that were obtained with GC were readily reproduced in SFC; no compromise in time of analysis or separation was noted.

Many reports (78-84) investigated the differences in packed and capillary supercritical fluid chromatography. Unfortunately, the rift between packed and capillary column users of SFC impeded the development of the science. This rift is a likely cause of the current low interest in SFC. Ideally, the unique features of the mobile phase is the area of scientific exploration that should be exploited. Choice of column size or type should be dependent upon the analytical problem to be solved.

Each of the individual aspects of either packed or capillary column supercritical fluid chromatography demonstrate significant advantages while coping with typical disadvantages of both column diameters known from other chromatographic techniques. More importantly, comparison studies of both packed and capillary supercritical fluid chromatography versus liquid or gas chromatography are more appropriate, and address trade-offs between mobile phases (85-86). Results of comparison studies with LC indicated that faster analyses, lower detection limits and greater injection to injection reproducibility were obtained with packed column SFC. No appreciable difference in the linearity of response between packed or capillary SFC or liquid chromatography techniques was noted. The mild conditions required to maintain carbon dioxide, the most commonly used supercritical fluid mobile phase, in the supercritical state, makes it exceptionally attractive for the separation of thermally labile compounds. Schwartz documented that capillary SFC afforded greater efficiencies and higher resolution, as would be expected for a smaller internal diameter column with any mobile phase type (86).

With the instrument developments of the late 1980's, the combination of gradient elution, density or pressure programming and, to a lesser extent, temperature programming, affords SFC a wide variety of method development techniques for the separation scientist. If gas and liquid chromatography had not been previously developed and extensively used, supercritical fluid chromatography might enjoy a more prestigious position than it currently does. SFC is no more difficult to use, interfaces with a wide assortment of detectors from both LC and GC and can readily separate those compounds which are routinely partitioned by liquid or gas chromatography. Albeit, LC and GC enjoy the position of having been developed first. As the use of supercritical fluids via extraction technology becomes routine in the analytical laboratory, the use of supercritical fluids in chromatography will most likely be successfully re-visited. Detection technology, still an Achilles heel in SFC in terms of lower limits of detection, will hopefully continue to be pursued.

The strength of the mobile phase has caused part of the slow development of SFC. Initially, the fact that we could easily change fluid density and thus solvent strength was a novel concept to the analytical chemist. Studies have been conducted to relate the strengths or weaknesses of these fluids in terms of polarity and solubilizing power to the commonly known solubilizing strengths of liquids. Theoretical studies have demonstrated the examination of solvent strengths using solvatochromic dyes for the purpose of comparing relative scales of carbon dioxide with different modifiers with solvating powers of normal liquids (87-89). Tehrani et al. have developed a software program available through Isco, Inc. (Lincoln, NE) to predict Hildebrand solubility parameters (see chapter 16 of this text for a more complete description). These calculated values, not uncommon to chemical engineers working with supercritical fluids, increase the analytical scientists familiarity with the mobile phase.

An understanding of chromatographic interactions of solute molecules with the supercritical fluid chromatographic system (the mobile phase and the stationary phase) has been conducted in a series of papers. Fields and Grolimund reported on basicity limits when carbon dioxide was used as the mobile phase (90). The objective was to develop a relationship between the basicity of amines and compatibility with carbon dioxide. A basicity limit of $pK_b = 9$ was proposed by the studies of Francis (91) and Dandge et al. (92) but did not hold for the tertiary alkyl amines investigated by Fields and Grolimund.

Wheeler and McNally reported several investigations of chromatographic retention shifts with changes in modifier composition and systematic alteration of the functional moieties on some basic molecular structures (93,94). Few models have been developed for the chromatographic theory in SFC, but Martire published results of the unified theory of adsorption chromatography with gas, liquid and supercritical fluid mobile phases (95). Many more theoretical models have been developed for supercritical fluid extraction processes than for chromatographic investigations. The utilization of these models to SFC is forthcoming.

Mobile fluid interaction with the stationary phase in SFC was investigated with mass spectrometric tracer pulse chromatography (96). Using capillary supercritical fluid chromatography, the effect of methanol as an additive was studied on the partition behavior of n-pentane into 5% phenylmethylsilicone stationary phase. The results showed that the mobile fluid uptake by the stationary phase decreased with increasing temperature and pressure. Thus suggests that stationary phase swelling, may occur in SFC.

Supercritical fluid chromatography offers some unique advantages in the separation field. To date, its use and therefore advantages have yet to be fully exploited.

Applications of Supercritical Fluids in Analytical Extraction

Supercritical fluid extraction (SFE) has the potential to change the analytical laboratory and its current extraction procedures dramatically. Predictions on the monetary exploitation of this technique are astronomical. Subsequently, the amount of research, development and applications in SFE is growing rapidly.

The theoretical principles of supercritical fluid extraction have been widely explored by chemical engineers. The most concise and comprehensive presentation of supercritical fluid phase diagrams to predict supercritical fluid extraction processes is the text by McHugh and Krukonis (97). Phase behavior, methods for determining solubilities in supercritical fluids and a wide variety of traditional supercritical fluid extraction application processes are outlined. This text provides a comprehensive view of supercritical phenomena for the novice as well as the experienced analyst. The optimization of supercritical fluid extraction processes was detailed for utility and cost by Cygnarowicz and Seider (98). Emphasis was placed on the development of strategies for optimizing steady-state flowsheets of supercritical fluid extraction processes. The cost of the supercritical fluid high pressure process equipment as well as utility and operational costs were compared with conventional separation processes. In general, the theoretical data available through the engineering research journals is process scale and only a stepping stone to the required information needed to initiate analytical scale extraction processes (99). The systems that have been investigated are generally simpler than those encountered in the analytical scale sample matrix. The components of interest are generally more complex and may not be present as one single solute of interest. The mobile phases most commonly employed by the analytical chemist are not the fluids that have been explored by the chemical engineers.

On an analytical scale, King reported the knowledge of four basic parameters of supercritical fluid extraction as an aid in understanding solute behavior in compressed gas media (100). These four parameters are the miscibility or threshold pressure, the solubility maxima, the fractionation pressure range and the cumulative physical properties of the solute. Specifically, the *miscibility or threshold pressure* is the pressure at which the solute starts to dissolve in the supercritical fluid. This pressure is technique dependent and varies with the analytical method sensitivity used to measure the solute concentration in the supercritical fluid phase. *Solubility maxima* is the pressure at which the solute attains its maximum solubility in the supercritical fluid. When the solvent strength of the extraction solvent is matched to the solute, the system is at the solubility maxima. The pressure range between the miscibility pressure and the solubility maximum is the *fractionation pressure* range. In this range, enrichment of one component over another in the supercritical fluid is possible by varying the pressure or density. This regulation of solute solubility in the supercritical solvent is still at the primitive stages. The fractionation is enhanced by differences in the solute physical properties. Two solutes of widely different vaporization or melting points can be more easily separated by temperature changes in the SFE process. Other solutes, with common properties are not as easily fractionated.

Environmental applications of SFE appear to be the most widespread in the literature. A typical example is the comparison of extraction efficiency for 2,3,7,8 - tetrachlorodibenzo-p-dioxin   (2,3,7,8-TCDD)   from   sediment   samples   using supercritical fluid extraction and five individual mobile phases with Soxhlet extraction was made (101). The mobile phases, carbon dioxide, nitrous oxide, pure and modified with 2% methanol as well as sulfur hexafluoride were examined. Pure nitrous oxide, modified carbon dioxide and modified nitrous oxide systems gave the recoveries in the acceptable range of 80 to 100%. Carbon dioxide and sulfur hexafluoride showed recoveries of less than 50% under identical conditions. Classical Soxhlet recoveries by comparison illustrated the poorest precision with average extraction efficiencies of less than 65%. Mobile phase choice, still as yet a major question in the science of supercritical fluid extraction, seems to be dependent upon several factors: polarity of the solute of interest, stearic interactions, as well as those between the matrix and the mobile phase. Physical parameters of the solute of interest, as suggested by King, must also be considered. Presently, the science behind the extraction of analytes of interest from complex matrices is not completely understood.

The matrix appears to control extractability in a wide variety of samples, the effect of water content in the matrix has been reported (101,102). Onuska and Terry reported decreases in extraction efficiency with wet and dry sediment, differences in weight to weight percent water were 0.3% to 19.8%. McNally and Wheeler suggested the opposite; increased extraction efficiencies with the addition of water as a modifier for the extraction of sulfonylureas molecules from a variety of soils. The differences in these two studies was principally the polarity of the solute molecule and the age of the sample. Aged environmental samples, especially those that have been field weathered, present more difficult extraction problems. Solutes of interest become trapped in the interstitial volumes of the sample matrix. When these environmental matrices have polar or ionic character and the solute also shares these properties, chemical interactions can occur. Non-polar solutes, such as those extracted in the Onuska study, will have less covalent and ionic bonding potential with the polar matrix. The kinetic movement of the solute of interest out of the matrix becomes dependent on the ability of the extraction solvent to break these bonds if they have formed as well as solubilize the solute (103).

Supercritical fluid extraction conditions were investigated in terms of mobile phase modifier, pressure, temperature and flow rate to improve extraction efficiency (104). High extraction efficiencies, up to 100%, in short times were reported. Relationships between extraction efficiency in supercritical fluid extraction and chromatographic retention in SFC were proposed. The effects of pressure and temperature as well as the advantages of static versus dynamic extraction were explored for PCB extraction in environmental analysis (105). High resolution GC was coupled with SFE in these experiments.

Coupled investigations of supercritical fluid extraction and various methods of chromatography abound in the literature. Hawthorne and Miller have used SFE coupled with GC in a simple procedure which can be utilized with a wide variety of capillary gas chromatography detection techniques (106). No sample handling was employed between the two steps, quantitative extraction and recovery of the target analytes was achieved. Quantitative transfer of the analytes of interest between the extraction and the detection steps was obtained to facilitate the lowest minimum detectable limit with the smallest original extraction sample size. The chromatographic peak shapes that were obtained were comparable to those that could be expected with more conventional sample introduction techniques. Ultimately, the methods were developed so that the SFE step and chromatographic analysis are complete in one hour. Successful examples included: the extraction of lemon and lime peel, basil and eucalyptus leaves for principle flavor components, residential chimney effluent for hardwood smoke phenolics which had been adsorbed onto a PUF sorbent plug and the quantitative extraction of polynuclear aromatic hydrocarbons from urban dust.

Levy and co-workers have reported the extraction of the components of gasoline followed by multi-dimensional supercritical fluid chromatography (107). Saturates, aromatics, heavy hydrocarbons and olefins were successfully analyzed with column switching, heart cutting and backflushing. Lee et al. used SFE-SFC with fraction collection to determine ouabain, $C_{29}H_{44}O_{22}$, a steroid-derived polar compound with eight hydroxyl groups (108). It belongs to a group of drugs used for the treatment of heart failure. A cryogenic trap was required to achieve acceptable recoveries and analytical detection.

Liquid chromatography with electrochemical detection was used to analyze menadione from animal feed after extraction by SFE (109). Average recovery declined with decreasing amounts of menadione spiked onto rat chow. At 0.1 to 1.5 mg of menadione spiked onto 0.5 g samples, recoveries ranged from 79.7% to

95.8%. At spiking levels of 0.02 and 0.05 mg, the recoveries dropped to 42 and 65%, respectively. These low recoveries were attributed to a low unextractable level of menadione that remained adsorbed on a component of the feed matrix. Pure carbon dioxide at 8000 psi and 60 °C for 20 minutes was used for these extractions. No mention of the use of modifiers to enhance extraction efficiency at these unextractable levels was made. These analyses were conducted off-line.

Fat tissue has been extracted from meat products with supercritical fluid carbon dioxide (110). The time required to perform these extractions depended on the 1) pressure and temperature at which the extraction was conducted, 2) the quantity of lipid matter in the sample, 3) the weight of the sample taken for the extraction and 4) the flow rate of the extraction gas. A high flow rate of the extraction gas to ensure rapid depletion of the fat phase from the protein matrix was reported. Percentages of fat ranged from 4.4 to 21.5%, in all cases extraction recoveries achieved were greater than 96%. Samples which contained high percentages of water, i.e. imported ham had a 73.4% water by weight, were dehydrated prior to extraction with carbon dioxide.

Aqueous samples were extracted for phenol and 4-chlorophenol using pure carbon dioxide in a specially designed phase separator apparatus (111). The extraction efficiency for these phenols was reported to be over 85%, with a RSD of 8% for eight samples. Additional liquid sample extractions have been investigated for the extraction of phenol from a 6M sulfuric acid solution as well as the extraction of the components of commercial soft drinks and orange juice (112-113). In all cases, specifically designed extraction vessels were utilized.

Supercritical fluid extraction has thus far demonstrated a wide variety of applications, more can be expected. Theoretical considerations, as of yet not completely understood, are currently being investigated. The promise of this technique is broad with many areas of investigation being actively pursued.

Conclusions

This chapter provides a brief introduction and necessary background for the remaining chapters of this volume. The fundamental studies using state-of-the-art spectroscopic techniques are presented first, followed by the chapters on modern application of supercritical science and technology (SFC and SFE).

Literature Cited

1. Reid, R.C.; Prausnitz, J.M.; Poling, B.E. *The Properties of Gases and Liquids*; 4th ed.; McGraw-Hill: New York, NY, 1987.
2. Bruno, T.J.; Ely, J.F. *Supercritical Fluid Technology: Reviews in Modern Theory and Applications*; CRC Press: Boca Raton, FL, 1991.
3. Paulaitis, M.E.; Krukonis, V.J.; Kurnik, R.T.; Reid, R.C. *Rev. Chem. Eng.* **1983**, 1, 179.
4. Paulaitis, M.E.; Kander, R.G.; DiAndreth, J.R. *Ber. Bunsenges. Phys. Chem.* **1984**, 88, 869.
5. Brennecke, J.F.; Eckert, C.A. *AIChE J.* **1989**, 35, 1409.
6. Klesper, E. *Angew. Chem. Int. Ed. Engl.* **1978**, 17, 738.

7.   Novotny, M.V.; Springston, S.R.; Peaden, P.A.; Fjeldsted, J.C.; Lee, M.L. *Anal. Chem.* **1981**, 53, 407A.
8.   *Supercritical Fluid Chromatography*; Smith, R.M., Ed.; Royal Society of Chemistry Monograph; Royal Society of Chemistry: London, UK, 1988.
9.   Smith, R.D.; Wright, B.W.; Yonker, C.R. *Anal. Chem.* **1988**, 60, 1323A.
10. Brunner, G. *Ion. Exch. Solvent Extr.* **1988**, 10, 105.
11. Eckert, C.A.; Van Alsten, J.G. *Environ. Sci. Technol.* **1986**, 20, 319.
12. *Supercritical Fluids - Chemical Engineering Principles and Applications*; Squires, T.G.; Paulaitis, M.E., Eds. ACS Symp. Ser. 329; ACS: Washington, DC, 1987.
13. van Wasen, U.; Swaid, I.; Schneider, G.M. *Angew. Chem. Int. Ed. Engl.* **1980**, 19, 575.
14. Aaltonen, O.; Rantakyla, M. *CHEMTECH* **1991**, 21, 240.
15. Eckert, C.A.; Ziger, D.H.; Johnston, K.P.; Ellison, T.K. *Fluid Ph. Equil.* **1983**, 14, 167.
16. Eckert, C.A.; Ziger, D.H.; Johnston, K.P.; Kim, S. *J. Phys. Chem.* **1986**, 90, 2738.
17. Kim, S.; Johnston, K.P. *Ind. Engr. Chem. Res.* **1987**, 26, 1206.
18. Kim, S.; Johnston, K.P. *AIChE J.* **1987**, 33, 1603.
19. Brennecke, J.F.; Eckert, C.A. In *Proceeding of the International Symposium on Supercritical Fluids*; Perrut, M., Ed.; Nice, France, 1988, p. 263.
20. Johnston, K.P.; Kim, S.; Combs, J. In *Supercritical Fluid Science and Technology*; Johnston, K.P.; Penninger, J.M.L., Eds.; ACS Symp. Ser. Vol. 406; ACS: Washington, DC, 1989, Chapter 5.
21. Hyatt, J.A. *J. Org. Chem.* **1984**, 49, 5097.
22. Sigman, M.E.; Lindley, S.M.; Leffler, J.E. *J. Am. Chem. Soc.* **1985**, 107, 1471.
23. Sigman, M.E.; Leffler, J.E. *J. Phys. Chem.* **1986**, 90, 6063.
24. Yonker, C.R.; Frye, S.L.; Kalkwarf, D.R.; Smith, R.D. *J. Phys. Chem.* **1986**, 90, 3022.
25. Smith, R.D.; Frye, S.L.; Yonker, C.R.; Gale, R.W.J. *J. Phys. Chem.* **1987**, 91, 3059.
26. Yonker, C.R.; Smith, R.D. *J. Phys. Chem.* **1988**, 92, 2374.
27. Johnston, K.P.; Kim, S.; Wong, J.M. *Fluid Ph. Equil.* **1987**, 38, 39.
28. Yonker, C.R.; Smith, R.D. *J. Phys. Chem.* **1988**, 92, 235.
29. Okada, T.; Kobayashi, Y.; Yamasa, H.; Mataga, N. *Chem. Phys. Lett.* **1986**, 128, 583.
30. Kajimoto, O.; Futakami, M.; Kobayashi, T.; Yamasaki, K. *J. Phys. Chem.* **1988**, 92, 1347.
31. Brennecke, J.F.; Eckert, C.A. In *Supercritical Fluid Science and Technology*; Johnston, K.P.; Penninger, J.M.L., Eds.; ACS Symp. Ser. Vol. 406; ACS: Washington, DC, 1989, Chapter 2
32. Hrnjez, B.J.; Yazdi, P.T.; Fox, M.A.; Johnston, K.P. *J. Am. Chem. Soc.* **1989**, 111, 1915.

33. Betts, T.A.; Bright, F.V. *Appl. Spectrosc.* **1990**, 44, 1196.
34. Betts, T.A.; Bright, F.V. *Appl. Spectrosc.* **1990**, 44, 1204.
35. Debenedetti, P.G. *Chem. Eng. Sci.* **1987**, 42, 2203.
36. Debenedetti, P.G.; Kumar, S.K. *AIChE J.* **1988**, 34, 645.
37. Debenedetti, P.G.; Mohamed, R.S. *J. Chem. Phys.* **1989**, 90, 4528.
38. Cochran, H.D.; Lee, L.L.; Pfund, D.M. *Fluid Ph. Equil.* **1988**, 39, 161.
39. Shing, K.S.; Chung, S.T. *J. Phys. Chem.* **1987**, 91, 1674.
40. Gitterman, M.; Procaccia, I. *J. Chem. Phys.* **1983**, 78, 2648.
41. Cochran, H.D.; Pfund, D.M.; Lee, L.L. *Sep. Sci. Tech.* **1988**, 23, 2031.
42. Petsche, I.B.; Debenedetti, P.G. *J. Chem. Phys.* **1989**, 91, 7075.
43. Combes, J.R.; Johnston, K.P.; O'Shea, K.E.; Fox, M.A. In this volume.
44. Brennecke, J.F.; Tomasko, D.L.; Peshkin, J.; Eckert, C.A. *Ind. Eng. Chem. Res.* **1990**, 29, 1682.
45. Brennecke, J.F.; Tomasko, D.L.; Eckert, C.A. *J. Phys. Chem.* **1990**, 94, 7692.
46. Olsen, E.D. *Modern Optical Methods of Analysis*; McGraw-Hill: New York, NY, 1975.
47. Ingle, J.D., Jr.; Crouch, S.R. *Spectrochemical Analysis*; Prentice Hall: Englewood, NJ, 1988.
48. Lakowicz, J.R. *Principles of Fluorescence Spectroscopy*; Plenum Press: New York, NY, 1983.
49. Demas, J.N. *Excited State Lifetime Measurements*; Academic Press: New York, NY, 1983.
50. Kamlet, M.J.; Abboud, J.-L.M.; Abraham, M.H.; Taft, R.W. *J. Org. Chem.* **1983**, 48, 2877.
51. Kamlet, M.J.; Abboud, J.-L.M.; Taft, R.W. *Prog. Phys. Org. Chem.* **1980**, 13, 485.
52. Kamlet, M.J.; Doherty, R.M.; Taft, R.W.; Abraham, M.H. *J. Am. Chem. Soc.* **1983**, 105, 6741.
53. Abraham, M.H.; Kamlet, M.J.; Taft, R.W.; Weathersby, P.K. *J. Am. Chem. Soc.* **1983**, 105, 6797.
54. Taft, R.W.; Abboud, J.-L.M., Kamlet, M.J. *J. Org. Chem.* **1984**, 49, 2001.
55. Kamlet, M.J.; Abboud, J.-L.M.; Taft, R.W. *J. Am. Chem. Soc.* **1977**, 99, 6027.
56. Bayliss, N.S.; McRae, E.G. *J. Phys. Chem.* **1954**, 58, 1002.
57. McRae, E.G. *J. Phys. Chem.* **1957**, 61, 562.
58. Onsanger, L. *J. Am. Chem. Soc.* **1936**, 58, 1486.
59. Buback, M. *Angew. Chem. Int. Ed. Engl.* **1991**, 30, 641.
60. Blitz, J.P.; Yonker, C.R.; Smith, R.D. *J. Phys. Chem.* **1989**, 93, 6661.
61. Beckman, E.J.; Smith, R.D. *J. Phys. Chem.* **1990**, 94, 345.

62.   Kaler, E.W.; Billman, J.F.; Fulton, J.L.; Smith, R.D. *J. Phys. Chem.* **1991**, 95, 458.
63.   Tingey, J.M.; Fulton, J.L.; Smith, R.D. *J. Phys. Chem.* **1990**, 94, 1997.
64.   Smith, R.D.; Fulton, J.L.; Blitz, J.P.; Tingey, J.M. *J. Phys. Chem.* **1990**, 94, 781.
65.   Yazdi, P.; McFann, G.J.; Fox, M.A.; Johnston, K.P. *J. Phys. Chem.* **1990**, 94, 7224.
66.   Lemert, R.M.; Fuller, R.A.; Johnston, K.P. *J. Phys. Chem.* **1990**, 94, 6021.
67.   McFann, G.J.; Johnston, K.P. *J. Phys. Chem.* **1991**, 95, 4889.
68.   Zerda, T.W.; Song, X.; Jonas, J. *Appl. Spectrosc.* **1986**, 40, 1194.
69.   Lee, M.; Holtom, G.R.; Hochstrasser, R.M. *Chem. Phys. Lett.* **1985**, 118, 359.
70.   Schroeder, J.; Thoe, J. *Chem. Phys. Lett.* **1985**, 116, 453.
71.   Otto, B.; Schroeder, J.; Troe, J. *J. Chem. Phys.* **1984**, 81, 202.
72.   Eastoe, J.; Robinson, B.H.; Visser, A.J.W.G.; Steytler, D.C. *J. Chem. Soc. Faraday Trans.* **1991**, 87, 1899.
73.   Klesper, E.; Corwin, A.H.; Turner, D.A. *J. Org. Chem.* **1962**, 27, 700.
74.   Conaway, J.E.; Graham, J.A.; Robers, L.B. *J. Chromatogr. Sci.* **1978**, 16, 102.
75.   Gere, D.R.; Board, R.; McManigill, D. *Anal. Chem.* **1982**, 54, 736.
76.   Gemmel, B.; Schmitz, F.P.; Klesper, E. *J. Chromatogr.* **1988**, 455, 17.
77.   Berry, V. *LC-GC Mag. Liq. Gas Chromatogr.* **1986**, 4(5), 471.
78.   Shoenmakers, P.J. *HRC & CC. J. High Res. Chromatogr. & Chromatogr. Comm.* **1988**, 11, 278.
79.   White, C.M.; Houck, R.K. *HRC & CC. J. High Res. Chromatogr. & Chromatogr. Comm.* **1986**, 9, 3.
80.   Fjeldsted, J.C.; Lee, M.L. *Anal. Chem.* **1984**, 56, 619A.
81.   Malik, A.; Jumaev, A.R.; Berezkin, V.G. *HRC & CC. J. High Res. Chromatogr. & Chromatogr. Comm.* **1986**, 9, 312.
82.   Jones, B.A.; Markides, K.E.; Bradshaw, J.S.; Lee, M.L. *Chromatography Forum* **1986**, May-June, 38.
83.   Schwartz, H.E.; Barthel, P.J.; Moring, S.E.; Lauer, H.H. *LC-GC Mag. Liq. Gas Chromatogr.* **1987**, 5(6), 490.
84.   Ettre, L.S. *HRC & CC. J. High Res. Chromatogr. & Chromatogr. Comm.* **1987**, 10, 637.
85.   Wheeler, J.R.; McNally, M.E. *J. Chromatogr.* **1987**, 410, 343.
86.   Schwartz, H.E. *LC-GC, Mag. Liq. Gas Chromatogr.* **1987**, 5, 14.
87.   Levy, J.M.; Ritchey, W.M. *J. Chromatogr. Sci.* **1986**, 24, 242.
88.   Levy, J.M.; Ritchey, W.M. *HRC & CC, J. High Res. Chromatogr. & Chromatogr. Comm.* **1987**, 10, 493.
89.   Deye, J.F.; Berger, T.A.; Anderson, A.G. *Anal. Chem.* **1991**,
90.   Fields, S.M.; Grolimund, K. *HRC & CC, J. High Res. Chromatogr. & Chromatogr. Comm.* **1988**, 11, 727.
91.   Francis, A.W. *J. Phys. Chem.* **1954**, 58, 1099.

92. Dandge, D.K.; Heller, J.P.; Wilson, K.V. *Ind. Eng. Chem. Prod. Res. Dev.* **1985**, 24, 162.

93. Wheeler, J.R.; McNally, M.E. *Fres. Z. Anal. Chem.* **1988**, 330, 237.

94. McNally, M.E.; Wheeler, J.R.; Melander, W.R. *LC-GC Mag. Liq. Gas Chromatogr.* **1988**, 6(9), 816.

95. Martire, D.E. *J. Chromatogr.* **1988**, 452, 17.

96. Strubinger, J.R.; Selim, M.I. *J. Chromatogr. Sci.* **1988**, 26, 579.

97. McHugh, M.A.; Krukonis, V.J. *Supercritical Fluid Extraction - Principles and Practice*; Butterworths: Boston, MA, 1986.

98. Cygnarowicz, M.L.; Seider, W.D. Societe Francaise de Chemie, International Symposium on Supercritical Fluid Proceedings, M. Perrut, Coordinator, Oct. 17-18th, **1988**, Nice, France.

99. Johnston, K.P.; Penninger, J.M.L. *Supercritical Fluid Science and Technology*; ACS Symposium Series, No. 406; ACS: Washington, DC, 1989.

100. King, J.W. *J. Chromatogr. Sci.* **1989**, 27, 355.

101. Onuska, F.I.; Terry, K.A. *HRC & CC, J. High Res. Chromatogr. & Chromatogr. Comm.* **1989**, 12, 357.

102. McNally, M.E. Pittsburgh Conference, New York, NY, **1990**, Paper No. 276.

103. Fahmy, T.M.; University of Delaware, Chemical Engineering Thesis, June, **1991**.

104. McNally, M.E.; Wheeler, J.R. *J. Chromatogr.* **1988**, 447, 53.

105. Onuska, F.I.; Terry, K.A. *HRC & CC, J. High Res. Chromatogr. & Chromatogr. Comm.* **1989**, 12, 527.

106. Hawthorne, S.B.; Miller, D.J.; Krieger, M.S. *J. Chromatogr. Sci.* **1989**, 27, 347.

107. Levy, J.M.; Cavalier, R.A.; Bosch, T.N.; Rynaske, A.F.; Huhak, W.E. *J. Chromatogr. Sci.* **1989**, 27, 341.

108. Xie, Q.L.; Markides, K.E.; Lee, M.L. *J. Chromatogr. Sci.* **1989**, 27, 365.

109. Schneiderman, M.A.; Sharma, A.K.; Locke, D.C. *J. Chromatogr. Sci.* **1988**, 26, 458.

110. King, J.W.; Johnson, J.H.; Friedrich, J.P. *J. Agric. Food Chem.* **1989**, 37, 951.

111. Thiebaut, D.; Chervet, J.P.; Vannoort, R.W.; De Jong, G.J.; Brinkman, U.A. Th.; Frei, R.W. *J. Chromatogr.* **1989**, 477, 151.

112. Hedrick, J.L.; Taylor, L.T. *HRC & CC, J. High Res. Chromatogr. & Chromatogr. Comm.* **1990**, 13, 312.

113. Levy, J.M.; Suprex Corporation, Personal Communication.

RECEIVED December 5, 1991

SPECTROSCOPIC INVESTIGATIONS

Chapter 2

Raman Studies of Molecular Potential Energy Surface Changes in Supercritical Fluids

Dor Ben-Amotz, Fred LaPlant, Dana Shea, Joseph Gardecki, and Donald List

Department of Chemistry, Purdue University, West Lafayette, IN 47907

Raman spectroscopy is used to measure changes in the vibrational potential energy surfaces of molecules in supercritical fluids. The NN, CC, and CH stretching vibrations of nitrogen, methane, and ethane are studied as a function of pressure and temperature throughout the vapor, liquid, and supercritical fluid range. The results are used to test a recently developed perturbed hard sphere fluid model. Good agreement with model predictions is found both far from the critical point and under near critical conditions.

Current applications of supercritical fluids (SCFs) in natural product extraction and separation processes are well known (1,2). More recent studies suggest further applications of SCFs in chemical reactivity enhancement (1-6). Preliminary reactivity studies reveal marked changes in reaction rates and equilibrium constants, particularly in the near critical region. Theoretical efforts have focused on modeling SCF solvation and reactivity enhancement effects in terms of macroscopic equations of state or microscopic density fluctuation models (7-12). Our aim in this work is to apply new experimental techniques and theoretical modeling approaches to SCF solvation and reactivity. In particular we focus on experimentally probing SCF induced changes in molecular potential energy surfaces and testing the applicability of perturbed hard sphere liquid theories to SCFs.

The effects of solvation on molecular potential surfaces are interrogated by measuring molecular vibrational frequency shifts in SCFs. Thus SCF solvation induced changes in vibrational potential energy surfaces are used to test models for solvent effects on more general isomerization and reaction potentials. We use Raman scattering to measure solute vibrational frequencies as a function of solvent density and temperature. Although these studies are not the first examples SCF Raman scattering (13-22), they are perhaps the first to critically test perturbed hard sphere liquid theories for SCF solvation induced changes in molecular potentials.

0097–6156/92/0488–0018$06.00/0

Perturbed hard sphere liquid theories have previously been used to model molecular vibrational potential surface changes in high pressure liquids (*23-28*). Our aim in this work is to test these theories over a wider density and temperature range, with particular emphasis on the near critical regime. Our results reveal surprisingly good agreement between measured and predicted vibrational frequency shifts in pure nitrogen, methane and ethane over a wide density and temperature range. The relevance of our findings to solvation and reactivity enhancement in SCFs is discussed.

Experimental Procedure

The Raman scattering instrument design is shown in Figure 1. Unique features of this design include; (i) The use of a commercial camera lens (Olympus, 50 mm, f/1.2) to both focus the laser (Coherent, Inova 70-5 argon ion laser) into the SCF cell and collect the back-scattered Raman signal; (ii) Simultaneous collection of light from a neon pen lamp (Pen Ray) for frequency calibration; (iii) The use of a spatial filter to enhance the rejection of fluorescence and stray light scattering emanating from outside the 5×10^{-7} cm^3 laser focal region inside the SCF sample. Since we are only concerned with frequency shifts, the polarization of the scattered light is not analyzed in these studies. A remotely scanned Spex 1400 double monochromator (f/8) with 1200 grooves/mm gratings and a cooled EMI 9558A photomultiplier tube is used to disperse and detect the Raman scattering signal. The 300 μm input and exit slits are used for maximum collection efficiency. Although this represents a spectral width of about 4 Å, it is sufficient for peak frequency measurement with better than ± 0.2 cm^{-1} accuracy. After preamplification and discrimination, the photon pulses from the photomultipier are counted using a Macintosh IIcx computer (National Instruments NB-MIO-16H-9 board and Labview software).

A schematic of the pressure cell used in these studies is shown in Figure 2. This small volume cell design has been tested at up to 20,000 PSI. The primary pressure seal consists of a Viton o-ring seated in the stainless steel cell body, in contact with the 3/8" thick sapphire cell windows. Although two windows are not strictly necessary for the backward Raman scattering geometry used in this study, the second window offers advantages in ease of alignment, rejections of stray laser light, and compatibility with transmission spectroscopies. The entire cell is immersed in a sealed optically transparent Lexan housing. Temperature control is provided by circulating water from a temperature regulated water bath (Lauda, RMT-6) through the Lexan housing. The input laser light, and the output Raman signals, are transmitted through the housing and temperature bath fluid with little loss in collection efficiency or increase in background signal.

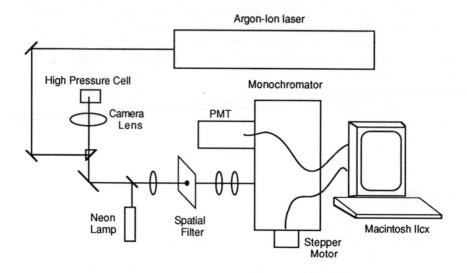

Figure 1. Spectrometer used for SCF Raman studies as a function of pressure and temperature.

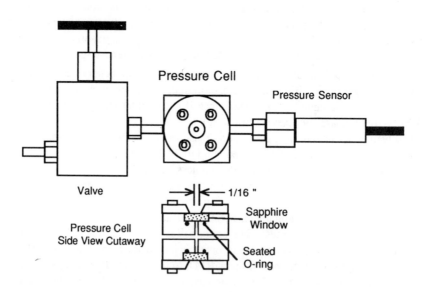

Figure 2. Detailed view of the high pressure cell used for SCF Raman studies.

The sample pressure is measured using a thin film pressure sensor (Omega PX603) attached to one arm of the cell. Internal calibration of this pressure gauge is achieved by comparing the temperature derivative of the sample pressure with tabulated equation of state values (*29,30,31*). This calibration results in an independent measure of both the pressure and the density of the sample, even in the highly compressible near critical regime. Representative pressure induced vibrational frequency shift results for nitrogen at 10 MPa (~100 atm) and 100 MPa (~1,000 atm) are shown in Figure 3.

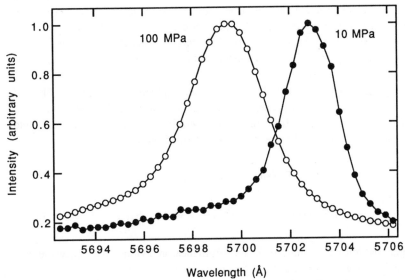

Figure 3. Nitrogen Raman scattering resonance observed using a 5017.16 Å laser wavelength.

Theoretical Modeling

The perturbed hard sphere fluid model used in this study is an extension of the the work of Chandler and coworkers (*23,32*). In this model the chemical potential of the solute molecule and the resulting Raman frequency shift are separated into two terms.

$$\Delta v = \Delta v_0 + \Delta v_a \tag{1}$$

The first term in this expression, Δv_0, represents the frequency shift resulting from short range repulsive packing forces. These many-body repulsive forces are in general expected to lead to a non-linear dependence of frequency on density. However, this complex non-linear behavior can be accurately modeled using a hard-sphere reference fluid, with appropriately chosen density, temperature and molecular diameters (see below).

In order to derive a practical approximation for the repulsive contribution to vibrational frequency shifts the excess chemical potential, $\Delta\mu_0$, associated with the formation of a hard diatomic of bond length r_{12} from two hard spheres at infinite separation in a hard sphere reference fluid is assumed to have the following form.

$$\Delta\mu_0(r_{12}) = -k_B T \{A + B\, r_{12} + C\, r_{12}^3 + D\, (1/r_{12})\} \qquad (2)$$

This functional form is derived from exact results in the dilute vapor and hydrodynamic solvent limits. The coefficients A, B, C and D used in modeling high density fluids are determined uniquely from the equation of state of the corresponding hard sphere reference system (33,34). This hard sphere fluid chemical potential model has been shown to accurately reproduce computer simulation results for both homonuclear and heteronuclear hard sphere diatomics in hard sphere fluids up to the freezing point density (35).

The repulsive frequency shift, $\Delta\nu_0$, is expressed explicitly in terms of the first and second derivatives of the excess chemical potential (equation 2) along with the vapor phase vibrational transition frequency, ν_{vib}, equilibrium bond length, r_e, and harmonic and anharmonic vibrational force constants, f and g (23,25,28).

$$\Delta\nu_0 = \nu_{vib}\,\frac{F}{f}\left\{\frac{-3}{2}\left(\frac{g}{f}\right) + \left(\frac{G}{F}\right)\right\} \qquad (3)$$

$$F = (\partial\Delta\mu_0/\partial r_{12})_{r_e} \text{ and } G = 1/2\,(\partial\Delta\mu_0/\partial r_{12})_{r_e} \qquad (4)$$

The vibrational force constants are obtained from measured vibrational frequencies and bond lengths along with extended Barger's rule correlations (36).

The attractive force frequency shift, $\Delta\nu_a$, is assumed to depend linearly on density solvent density (23)

$$\Delta\nu_a = C_a\,\rho \qquad (5)$$

This perturbative expression for the attractive force shift is derived from a van der Waals mean field approximation (23). Although the predictions of this model have been found to agree with numerous high pressure vibrational frequency shift measurements (23,25,28), a non-linear attractive force model has recently been suggested to be appropriate for some systems (26,27).

Equations 1 - 5 completely define the "hard fluid" model for solvent induced changes in the vibrational frequency of a diatomic (or pseudo-diatomic) solute. *The only adjustable parameter in this model is the coefficient* C_a *appearing in equation 5.* The other parameters, such as the diameters of the solute and solvent as well as the solvent density and temperature, are determined using independent measurements and/or parameter correlations (37). The value of C_a can be determined with a minimal amount of experimental data. In particular we use the frequency shift observed in going from the dilute gas to a dense fluid to fix the value of C_a. Having done this, the measured frequency shifts at all other densities can be used to test the hard fluid model predictions.

In order to apply this model to polyatomic solute/solvent systems a procedure for determining the effective diatomic(s) corresponding to the polyatomic solute of interest must be established. In the original work of Schweizer and Chandler (23), polyatomics were treated as a superposition of isolated diatomic bonds. This model completely ignores any shielding of solvent collisions by neighboring bonds in a polyatomic solute. Later studies have modeled the entire polyatomic solute as psuedo-diatomic whose "atoms" represents polyatomic units on either side of the bond of interest (25,38). This approach, which is the one we adopt here, is believed to more realistically represent the solvent induced forces along a bond in a polyatomic molecule.

The effective diameters for the pseudo-diatomic representing a given polyatomic solute vibrational normal mode depend on the nature of the normal mode. For example, the C–C stretch of ethane involves a concerted motion of the two -CH$_3$ groups relative to each other, and so the diameter of each "atom" in the pseudo-diatomic is taken to be that appropriate for a methyl group. The symmetric CH stretch of methane, on the other hand, is represented by a hetero-nuclear pseudo-diatomic with diameters appropriate for an –H and a -CH$_3$ group. Alternatively, the pseudo-diatomic for methane can, for example, be taken to a have a (-H)$_3$ and a CH group with little change in the results (as long as the force constants of the pseudo-diatomic are tripled to account for the three -H bonds between the two "atoms" of the pseudo-diatomic). The diameters of the "atoms" in this model are chosen by requiring that the total volume of the resulting psuedo-diatomic be equal to that of the hard sphere which best represents repulsive contributions to the PVT properties of the fluid(37). Table I contains the solvent and solute diameters and force constants used in these hard fluid calculations.

Table I. Molecular Parameters for SCF Studies

mol.	bond	T	v_{vib}[a]	r_e[a]	σ_s[b]	σ_1[b]	σ_2[b]	force constants[c]		Attractive force[d]
								$f r_e^2$	$g r_e^3$	parameter, C_a
		(°C)	(cm^{-1})	(Å)	(Å)	(Å)	(Å)	(dyne·Å)	(dyne·Å)	(cm^{-1}·l/mole)
N$_2$	N-N	23	2329.9	1.098	3.449	3.00	3.00	0.02766	-0.0770	-0.131
CH$_4$	C-H	23	2916.5	1.091	3.582	3.53	2.22	0.00601	-0.0136	-0.586
C$_2$H$_6$	C-C	23	994.8	1.543	4.237	3.63	3.63	0.00706	-0.0177	-0.279
	C-H	26	2953.7	1.110	4.234	4.23	2.22	0.00582	-0.0118	-0.962
		39	2953.7	1.102	4.222	4.22	2.21	0.00582	-0.0118	-0.967
		79	2953.7	1.102	4.187	4.18	2.19	0.00582	-0.0118	-0.990

a) Vibrational resonance frequencies, v_{vib}, and bond lengths, r_e, in the low density vapor phase (14,39).
b) Effective molecular and psuedo-atomic hard sphere diameters σ_s, σ_1 and σ_2 are estimated using molecular hard sphere volumes (37) and van der Waals volume increments (scaled to hard sphere volumes) (40).
c) Force constants are obtained from vibrational frequencies and extended Barger's rule correlations (36).
d) Attractive force parameters are obtained by fitting the hard fluid model to high density frequency shifts (see text).

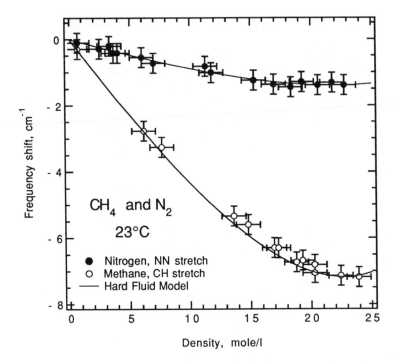

Figure 4. Vibrational frequency shifts as a function of density in supercritical fluid nitrogen and methane. The experimental and theoretical shifts are measured relative to their zero density vapor phase values (see table I).

Results

The frequency shifts observed in pure SCF nitrogen and methane at room temperature are plotted as a function of the density in Figure 4. The shifts are measured relative to their zero density values. These are found to red shift as density is increased through the vapor and near critical density range. The magnitude of this red shift is clearly much larger for methane than for nitrogen. This is consistent with the larger derivative of the polarizability with respect to the vibrational normal mode coordinate for the CH ($d\alpha/dQ = 2.08$ and 0.66 for CH_4 and N_2 respectively (41)).

A larger polarizability derivative tends to favor solvation of the excited vibrational state, and a corresponding decrease in curvature of the vibrational potential energy surface. Thus the stronger red shift for methane than nitrogen is a direct consequence of the enhanced solvation energy of the more polarizable excited vibrational state of methane. The solid lines through the data points in Figure 4 represent predictions of the hard fluid model (see Theoretical Modeling section). The one adjustable parameter (C_a) in this model is fixed using the observed frequency shift at high density. The magnitudes of the attractive force shift parameters are given in Table I, along with the corresponding input parameter values.

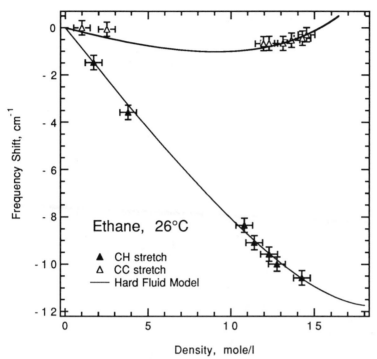

Figure 5. Vibrational frequency shifts as a function of density for two different vibrational modes in sub-critical ethane. The experimental and theoretical shifts are measured relative to their zero density vapor phase values (see table I).

The CH and CC vibrational modes of ethane were studied as a function temperature and pressure in the liquid, vapor, and SCF region. This system offers an opportunity to probe near critical solvation forces and their effects on different internal molecular coordinates within the same solute molecule. The room temperature frequency shifts values for the CC and symmetric CH stretch vibrations are shown in Figure 5.

The density gap in Figure 5 is due to a vapor/liquid phase transition. The solid lines through the data points represent predictions of the hard fluid model, using experimental high density frequency shifts to fix the attractive force parameter, C_a, in equation 5. Again, the more polarizable CH vibration is found to experience a significantly larger red shift than the less polarizable CC vibration ($d\alpha/dQ = 2.65$ and 0.41 for CH and CC, respectively (*41*)). These results offer striking evidence for the marked difference in solvation forces along different coordinates of motion within a given polyatomic solute molecule. This is particularly pertinent to the consideration of branching chemical reactions. Our frequency shift results suggest that, in general, one may expect to observe large pressure dependent changes in the relative product yields for branching chemical reactions in SCFs.

In order to further probe potential energy surface changes in near the critical region, CH stretch frequency shift measurements were performed over the entire vapor/liquid density range in supercritical ethane at 39 °C and 79 °C ($T_c \sim 32$ °C, ρ_c = 7.3 mole/l). The resulting frequency shifts are shown in Figure 6. The dashed and solid lines through the data points represent hard fluid model fits to the 39 °C and 79 °C results, respectively. The density dependence of the frequency shifts observed at these two supercritical temperatures are essential identical to those found along the sub-critical 26 °C isotherm (Figure 5).

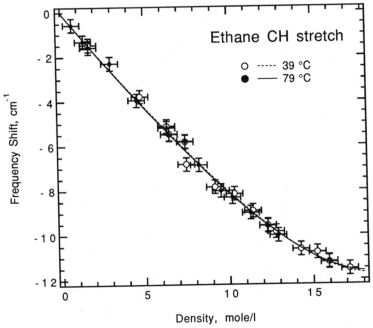

Figure 6. Vibrational frequency shift for the CH stretch of ethane at two supercritical temperatures. The experimental and theoretical shifts are plotted relative to their zero density vapor phase values.

The temperature independence of the CH frequency shifts is also reflected in the nearly constant attractive force parameters (see Table I). In fact, the frequency shifts predicted using the average attractive force parameter, $C_a = 0.973$, reproduce the experimental results to within 3% throughout the experimental density and temperature range. It thus appears that the attractive force parameter may reasonably be treated as a temperature and density independent constant. This behavior is reminiscent of that found for attractive force parameters derived from high pressure liquid equation of state studies using a perturbed hard sphere fluid model (37).

Discussion

It is clear from the representative Raman spectra shown in Figure 3 that both the mean frequency and the width of vibrational bands may be sensitive to solvent density and temperature. Changes in peak frequency are directly related to the potential of the mean force, while band width changes may depend in a more complex way on homogeneous and inhomogeneous broadening mechanisms (*23*). On the other hand, our results, like those of previous near critical band width studies (*18*), do not indicate any anomalous changes in vibrational band width near the critical point. We have therfor restricted our analysis to the interpretation of density and temperature dependent changes in the mean vibrational frequency of molecules dissolved in SCFs.

The hard fluid model is found to quantitatively reproduce observed vibrational frequency shifts in supercritical N_2, CH_4 and near critical C_2H_6. In nitrogen and methane at room temperature T/T_c is equal to 2.3 and 1.5, respectively. At such high reduced temperatures repulsive forces are expected to exert a predominant influence on fluid structure. Thus it is perhaps not surprising that the hard fluid model is successful in reproducing the observed frequency shifts in these two fluids.

On the other hand, the success of the hard fluid models in describing the frequency shifts in ethane is perhaps more significant. In the near critical region attractive force interactions are in general expected play a more important role. These can lead to long range density correlation, or cluster formation, and as a result to strong perturbations in fluid structure. Thus attractive force effects in near-critical systems tend to weaken the basic perturbative assumption that the fluid structure remains close to that of the hard sphere reference system. Nevertheless, the results in Figures 5 and 6 clearly do not indicate any significant failure of the hard fluid model.

Interestingly, our initial studies of ethane suggested more significant discrepancies between the observed frequency shifts and hard fluid model predictions along the 39 °C isotherm. These deviations were subsequently found to be artifacts resulting from pressure/density calibration errors in the highly compressible near critical region. After correction of the densities and pressures using measurements of the temperature derivative of the pressure (at constant density) the apparent deviations from hard fluid model predictions were found to essentially disappeared. These results underscore the difficulty in quantitating spectroscopic evidence for cluster formation and anomalous near critical behavior in SCFs.

Previous experimental and theoretical studies have found what appears to be clear evidence for cluster formation, or local density enhancement, in near critical solutions (*7,12,42-45*). These include experimental optical absorption, fluorescence and partial molar volume measurements as well as theoretical simulation studies. These offer compelling evidence for local solvent density enhancement in near critical binary SCF systems. Theoretical models suggest that local density enhancement should be strongly dependent on the relative size and attractive force interactions strengths of the solute and solvent species as well as on bulk density and temperature (*7,44*).

Our vibrational frequency shift measurements in pure SCF systems can be viewed as studies of a special class of solutions in which the solute and solvent species are the same. The absence of anomalous near critical frequency shifts in such systems suggests the lack of significant local density enhancement. These results appear to be consistent with recent calculations by Petsche and Debenedetti (7,44) which predict a vanishing local density enhancement when the solvent/solute size and attractive force strength become equal. Further vibration and isomerization studies in two component SCF systems are planned. These will be used to further test hard fluid and local density enhancement models.

In summary, we have measured solvation induced perturbations of molecular vibrational potential energy surfaces in pure N_2, CH_4 and C_2H_6. The results reveal that more polarizable molecular vibrations are more strongly red shifted as a function of density. The observed perturbations are quantitatively reproduced by a model which approximates the structure of a SCF using a hard sphere reference fluid, attractive forces effects using a van der Waals mean field model and the vibration of a polyatomic molecule using a corresponding psuedo-diatomic. Further tests of this approach to modeling SCF solvation and its effects on molecular vibration, isomerization and reactivity are in progress.

Legend of Symbols

Δv	- Vibrational frequency shift, relative to low density vapor
Δv_0	- Repulsive contribution to the vibrational frequency shift
Δv_0	- Attractive contribution to the vibrational frequency shift
$\Delta \mu$	- Excess chemical potential of a solute, relative to ideal gas phase
$\Delta \mu_0$	- Repulsive contribution to the solute chemical potential
r_{12}	- Diatomic solute bond length
χ	- Reduced bond length of the solute
σ_s	- Solvent hard sphere diameter
σ_1 and σ_1	- Solute diatomic hard sphere diameters
σ_0	- Average solvent/solute atom contact diameter
A, B, C and D	- Coefficients for bond length dependence of chemical potential
v_{vib}	- Vapor phase solute vibrational frequency
r_e	- Vapor phase solute bond length
f and g	- harmonic and anharmonic solute force constants
F and G	- linear and quadratic hard sphere solvation force coefficients
C_a	- Attractive force frequency shift parameter
ρ	- Solvent number density
α	- Solute polarizability
Q	- Solute vibrational normal mode coordinate
T_c	- Critical temperature
ρ_c	- Critical density

Acknowledgments

This work was supported in part by the Exxon Education Foundation and the Purdue Research Foundation.

Literature Cited

1. Johnston, K. P.; Penninger, J. M. L. *American Chemical Society Symposium Series 406;* American Chemical Society; Washington DC, 1989.
2. Squires, T. G.; Paulaitis, M. E. *American Chemical Society Symposium Series 329;* American Chemical Society; Washington DC,1987.
3. Flarsheim, W. M.; Bard, A. J.; Johnston, K. P. *J. Phys. Chem.*1989, **93**, 4234.
4. Paulaitis, M. E.; Alexander, G. C. *Pure and Appl. Chem.*1987, 59, 61.
5. Townsend, S. H.; Abraham, M. A. ; Huppert, G. L ; Klein, M. T. ; Paspek, S. C. *Ind. Eng. Chem. Res.*1988, 27, 143 .
6. Subramaniam, B.; Mchugh, M. A. *Ind. Eng. Chem. Process Des. Dev.*1986, **25**, 1.
7. Petche, I. B.; Debenedetti, P. G. *J. Phys. Chem.* 1991, **95**, 386.
8. Peck, D. G.; Mehta, A. J.; Johnston, K. P. *J. Phys. Chem.*1989, **93**, 4297.
9. Saim, S.; Ginosar, D. M. ; Subramaniam, B. in *American Chemical Society Symposium Series 406;* Johnston, K. P. and Penninger, J. M. Ed.; American Chemical Society; Washington DC, 1989, 301.
10. Saim, S. ; Subramaniam, B. *Chem. Eng. Science* 1988, **43**, 1837 .
11. Debenedetti, P. G. ; Mohamed, R. S. *J. Chem. Phys.*1988, **90**, 4528.
12. Kim, S.; Johnston, K. P. *Ind. Eng. Chem. Res.* 1987, **26**, 1206.
13. Kroon, R.; Baggen, M.; Lagendijk, A. *J. Chem. Phys.* 1989, **91**, 74.
14. Lavorel, B.; Chaux, R.; Saint-Loup, R.; Berger, H. *Opt. Comm.* 1987, **62**, 25.
15. Zinn, A. S.; Schiferl, D.; Nicol, M. F. *J. Chem. Phys.* 1987,**87**, 1267.
16. Hacura, A.; Yoon, H. J., Baglin, F. G. *J. Raman Spectrosc.* 1987, **18**, 377.
17. Zerda, T. W.; Song, X. ; Jonas, J. *Appl. Spectrosc.*1986b, **40**, 1194 .
18. Wood, K. A.; Stauss, H. L. *J. Chem. Phys.* 1981, **74**, 6028.
19. Garrabos, Y.; Tufue, R.; Le Neindre, B.; Zalczer, G.; Beysens, D. *J. Chem. Phys.* 1980, **72**, 4637.
20. Wang, C. H.; Wright, R. B. *J. Chem. Phys.* 1974, **59**, 1706.
21. Wright, R. B.; Wang, C. H. *J. Chem. Phys.* 1973, **58**, 2893.
22. Wang, C. H.; Wright, R. B. *Chem. Phys. Lett.* 1973, **23**, 241.
23. Schweizer, K. S.; Chandler, D. *J. Chem. Phys.*1982, **76**, 2296 .
24. Pratt, L. R.; Chandler, D. *J. Chem. Phys.* 1980, **72**, 4045 .
25. Zakin, M. R.; Herschbach, D. R. *J. Chem. Phys.*1986, **85**, 2376.
26. Zakin, M. R.; Herschbach, D. R. *J. Chem. Phys.*1988, **89**, 2380.
27. LeSar, R. *J. Chem. Phys.*1987, **86**, 4238.
28. Ben-Amotz, D.; Zakin,M. R.; King, H. E. Jr.; Herschbach, D. R. *J. Phys. Chem*1988, **92**, 1392.
29. Agnus, S; Armstrong, B.; K. M. de Reuck *International Thermodynamic Tables of the Fluid State - 5. Methane;* Pergamon Press; New York, 1976.
30. Agnus, S; Armstrong, B.; K. M. de Reuck *International Thermodynamic Tables of the Fluid State - 6. Nitrogen;* Pergamon Press; New York, 1979.
31. Landolt-Bornstein *Zahlenwerte und Funktionen, Band II, Teil 1;* Springer-Verlag; Berlin, 1971.
32. Pratt, R.; Hsu, C. S. ; Chandler, D. *J. Chem. Phys.*1978, **68**, 4202.

33. Mansoori, G. A.; Carnahan, N. F.; Starling, K. E.; Leland,T. W. Jr. *J. Chem. Phys.*1971, **54**, 1523.
34. Grundke, E. W. ; Henderson, D. *Mol. Phys* 1972, **24** .
35. A detailed description of the hard fluid model is in preparation by D. Ben-Amotz and D. R. Herschbach. The model is closely related to that described in references 23, 25, 28, and 32.
36. Herschbach, D. R.; Laurie, V. W. *J. Chem. Phys.*1961, **35**, 458 .
37. Ben-Amotz, D.; Herschbach, D. R. *J. Phys. Chem.* 1990, **94**, 1038.
38. Myers, A. B.; Markel, F. *Chem. Phys.* 1990, **149**, 21.
39. Sverdlov, L. M.; Kovner, M. A.; Krainov, E. P. *Vobrational Spectra of Polyatomic Molecules;* John Wiley and Sons; New York, 1974.
40. Bondi, A. *J. Phys. Chem.* 1964, **68**, 441.
41. Murphy, W. F.; Holzer, W.; Bernstein, H. J. *Appl. Spectrosc.* 1969, **23**, 211.
42. Shim, J.-J.; Johnston, K. P. *J. Phys. Chem.* 1991, **95**, 353.
43. Brennecke, F. F.; Tomasko, D. L.; Peshkin, J.; Eckert, C. A. *Ind. Eng. Chem. Res.* 1990, **29**, 1682.
44. Petsche, I. B.; Debenedetti, P. G. *J. Chem. Phys.* 1989, **91**, 7075.
45. Debenedetti, P. G. *Chem. Eng. Sci.*1987, **42**, 2203.

RECEIVED November 25, 1991

Chapter 3

Influence of Solvent–Solute and Solute–Solute Clustering on Chemical Reactions in Supercritical Fluids

J. R. Combes, K. P. Johnston, K. E. O'Shea, and M. A. Fox

Departments of Chemical Engineering and Chemistry, University of Texas, Austin, TX 78712

In a highly compressible supercritical fluid(SCF), local densities of the solvent and solute about a solute molecule are augmented over bulk values. The influence of this solvent-solute clustering on reactions is examined based on the photolysis of 1,3-diphenylacetone and a new interpretation of the photolysis of iodine. Together these studies indicate that solvent-solute clustering causes the solvent cage effect to be larger than expected based on bulk properties, but smaller than in liquid solvents. New experimental results indicate that the rate of cyclohexenone photodimerization and the regioselectivity to the more polar head-to-head versus the less polar head-to-tail dimer increase sharply as pressure is decreased to the critical point. This unusual result is attributed to solute-solute clustering, which increases the local polarity and the number of encounters between reacting species. Solute-solute clustering is shown to occur in a single phase region, where volume fluctuations are large, just prior to the onset of nucleation and growth of a condensed phase.

At an ACS symposium on supercritical fluids, Kim and Johnston (1) presented spectroscopic evidence that the local density of a supercritical fluid solvent about a solute is augmented in excess of the bulk density. This phenomenon, commonly called solvent-solute clustering is due to the large isothermal compressibility of a fluid, as described by Kirkwood-Buff solution theory (1-3). According to the theory of volume fluctuation thermodynamics (4), the phase stability of a compressible multicomponent mixture is always less than that of the corresponding incompressible solution. Through volume fluctuations, solute and solvent molecules are able to explore other regions in configuration space to form inhomogeneous regions, which can become phase separated.

0097–6156/92/0488–0031$06.00/0
© 1992 American Chemical Society

Brennecke et al. (5-6) proposed another type of local density augmentation, solute-solute clustering, based on a study of pyrene excimer formation. In a more detailed study of this system, Betts and Bright (7) did not observe evidence of this type of clustering, for reasons unknown to us. Possibly, the lack of clustering may be explained by the extremely small concentrations. Wu et al. (8) characterized solute-solute clustering with integral equation theories and speculated that these clusters can influence the macroscopic properties in SCF solution, e.g. dimerization reactions. In another study in this book, Chateauneuf et al. (9) propose that a third type of clustering occurs, co-solvent - solute clustering, based on the rate of hydrogen abstraction from excited state benzophenone to form a ketyl radical.

The primary objective of this work is to determine to what degree, if any, solute-solute clustering influences chemical reaction rate constants and selectivities in supercritical fluids. To put this objective in context, we begin with a summary of the effects of solvent-solute clustering on reactions, including a new interpretation of the classic study of the photolysis of iodine in the gaseous, supercritical fluid, and liquid states. Next, we describe how our results on the photodimerization of cyclohexenone in supercritical ethane suggest that solute-solute clustering is occurring. This type of clustering is examined as a function of pressure near the critical point in order to determine how it influences both the rate of dimerization and the product selectivity.

Transition State Theory for Reactions in Supercritical Fluids

Before discussing clustering, it is instructive to examine thermodynamic effects on reactions in supercritical fluids based on transition state theory. The activation volume is a quantity defined as the negative of the pressure derivative of the logarithm of the rate constant multiplied by RT. Another term must be added if the rate constant is in pressure dependent concentration units (10). Equivalently, according to transition state theory, the activation volume is equal to the difference in partial molar volumes between the transition state and the reactants. In a highly compressible supercritical fluid, several studies have shown that pressure can have a large effect on a reaction rate constant with pronounced activation volumes of thousands of mL/mol positive or negative, for example in the unimolecular decomposition of α-chlorobenzylmethyl ether (11). For dilute systems, the activation volume is a function of (1) the differences in the repulsive and attractive interactions (van der Waals forces) of the reactants and transition state with the solvent (described by $\partial P/\partial n_i$ factors), and (2) the isothermal compressibility of the pure solvent (11). As the compressibility becomes large it can magnify the influence of either repulsive (indicative of size changes in the repulsive parameters in the equation of state) or attractive interactions, leading to large pressure effects. Simply put, small changes in pressure cause large changes in density, and thus certain rate constants. If the activation volume of a reaction in an SCF is normalized by the fluid's isothermal compressibility, the resulting value is comparable to that in a liquid solvent, and may be used to help elucidate reaction mechanisms.

Influence of Solvent-solute Clustering on Reactions Due to Augmentation in Local Polarity

Although various reactions have been investigated in supercritical fluids, only a few studies have addressed the effect of solvent-solute clustering on a rate constant or equilibrium constant. Johnston and Haynes (11) found that this clustering increases the rate constant for the unimolecular decomposition of α-chlorobenzylmethyl ether because of the augmentation of the local density and local dielectric constant. For the reversible tautomerization between 2-hydroxypyridine and 2-pyridone in SCF 1,1-difluoroethane, the large volume change on reaction indicates that more solvent molecules are attracted to the more polar tautomer (12). Flarsheim et al. (13) compared the degree of clustering of supercritical water about I^- versus I_2 based on electrochemical measurements of the volume change on reaction as interpreted with a modified Born equation. The electrostriction about I^- (density augmentation) becomes pronounced towards the critical point.

Influence of Solvent-solute Clustering on Cage Effects in Photolytic Reactions

In this section, we describe how solvent-solute clustering influences the dynamics of the solvent cage effect for photodissociation reactions. This dynamic clustering effect is a totally different phenomena than the effect of clustering on local polarity described above. Since these photodissociation reactions proceed via free-radical mechanisms, solvent polarity effects are expected to be negligible. In these free radical dissociations, the diffusive separation of the geminate radical pair resulting from the bond scission is inhibited by recombination to starting material within the solvent cage. This cage effect is thought to play a role in the oxidation of organics in supercritical water (14), the cracking of hydrocarbons (15), and the free radical dissociation of benzyl phenyl ether in supercritical toluene (16).

We begin with a brief summary of a recent study of the photolysis of dibenzylketones in carbon dioxide and ethane (17). As DBK photodissociates via a two bond scission, it does not exhibit cage effects in liquid solvents (18). An organized molecular assembly such as a micelle is required for a cage effect. Thus, this reaction may be used to determine if a cluster in a supercritical fluid is considerably more rigid than a cage in a liquid solvent. Next, we revisit the classic study of the photolytic cage effect for iodine in compressed fluids (19). The objective is to give an alternative interpretation of solvent effects in the critical region in light of the recent characterization of solvent-solute clustering. In contrast to DBK photolysis, cage effects are prevalent for the photolysis of iodine in SCF ethane, and we will present a model to describe the cage effect with and without solvent-solute clustering.

Photolysis of Dibenzylketones. Despite the clear evidence for the existence of solvent-solute clusters, their rigidity and lifetime are still unclear. One means to investigate the structural integrity of a cluster is to probe the existence of cage effects in a SCF. The free radical dissociation of an unsymmetrically substituted dibenzyl ketone, Figure 1, is a classical photoactivated system known to produce observable cage effects in an organized molecular assembly such as a micelle, but not in liquid solvents such as alkanes (18, 20). If the solvent cage is capable of maintaining structural integrity even after decarbonylation, then the concentration of the asymmetric photoproduct (AB) should be well above 50%. The percentage of the AB photoproduct ranges from 49.3 to 50.7 which is within the experimental error of the distribution expected in the absence of cage effects, see Figure 2. The statistical distribution of photoproducts indicates random radical recombination characteristic of the absence of a cage effect, even in the highly compressible near-critical region where clustering is believed to be maximum. Hence, solvent-solute clusters do not present any unusually rigid cage effects to this photolysis compared with liquid solvents.

Photolysis of iodine. Unlike DBK, molecular iodine photodissociates via a one bond scission. Also the wavelength of the laser pulse used by Otto et al. (19) results in a relatively mild, low energy separation of the photogenerated radicals. Therefore I_2 photodissociation should be a more sensitive probe for the characterization of solvent cage effects.

In the following, we present a new interpretation of a study of the laser pulse photolysis of molecular iodine (19), to consider the effect of solvent-solute clustering on the cage effect in the near critical region. A simplified mechanism in various gases and fluids may be expressed in terms of three elementary steps. Although the detailed mechanism is considerably more complicated (21), this crude mechanism takes into account the key features of the solvent cage effect. An extensive study is in progress to consider the effects of supercritical fluids on the complex processes which occur in the excited state (22).

The initiation step is laser photoexcitation of the molecule to yield a caged geminate pair of atomic iodine radicals. The two competing steps open to the caged pair are: 1) separative diffusion to yield atomic iodine radicals and 2) recombination to give molecular iodine.

$$I_2 + h\nu \longrightarrow \left[I\cdot\cdot I\right]^* \tag{1}$$

$$\left[I\cdot\cdot I\right]^* \xrightarrow{\ k_{-1}\ } I_2 \tag{2}$$

$$\left[I\cdot\cdot I\right]^* \xrightarrow{\ k_d\ } 2I\cdot \tag{3}$$

The quantum yield of iodine radicals produced from the laser pulse, Φ, is

$$\Phi = \frac{k_d}{k_d + k_{-1}} \tag{4}$$

The rate of diffusive separation, k_d, was determined from separate experimental measurements of iodine radical diffusion rates in the high pressure diffusion limited regime (19). The rate of excited state deactivation, k_{-1}, was calculated from the measured quantum yields at high densities where $\Phi = k_d/k_{-1}$ (18). It was assumed that k_{-1} is proportional to the inverse diffusion coefficient, D^{-1} (19, 23) as both properties are related to the collision frequency.

Although the proposed mechanism is consistent for photolysis of iodine in helium, nitrogen and methane (24), substantive deviations were present at low densities and especially near the critical point of ethane. As Figure 3 shows, the quantum yields at these low densities are consistently below one, the value expected in this high diffusivity regime where $k_d \gg k_{-1}$.

These deviations were attributed to chemical complexation between the iodine and ethane in the ground state, so that after excitation, the excited species fully relaxes (19, 25). It is suggested that only the free I_2 has the potential to dissociate, hence the reduction in Φ is described by the following mechanism

$$I_2 + nM \;\rightleftharpoons\; I_2M_n \tag{5}$$

$$K_c = \frac{[I_2M_n]}{[I_2][M]^n} \tag{6}$$

$$I_2M_n + h\nu \;\longrightarrow\; \left(I_2M_n\right)^* \;\longrightarrow\; I_2 + nM \tag{7}$$

where M denotes a solvent molecule and n is a coordination number. We combine steps shown in Equations 1 through 3 with Equations 5-7 to yield the following expression:

$$\Phi = \frac{1}{1 + K_c[M]^n} \left\{ \frac{k_d}{k_d + k_{-1}} \right\} \tag{8}$$

At low densities $k_d \gg k_{-1}$, and the bracketed term is unity. The value of n, which was regressed in the density range of 0 to 10^{-4} mol/cm^3, is 0.38.

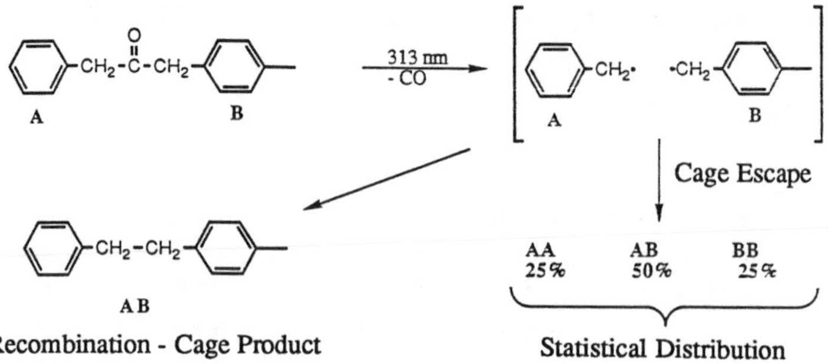

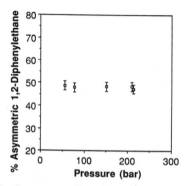

Figure 1. Mechanism of the cage effect for the photolysis of 1-(4-methylphenyl)-3-phenyl-2-propanone.

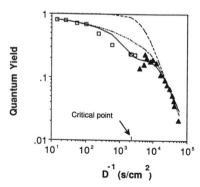

Figure 2. Fraction (%) of cross-coupling photoproduct AB obtained by photolysis of 1-(4-methylphenyl)-3-phenyl-2-propanone in supercritical ethane as a function of pressure. T=35°C, 39mM reactant, 2.5 min exposure.

Figure 3. Quantum yield for the laser pulse photolysis of iodine in ethane at gaseous, supercritical fluid, and liquid densities.
  ----- model of Otto et al. (1984).
  --·-- model including correction for chemical complexation.
  —— model including chemical complexation and clustering correction.
    □ - Gas phase data of Otto et al. (1984) T=41°C.
    ▲ - Liquid phase data of Otto et al. (1984) T=25°C.

The model in Equation (8) does not fit the data over all densities. At moderate densities where the fluid is highly compressible, there are still large deviations between theory and experiment as shown in Figure 3. These deviations appear to be largest near the critical point where the compressibility is largest and diminish with either an increase or decrease in density. We propose that these deviations are due to solvent-solute clustering. This type of clustering is characterized by physical local density augmentation, and is thus different than the stoichiometric chemical complexation just described.

The same type of behavior has been observed for deviations between experimentally measured solvatochromic shifts and values calculated with an Onsager reaction field for a continuous fluid as described in a recent review (3). These deviations are attributed to augmentations in the local density (solvent-solute clusters) and tend to be largest in the near-critical region where the compressibility is large. Examples include shifts in the absorption maximum for phenol blue in ethylene (1), of 2-nitroanisole in CO_2 (1, 26), shifts in the emission maximum for (N,N-dimethylamino)benzonitrile in CHF_3 (27) and changes in the ratio of fluorescence intensity for two pyrene emission peaks (5, 6). These studies indicate that solvent-solute clustering is related to the isothermal compressibility. The shape of the isothermal compressibility versus density may be approximated with a simple Gaussian function. Therefore, we assume that the local density augmentation of solvent about solute also may be approximated with a Gaussian function, although this assumption is somewhat over-simplified based on a recent study (28).

The local density augmentation caused by the large isothermal compressibility of the fluid may conceivably influence k_{-1} or k_d. We assume that the lifetime of the clusters is extremely short and thus there is no effect on k_d, based on the molecular dynamics study of Petsche and Debenedetti (29) and experimental measurements of binary diffusion coefficients near the critical point. It seems more likely that a higher local density would affect k_{-1} due to an increase in the number of collisions which may lead to vibrational deactivation of the radical pair, $[I \cdot\cdot I]^*$ The relationship between k_{-1} and density has already been established in regions where local density augmentation is minimal (19). Therefore it is reasonable to assume a Gaussian clustering correction to k_{-1} of the following form:

$$k_{-1} = (k_{-1})_0 \frac{D^{-1}}{(D^{-1})_0} + A \exp\left\{ \frac{-(\rho - \rho_c)^2}{\sigma^2} \right\} \tag{9}$$

where A and σ are adjustable parameters in the Gaussian expression. The value of A is dependent on the size of the cluster (degree of local density augmentation) for a particular system. The variance, σ, determines the breadth of densities over which the clustering is significant. It should increase as temperature increases from the critical point based on the behavior of the isothermal compressibility. Although these parameters were adjusted to fit Φ, it should be possible to determine how the parameters vary based on solvatochromic and other spectroscopic measurements, although it is not yet known to what degree the clustering phenomena varies from system to system.

The final expression for the quantum yield, which includes the effect of clustering on k_{-1} is

$$\Phi = \frac{1}{1 + K_c[M]^n} \left\{ \frac{k_d}{k_d + (k_{-1})_0 \dfrac{D^{-1}}{(D^{-1})_0} + A \exp\left\{ \dfrac{-(\rho - \rho_c)^2}{\sigma^2} \right\}} \right\} \qquad (10)$$

As shown in Figure 3, the data are correlated much more accurately with this model than the basic model in Equation 4. If a system could be found where Φ is unity in the low pressure limit ($K_c = 0$), it would be possible to focus more clearly on the solvent-solute clustering effect . However, a comparison between the curves with and without the Gaussian clustering correction suggests that there is a strong clustering effect. The clustering causes the solvent cage effect to be larger than expected based on the bulk density. This excess cage effect should be even larger for more polarizable solutes, in which the degree of clustering is more significant. Although the data are presented over an extremely wide density range from the dilute gas to dense liquid state, relatively few data were obtained in the near supercritical region. Additional study in this region is warranted, both experimentally and theoretically.

Effect of Solute-solute Clustering on the Photodimerization of Cyclohexenone

The purpose of this section is to demonstrate the reaction rate constant and selectivity of a bimolecular reaction may be influenced by solute-solute clustering as the pressure is lowered towards the critical point. The solvent sensitive photodimerization of 2-cyclohexen-1-one is an ideal candidate as both the rate of dimerization and the product selectivity may be monitored.

The [2+2] cycloaddition of 2-cyclohexen-1-one is believed to proceed according to the mechanism in Figure 4. The cyclohexenone is excited to the singlet and undergoes intersystem crossing rapidly to the triplet. The triplet may be deactivated by the solvent with a rate constant k_2, or react with ground state cyclohexenone to form the dimer, with a rate constant of k_3. It is straightforward to show that the quantum yield for this mechanism, assuming no exciplex formation (30), is

$$\Phi = \Phi_{isc}\left[\frac{k_3A}{k_3A + k_2}\right] \qquad (11)$$

The relative linearity of a Stern-Volmer plot of $1/\phi$ versus $1/[A]$ for the photoreaction of cyclohexenone supports this mechanism and suggests that only one excited state is involved (30). The [2+2] addition of the triplet to a ground state molecule is not concerted, and there is a large solvent polarity effect on the regioselectivity for the head-to-head (HH) versus head-to-tail (HT) photodimers, with the more polar head-to-head isomer being favored in polar media. The polarity effect is attributed to the large difference in the dipole moments of the transition states leading to the products.

Experimental. All photodimerizations were carried out in a stainless steel fixed volume cell (1.75 cm ID with a 1.0 cm path length) with sapphire windows under the irradiation of a Hanovia medium pressure mercury lamp filtered through water and Pyrex for a 13.5 hour exposure. The cell and lamp assembly have been described previously (31). For selected runs a custom built 0.9 mL variable-volume pump was connected to the cell and the pressure was varied to determine the exact location of the phase boundary, based on light scattering measured in a Cary 2290 UV-Vis spectrophotometer (Varian Inst.). The spectrophotometer was also used to measure the concentrations of the monomeric cyclohexenone before and after reaction.

The 2-cyclohexen-1-one (Aldrich, 97%) was used as received. It was loaded into the cell to yield a reaction concentration of 29 mM, which was verified spectrophotometrically. Ethane (CP grade, Big Three Gases) was passed through an in-line activated carbon filter and an Oxy-Trap (Alltech Associates Inc.) prior to use to remove impurities and oxygen, respectively.

Temperature control was achieved through the use of two cartridge heaters (Gaumer Inc., 1 inch, 80W) connected to an Omega RTD temperature controller to maintain a constant value of $35 \pm 0.1°C$. A digital pressure transducer (Autoclave $\pm$ 20 psi) was used to record pressure during filling and a syringe pump with a vernier scale (60 ml, High Pressure Equipment Inc.) was used to load the solvent at a known density.

After exposure the cell was vented through 10 ml of hexane. Following depressurization the cell was rinsed several times with the same 10 ml solution. A GC analysis (HP 5890A gas chromatograph with a 25m polydimethylsiloxane capillary column, FID detector) of the hexane extract solution revealed the product isomer distribution.

In order to ensure that cyclohexenone did not undergo condensation on the walls of the cell, the phase behavior of 0.3 mole % cyclohexenone in ethane was studied using a variable-volume view cell. Further verification of a single fluid phase was provided by attaching the 0-0.9 ml variable-volume pump to the cell and spectroscopically detecting a dew point at a density below that at which the experiment was performed. At the dew point, the absorbance increased strongly due to light scattering. This procedure was repeated before and after exposure for several of the low density near-critical experiments and selected high density experiments as well. In each instance the dew point was detected at lower densities than those employed in the experiment.

Quantum yield. The relative quantum yield for the photoaddition of cyclohexenone is shown in Figure 5. It is given in terms of measured cyclohexenone concentrations by the relationship

$$\Phi/\Phi_0 = \frac{\dfrac{\left([A]_{initial} - [A]\right)}{[A]_{initial}}}{\dfrac{\left([A_o]_{initial} - [A_o]\right)}{[A_o]_{initial}}} \tag{12}$$

where the subscript o refers to a reference pressure, in this case 55 bar. All experimental runs are of the same exposure and contain equivalent concentrations of the enone.

As pressure is lowered towards the critical pressure, there is an abrupt increase in the the quantum yield of dimerization, by a factor of 10. There are two primary possibilities for this unusual result. It may be caused by less deactivation of the triplet due to fewer collisions with the solvent, that is a decrease in k_2. Alternatively, it may be caused by an increase in k_3A resulting from a larger number of encounters between A and $^3A^*$ facilitated by solute-solute clustering. According to the theoretical study by Wu et al. (8), solute-solute clustering is minimal at higher pressures, but increases sharply as a function of the compressibility near the critical pressure.   To distinguish between the effects of k_2 and k_3 on Φ, additional information is required, and fortunately for this reaction, the regioselectivity provides a useful probe.

Regioselectivity. The regioselectivity of HH to HT cyclohexenone dimers increases with solvent polarity in a series of liquid solvents, Figure 6. On the basis of the slope of this plot, it may be anticipated that the regioselectivity would increase in SCF ethane only slightly, approximately 20%. In this region the dielectric constant of the SCF ethane solvent increases from 1.3 to 1.6, a relatively small change in solvent polarity. Interestingly, the regioselectivity ratio actually decreases with increasing solvent density and dielectric constant, a counterintuitive result based on the results in liquid solvents. The regioselectivity changes little from 250 bar to 65 bar, but then increases sharply at lower pressures as shown in detail in Figure 7.

It seems unlikely that deactivation of the HT transition state would be much different than that of the HH isomer. Clearly, the sharp change in regioselectivity must be due to factors other than changes in solvent deactivation. Because the sizes of the regioisomers are similar, this effect can not be attributed to the repulsive contribution to the equation of state in transition state theory. However, in this near-critical region, large solute-solute fluctuations (i.e. solute-solute clusters) must

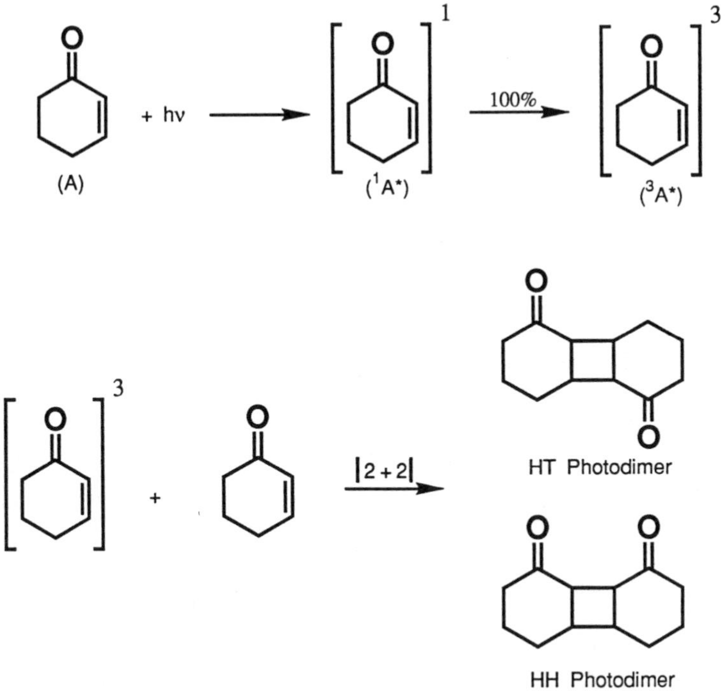

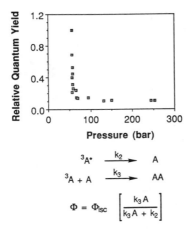

Figure 4. Photochemistry of 2-cyclohexen-1-one (A)

$$^3A^* \xrightarrow{k_2} A$$

$$^3A + A \xrightarrow{k_3} AA$$

$$\Phi = \Phi_{isc} \left[\frac{k_3 A}{k_3 A + k_2} \right]$$

Figure 5. Relative quantum yields for cyclohexenone dimerization in SCF ethane. T=35°C, 29 mM cyclohexenone.

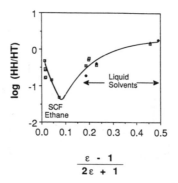

$$\frac{\varepsilon - 1}{2\varepsilon + 1}$$

Figure 6. Regioselectivity for cyclohexenone photodimers.
♦   - Data from Lam et al., 1967.
⊡   - This work.

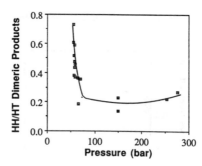

Figure 7. Regioselectivity for cyclohexenone photodimers versus pressure in SCF ethane. T=35°C, 29 mM cyclohexenone.

occur. These concentration fluctuations increase the number of solute molecules in the environment of the excited triplet. As the microscopic environment surrounding the solute molecules becomes more concentrated with other solute molecules (relative to the bulk concentration) the local dielectric constant increases leading to an increase in regioselectivity. In the extreme limit of neat cyclohexenone, the HH to HT ratio is 1.0 (30), consistent with the dielectric constant of cyclohexenone.

Additional evidence suggests that small clusters of solute molecules may be formed at the lowest pressures studied in Figure 7. The onset of condensation was studied in both the fixed volume cell in which we conducted the reactions (with the 0.9 ml variable volume pump attached) and in a variable volume view cell. The dew and bubble points were observed visually with the variable volume view cell as shown in Figure 8. At 35°C, notice that the dew point (53 bar) is well below the minimum pressure of the photodimerization experiments, (55 bar). We were careful in verifying that the data in Figure 7 were still in the one-phase region as discussed above. Since small decreases in pressure cause large decreases in the solvent density and solute solubility, the regioselectivity and the phase behavior data suggest that it is likely that small solute clusters containing a few molecules are present in this region, just prior to nucleation and growth of a new condensed phase.

In order to verify that higher cyclohexenone concentrations increase the HH/HT ratio in SCF solvents as it does in liquids, the photodimerization was performed in high density SCF ethane at different concentrations. In this regime (P = 124 bar and T = 35°C) clustering is presumably minimal. The HH/HT ratio increases with monomer concentration, which is consistent with the trend observed in liquid solvents (30), see Table I. As bulk solute concentration increases by a factor of 3, the increase in HH/HT is relatively small compared with the HH/HT increases portrayed in Figure 7. This comparison suggests that the local solute concentrations are increased dramatically, perhaps by even an order of magnitude or more, in solute-solute clusters.

Table I. Regioselectivity of 2-cyclohexen-1-one dimers as a function of concentration. T=35°C, P=124 bar

Reactant Concentration (mM)	Head-to-head/Head-to-tail Ratio
30	0.120
50	0.136
100	0.259

The unusual increase in regioselectivity with a decrease in pressure is not present for the photodimerization of a similar enone, isophorone, in SCF CO_2 (31). The difference for the two enones is a result of differences in their phase diagrams and differences in the phase region in which the experiments were performed. As shown in Figure 8, the critical point for the cyclohexenone-ethane mixture is close to that of pure ethane. Hence the compressibility is large. In contrast, the isophorone-CO_2 system exhibits a dew point curve which merges with a LL immiscibility curve far from the critical point of pure CO_2 (31). Consequently the large compressibility necessary at the mixture critical point for solute-solute clustering to occur is not accessible. Future work will compare the photodimerizations of both enones in SCF CHF_3, as well as the phase behavior.

Stereoselectivity. The SCF can influence the stereoselectivity as well as the regioselectivity of a cyclic enone dimerization. There are both syn and anti stereoisomers of the head-to-tail dimers of both isophorone and cyclohexenone. The differences in alignment of the cyclohexyl rings was portrayed schematically by Hrnjez et al. (31). Both HT photodimers for isophorone have similar dipole moments. Hence, a variation in solvent polarity is not expected to influence the anti/syn ratio and, indeed, these stereoisomers are formed in equal amounts regardless of the dielectric constant in liquid solvents (32). For the cyclohexenone HT stereoisomers, the anti configuration dominates in acetonitrile and benzene, but there has been limited study of stereoselectivity in other liquids.

It is therefore interesting to note that the ratio of HT anti/HT syn for the HT cyclohexenone photodimers increases with solvent density, see Figure 9. An identical trend in stereoselectivity for the HT anti/syn ratio was observed for the HT photodimers of isophorone (31). Two explanations are offered for this observation. The transition state leading to the HT syn configuration requires greater "desolvation" of nearest neighbor solvent molecules. Also, it has a smaller solvent accessible area and thus occupies more volume in solution. This desolvation is understandably inhibited at higher fluid densities, hence the increased selectivity of the HT anti stereoconfiguration.

Conclusions

In the last few years, it has been shown that solvent-solute clustering in a SCF influences reactions due to local polarity effects. The new model for the photolysis of iodine indicates that solvent-solute clustering can influence a dynamic property, the solvent cage effect. The cage effect is stronger than expected based on the bulk properties, but not unusually strong compared with liquid solvents, as evidenced by the lack of a cage effect for the photolysis of dibenzylketones. The regioselectivity and the quantum yield for the photoaddition of cyclohexenone provide experimental evidence for the first time that solute-solute clustering can increase reaction rate constants and selectivities near the critical point. The regioselectivity and phase behavior results suggest that solute-solute clustering is a precursor to nucleation and growth of a new condensed phase. The formation of inhomogeneous regions enriched in solute, in the single phase region, is to be expected in a highly compressible fluid, based on the theory of volume fluctuation thermodynamics (4). Although this study provides evidence that solvent-solute and solute-solute clustering influence chemical reactions, further studies, e.g. time resolved studies,

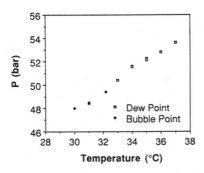

Figure 8. Phase behavior for 0.3 mole % cyclohexenone in ethane.

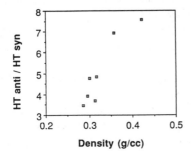

Figure 9. Stereoselectivity for the head-to-tail photodimers of cyclohexenone in ethane. T=35°C, 29 mM cyclohexenone.

are required to determine the generality of this effect and its relevance to practical applications.

Acknowledgment

Acknowledgement is made to the Separations Research Program at the University of Texas, the State of Texas Energy Research in Applications Program, and the Camille and Henry Dreyfus Foundation for a Teacher-Scholar Grant (to KPJ).

Literature Cited

1.   Kim, S.; Johnston, K.P. In *Supercritical Fluids* **1987**, Squires, T.G.; Paulaitis, M.E., Eds.; ACS Symposium Series 329; Washington, D.C.; p. 42.
2.   Debenedetti, P.G. *Chem. Eng. Sci.* **1987**, *42*, 2203.
3.   Johnston, K.P.; Kim, S.; Combes, J. In *Supercritical Fluid Science and Technology*, **1989**, Johnston, K. P.; Penninger, M. L., Eds.; ACS Symposium Series 406; Washington, D.C., p. 52.
4.   Sanchez, I. *Macromolecules*, **1991**, 24, 908.
5.   Brennecke, J.F.; Eckert, C.A. In *Supercritical Fluid Science and Technology* **1989**, Johnston, K.P.; Penninger, M.L., Eds.; ACS Symposium Series 406; Washington, D.C., p. 14.
6.   Brennecke, J.F.; Tomasko, D.L.; Peshkin, J.; Eckert, C.A. *Ind. Eng. Chem. Res.* **1990**, *29*, 1682.
7.   Betts, T.A.; Bright, F.V. *ACS Symposium Series* **1991**, this issue.
8.   Wu, R.-S.; Lee, L.L.; Cochran, H.D. *Ind. Eng. Chem. Res.* **1990**, 29, 977.
9.   Chateauneuf, J.E.; Roberts, C.B.; Brennecke, J.F. *ACS Symposium Series* **1991**, this issue.
10.   Eckert, C.A. *Ann. Rev. Phys. Chem* **1972**, *23*, 239.
11.   Johnston, K.P.; Haynes, C. *AIChE J.* **1987**, *33*, 2017.
12.   Peck, D.G.; Mehta, A.J.; Johnston, K.P. *J. Phys. Chem.* **1989**, *93*, 4297.
13.   Flarsheim, W.M.; Bard, A.J.; Johnston, K.P. *J.Phys. Chem.* **1989**, *93*, 4234.
14.   Yang, H.H.; Eckert, C.A. *Ind. Eng. Chem. Res.* **1988**, *27*, 2009.
15.   Paspek, S.C. "Conversion of High Boiling Liquid Organic Materials to Lower Boiling Materials," **1989**, U.S. Patent Number 4,840,725.
16.   Wu, B.C.; Klein, M.T.; Sandler, S.I. *The Influence of Diffusion on Reactions in Supercritical Fluid Solvents*, Paper presented at AIChE National Meeting, Orlando, Fla.; March 18-21, **1990**.
17.   O'Shea, K.E.; Combes, J.R.; Fox, M.A.; Johnston, K.P. *Photochem. and Photobiol* **1991**, 54, 571.
18.   Robbins, W .K.; Eastman, R.H. *J. Amer. Chem. Soc.* **1970**, *92*, 6077.

19. Otto, B.; Schroeder, J.; Troe, J. *J. Chem. Phys.* **1984**, *81*, 202.
20. Turro, N.J.; Cox, G.S.; Paczkowski, M.A. *Pure Appl. Chem.* **1986**, *58*, 1219.
21. Harris, A.L.; Brown, J.K.; Harris, C.B. *Ann. Rev. Phys. Chem.* **1988**, *39*, 341.
22. Knopf, F.C.; Xu, X.; Yu, S.-C.; Lingle, R.; Hopkins, J.B. *Second Int. Symp. on Supercritical Fluids* **1991**, Boston, p. 154.
23. Chatelet, M.; Kieffer, J.; Oksengorn, B. *Chem. Phys.* **1983**, *79*, 413.
24. Hippler, H.; Luther, K.; Maier, M.; Schroeder, J.; Troe, J. In *Laser induced Processes in Molecules*; Kompa, K.L., Smith, S.D., Eds. **1979**; p. 286.
25. Troe, J. *J. Phys. Chem.* **1986**, *90*, 357.
26. Yonker, C.R.; Smith, R.D. *J. Phys. Chem.* **1988**, *92*, 235.
27. Kajimoto, O.; Futakami, M,; Kobayashi, T.; Yamasaki, K. *J. Phys. Chem.* **1988**, *92*, 1347.
28. Sun, Y.-P.; Fox, M.A.; Johnston, K.P. *J. Am. Chem. Soc.* **1991**, in press.
29. Petsche, I. B.; Debenedetti, P.G. *Time Dependent Behavior of Near-critical Mixtures: A Molecular Dynamics Investigation*, Paper presented at the AIChE Annual Meeting, San Francisco, Ca.; November, **1989**.
30. Lam, E.Y.Y.; Valentine, D.; Hammond, G.S. *J. Amer. Chem. Soc.* **1967**, *89*, 3482.
31. Hrnjez,B.J.; Mehta, A.J.; Fox, M.A.; Johnston, K.P. *J. Amer. Chem. Soc.* **1989**, *111*, 2662.
32. Chapman, O.L.; Nelson, P.J.; King, R.W.; Trecker, D.J.; Griswold, A.A. *Record Chem. Progr.* **1967**, *28*, 167.

RECEIVED December 6, 1991

Chapter 4

Elucidation of Solute–Fluid Interactions in Supercritical CF$_3$H by Steady-State and Time-Resolved Fluorescence Spectroscopy

Thomas A. Betts, JoAnn Zagrobelny, and Frank V. Bright

Department of Chemistry, Acheson Hall, State University of New York at Buffalo, Buffalo, NY 14214

Steady-state and multifrequency phase and modulation fluorescence spectroscopy are used to study the photophysics of a polar, environmentally-sensitive fluorescent probe in near- and supercritical CF$_3$H. The results show strong evidence for local density augmentation and for a distribution of cluster sizes. These results represent the first evidence for lifetime distributions in a "pure" solvent system.

Much progress in supercritical fluid science and technology has occurred during the past decade (1-6). Many experimental (7-17) and theoretical advances (18-28) have helped to improve our understanding of solute-fluid interactions in supercritical fluids. However, progress is impeded somewhat because certain chemical information is lacking. Specifically, we lack a detailed understanding of many of the molecular processes occurring in the local solvation shells about the solute (i.e., the cybotactic region). To address this issue, our research group has set its sights on improving our understanding of the kinetics and mechanisms of solvation is supercritical media (29,30). From this new information, we believe chemists and engineers may begin to develop less empirical correlations and thus more accurate theories to predict fluid phase equilibria.

This paper reports on our most recent investigations of solvation processes in a pure, polar supercritical fluid, fluoroform (CF$_3$H). Fluoroform was chosen because it has a permanent dipole moment (1.60 D), easily accessible critical parameters (T$_c$ = 26.1 °C; P$_c$ = 48.6 bar), and dielectric properties are easily adjusted over a broad range with pressure. In this work, we employ the fluorescent probe PRODAN (Figure 1). PRODAN has been used previously to study dynamics in proteins (31) and normal liquids (32). PRODAN was also chosen because it is: 1) available in high purity, 2) effectively excited with our present laser-based systems, 3) strongly fluorescent in CF$_3$H, and 4) extremely sensitive to the local physicochemical properties of the solvent (e.g., refractive index and dielectric constant). Thus, as the properties of the fluid vary, PRODAN can be used to directly monitor and quantify the change.

0097–6156/92/0488–0048$06.00/0

Experimental. PRODAN was purchased from Molecular Probes and the purity checked by reverse phase HPLC. There were no detectable impurities. Stock solutions (1 mM) were prepared in absolute ethanol and stored in the freezer. CF$_3$H was purchased from Matheson and passed through a single adsorptive O$_2$ trap (Matheson) prior to entering the pumping system. According to the manufacturer this gives an O$_2$ level < 5 ppm.

The high-pressure cells and temperature control units are similar to the ones described by Betts and Bright (29). Samples for analysis were prepared by directly pipetting the appropriate amount of stock solution into the cell. To remove residual alcohol solvent, the optical cell was placed in a heated oven (60 °C) for several hrs. The cell was then removed from the oven, connected to the high-pressure pumping system (29), and a vacuum (50 μm Hg) maintained on the entire system for 10-15 minutes. The system was then charged with CF$_3$H and pressurized to the desired value with the pump (Isco, model SFC-500). Typically, we performed experiments at 10 μM PRODAN and there was no evidence for primary or secondary interfilter effects. HPLC analysis of PRODAN subjected to supercritical solvents showed no evidence of decomposition or additional components.

All steady-state absorbance measurements were made using a Milton-Roy model 1201 UV-Vis spectrophotometer. The majority of the steady-state and dynamic fluorescence measurements were performed using a modified multifrequency phase-modulation fluorometer (29) with parallel acquisition capabilities (33). One set of experiments was also carried out on a time-correlated single-photon counting instrument (see reference 32 for description). The results recovered by our frequency-domain instrumentation were extremely similar and indicate lack of artifacts or anomalies. For all experiments reported here, the excitation source was an argon-ion laser (351.1 nm; Coherent, model Innova 90-6) and emission was monitored using either long-pass or an ensemble of band-pass filters (Oriel). For all dynamic measurements, magic angle polarization was used to eliminate any polarization biases (34).

Dynamic data analysis was performed using a commercially available global analysis software package (Globals Unlimited). The details of this software and its features can be found elsewhere (35-37). Our main criteria for choosing one kinetic model over another are based on chi-squared (χ^2), randomness of the residual terms, and physical significance of the model. More details on these criteria can be found in recent review articles (38-40). Additional software for determining the emission and absorbance center of gravity, etc. were developed in house and written in BASIC.

The density- and temperature-dependent dielectric constant and refractive index for CF$_3$H were calculated using the Debye equation and the molar refractivity, respectively (41). The values obtained from this approach agree to within 5% with those determined by interpolation of existing data (42).

Results and Discussion

Steady-State Fluorescence. As mentioned in the Introduction section PRODAN is an environmentally-sensitive fluorescent probe. To illustrate this point, Figure 2

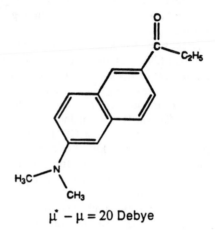

$\mu^* - \mu = 20$ Debye

Figure 1. PRODAN structure.

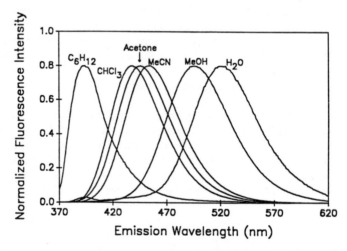

Figure 2. Normalized steady-state emission spectra for 10 μM PRODAN in normal liquids. $\lambda_{ex} = 351$ nm.

shows the normalized steady-state emission spectra for PRODAN in a series of liquid solvents at 25 °C. Clearly, as the polarity of the liquid increases the spectrum red shifts appreciably.

Figure 3 shows the effect of CO_2, N_2O, and CF_3H fluid density on the emission center of gravity for PRODAN. Several interesting trends are apparent in these data. First, for N_2O there is little influence of density on the emission spectra. (Experiments below a reduced density of 1.0 are difficult with N_2O because of background from the N_2O itself.) Second, there is a significant change in emission center of gravity with CO_2 density. We have not looked at this particular system in detail, but conclude that there is some degree of solute-solute or solute-fluid interaction (in CO_2) near a reduced density (ρ_r) of unity. Third, in CF_3H there is a significant shift in the PRODAN emission center of gravity with density. This is not too unexpected considering that CF_3H possesses a significant permanent dipole moment and its dielectric properties vary appreciably with density (42). This density-dependent spectral shift is a result of the significant increase in dielectric constant (ϵ) and the refractive index (n) of the medium. In CO_2 and N_2O these changes are modest (above $\rho_r = 1$) compared to CF_3H.

In order to investigate the observed spectral shifts in more detail, it is important to recognize that the solvent and its physicochemical properties influence the energy difference between the ground and excited states (43). These effects are most often described using the Lippert formalism (43):

$$\overline{v}_a - \overline{v}_f \approx \frac{2}{hc}\left(\frac{\epsilon - 1}{2\epsilon + 1} - \frac{n^2 - 1}{2n^2 + 1}\right)\frac{(\mu^* - \mu)^2}{a^3}, \qquad (1)$$

where h is Planck's constant, c is the speed of light, a is the radius of the cavity occupied by the fluorophore, v_a and v_f are the centers of gravity (in cm⁻¹) for the absorbance and fluorescence, respectively, and the term ($\mu^* - \mu$) is the difference between the excited- and ground-state dipole moments. The bracketed term that depends on the dielectric constant and refractive index is often referred to as the orientational polarizability. For PRODAN the cavity radius is about 4.2 Å and the dipole change is near 20 debye (31).

Figure 4 shows a plot of $v_a - v_f$ as a function of the orientational polarizability for PRODAN in supercritical CF_3H (points). In addition to this data, we show also the theoretical results expected, based on solvent dielectric constant, refractive indices, dipole change and cavity radius, for PRODAN in normal liquid solvents. Clearly, there are significant deviations at low density, but at higher density the two data sets begin to track one another. In addition it appears that the observed shift seen at low density is much greater than predicted by theory. That is, the local density about the probe is far greater that the bulk density of the fluid. This result is shown most clearly in Figure 5 and is a result of CF_3H molecules preferentially associating about the PRODAN in the highly compressible region near the critical point.

The spectral width of the fluorescence contour also gives insight into the immediate environment about the probe. Figure 6 shows that the spectral full-width-at-half maximum (FWHM) is initially wide at low density, goes through intermediate values as density is increased, and levels off at higher densities. These results are

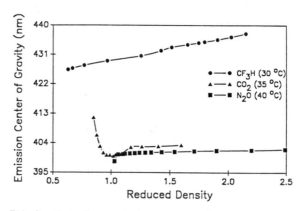

Figure 3. Density-dependent emission centers of gravity (in nm) for 10 μM PRODAN in CO_2, N_2O, and CF_3H.

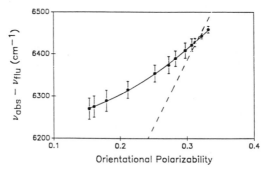

Figure 4. Lippert plot for 10 μM PRODAN in CF_3H. The solid points are the experimental data. The dashed line represents the theoretical prediction for PRODAN.

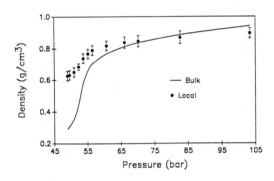

Figure 5. Local (points) and bulk fluid density versus pressure.

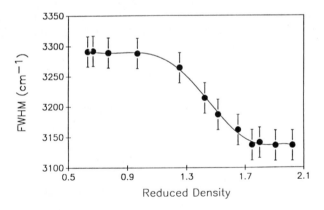

Figure 6. Influence of density on the spectral full-width-at-half maximum (FWHM) for 10 μM PRODAN in supercritical CF_3H.

consistent with PRODAN being in an ensemble of domains initially (at low density), the number of domains diminishing at intermediate densities, and the local environment about the PRODAN eventually coalescing into a more homogeneous environment at higher densities.

In an effort to learn more about probe-fluid interactions we investigated the effects of temperature on the PRODAN emission contour. Figure 7 presents the steady-state emission spectra for PRODAN in CF_3H as a function of temperature. These spectra are interesting because as one increases temperature one observes a systematic blue shift. In normal liquids, at mild temperatures (< 200 °C), one typically sees the opposite trend. At fairly high temperatures there is often a reversal in peak emission and spectra often begin to blue shift at higher temperatures. The explanation for this effect is that the thermal energy input into the system is large enough such that it precludes the solvent molecules from being able to reorganize about the solute during the excited-state lifetime (44,45). We propose the same mechanism here and attribute the effect to the much weaker interactions in supercritical media compared to normal liquids.

To this point it is important to recognize that steady-state fluorescence spectroscopy gives one only an average picture of the system under study. In order to determine the details of the photophysics, it is necessary to employ time-resolved methodologies.

Time-Resolved Fluorescence. To improve our understanding of the photophysics of the PRODAN-CF_3H system we performed a series of time-resolved fluorescence experiments at $\rho_r = 1.25$ as a function of temperature. Traditionally, one describes the time-resolved fluorescence intensity decay ($I(t)$) by a sum of (n) discrete exponentials of the form (43,46):

$$I(t) = \sum \alpha_i \exp(-t/\tau_i). \tag{2}$$

Here α_i is the pre-exponential (amplitude) factor corresponding to lifetime τ_i. The fractional intensity contribution (f_i) from component i is related to the pre-exponential factors by:

$$f_i = \frac{\alpha_i \tau_i}{\sum \alpha_i \tau_i}, \tag{3}$$

where $\Sigma f_i = 1$.

Recent reports from many groups, including our own, have shown (47-55) that systems originally described by discrete decay processes (Eqn. 2) are more accurately described using continuous distributions of decay times. In these situations, the fluorescence intensity decay can be written in integral form as:

$$I(t) = \int \alpha(\tau) \exp(-t/\tau)d\tau. \tag{4}$$

In the present work, we have found that a continuous Lorentzian lifetime distribution (two floating parameters) described by Eqn. 5, best modelled the experimental data:

$$\alpha(\tau) = A/\{1+[(\tau-\tau_c)/(W/2)]^2\}. \tag{5}$$

Here A is a constant from the normalization condition: $\int \alpha(\tau) = 1$, τ_c is the center value of the lifetime distribution, and W is the full-width-at-half maximum for the Lorentzian function.

Fits to single (one floating parameter) and double (three floating parameters) exponential decay laws are always poorer as judged by the χ^2 and residual traces. In the case where we assume that there is some type of excited-state process (e.g., solvent relaxation) we find that the spectral relaxation time is > 20 ns. This is much, much greater than any reasonable solvent relaxation process in supercritical CF₃H. For example, in liquid water, the solvent relaxation times are near 1 ps (56).

Figure 8 presents the recovered temperature-dependent lifetime distributions for PRODAN in CF₃H ($\rho_r = 1.25$). From these results it appears that, in the highly compressible region where these particular experiments were conducted, PRODAN is subjected to an ensemble of domains. To the best of our knowledge, however, such an observation in a pure "solvent" is unprecedented and merits further discussion.

In other media like micelles, cyclodextrin, binary solvent mixtures, and proteins (47-55), lifetime distributions are routinely used to model the decay kinetics. In all of these cases the distribution is a result of the (intrinsic or extrinsic) fluorescent probe distributing simultaneously in an ensemble of **different** local environments. For example, in the case of the cyclodextrin work from our laboratory (53-55), the observed lifetime distribution is a result of an ensemble of 1:1 inclusion complexes forming and coexisting. These complexes are such that the fluorescent probe is located simultaneously in an array of environments (polarities, etc.) in, near, and within the cyclodextrin cavity, which manifest themselves in a distribution of excited-state lifetimes (53-55). In the present study our experimental results argue for a unimodal lifetime distribution for PRODAN in pure CF₃H. The question then becomes, how can a lifetime distribution be manifest in a pure solvent?

We propose that the recovered lifetime distribution is a result of the different cluster sizes (i.e., domains) encountered by PRODAN in CF₃H. That is, if PRODAN were simultaneously distributed in clusters or aggregates with different solvation characteristics (sizes) one would anticipate a distribution of decay times. Thus, it appears that the observed lifetime distributions recovered here may in fact be a consequence of the actual distribution of cluster sizes in the highly compressible region of supercritical CF₃H.

Conclusions. The results from the present experiments lead to several conclusions and hints to the dynamics of solvation in supercritical CF₃H. [1] The local density about the PRODAN probe is far greater (in some cases 120% greater) than the bulk density. This is most likely a result of fluid aggregation (i.e., clustering, charm, charisma) about the solute in the highly compressible region. [2] Our steady-state fluorescence experiments show that the spectral FWHM is extremely sensitive to density. These particular observations are consistent with an ensemble of local domains about the solute. [3] The excited-state decay kinetics for PRODAN in CF₃H

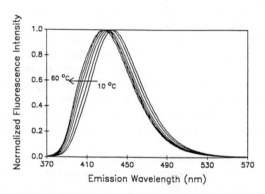

Figure 7. Temperature-dependent emission spectra for 10 μM PRODAN in CF$_3$H.

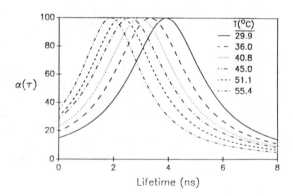

Figure 8. Temperature-dependent Lorentzian lifetime distributions for 10 μM PRODAN in CF$_3$H.

are not monoexponential.   In normal liquids like C_1-C_5 alcohols, CH_3CN, and water the fluorescence intensity decays are always well modelled by a nearly monoexponential decay law; only in supercritical CF_3H is the intensity decay modelled by a distribution of lifetimes. Because of our present time resolution (20 ps), it is possible that there is a much faster component that remains unresolved. The best fit model for describing the intensity decay is a simple unimodal Lorentzian lifetime distribution.   The evidence for a lifetime distribution has prompted us to propose that we are observing an ensemble of PRODAN molecules each being subjected to different local compositions of CF_3H.   That is, there is not a unique solvent cage about PRODAN; under our experimental conditions the cybotactic region is nonuniform and heterogeneous.

Presently, we are initiating density-dependent experiments on this system, elucidating the origin of the distribution, and determining if the observed results are probe dependent.

Acknowledgements.   This work has been generously supported by the United States Department of Energy (DE-FGO2-90ER14143).   We also thank the Tennessee Eastman Co. for sponsoring the ACS Analytical Division Fellowship (TAB) and SUNY-Buffalo for the Mark Diamond Research Award (TAB).   Special thanks are also extended to Gary Sagerman for his continued help with instrument construction. Finally, we thank Professor Mark Maroncelli for graciously allowing us time on his time-domain spectrometer to confirm our experimental observations.

Literature Cited

1.   van Wasen, U.; Swaid, I.; Schneider, G.M., *Angew. Chem. Int. Ed. Engl.* **1980**, *19*, 575.
2.   *Supercritical Fluid Chromatography*; Smith, R.M., Ed.; Royal Society of Chemistry Chromatography Monograph; The Royal Society of Chemistry: London, 1988.
3.   McHugh, M.; Krukonis, V. *Supercritical Fluid Extraction: Principles and Practice*; Butterworths: Boston, MA, 1986.
4.   *Supercritical Fluid Science and Technology*; Johnston, K.P.; Penninger, J.M.L., Eds.; American Chemical Society Symposium Series, No. 406; American Chemical Society: Washington, DC, 1989.
5.   Reid, R.C.; Prausnitz, J.M.; Poling, B.E. *The Properties of Gases and Liquids* 4ᵗʰ Ed.; McGraw-Hill: New York, NY, 1987.
6.   Brennecke, J.F.; Eckert, C.A. *AIChE J.* **1989**, *35*, 1409.
7.   Eckert, C.A.; Ziger, D.H.; Johnston, K.P.; Kim, S. *J. Phys. Chem.* **1986**, *90*, 2738.
8.   Lemert, R.M.; Johnston, K.P. *Fluid Ph. Equil.* **1989**, *45*, 265.
9.   Dobbs, J.M.; Wong, J.M.; Lahiere, R.J.; Johnston, K.P. *Ind. Eng. Chem. Res.* **1987**, *26*, 56.
10.   Kim, S.; Johnston, K.P. *AIChE J.* **1987**, *33*, 1603.
11.   Yonker, C.R.; Smith, R.D. *J. Phys. Chem.* **1988**, *92*, 235.
12.   Okada, T.; Kobayashi, Y.; Yamasa, H.; Mataga, N. *Chem. Phys. Lett.* **1986**, *128*, 583.

13. Hrnjez, B.J.; Yazdi, P.T.; Fox, M.A.; Johnston, K.P. *J. Am. Chem. Soc.* **1989**, *111*, 1915.
14. Kajimoto, O.; Futakami, M.; Kobayashi, T.; Yamasaki, K. *J. Phys. Chem.* **1988**, *92*, 1347.
15. Brennecke, J.F.; Eckert, C.A. In *Supercritical Fluid Science and Technology*; Johnston, K.P.; Penninger, J.M.L., Eds.; American Chemical Society Symposium Series, No. 406; American Chemical Society: Washington, DC, 1989, chapter 2.
16. Brennecke, J.F.; Tomasko, D.L.; Eckert, C.A. *J. Phys. Chem.* **1990**, *94*, 7692.
17. Johnston, K.P.; Kim, S.; Combs, J. In *Supercritical Fluid Science and Technology*; Johnston, K.P.; Penninger, J.M.L., Eds., American Chemical Society Symposium Series, No. 406; American Chemical Society: Washington, DC, 1989, chapter 5.
18. Cochrun, H.D.; Lee, L.L. In *Supercritical Fluid Science and Technology*; Johnston, K.P.; Penninger, J.M.L., Eds., American Chemical Society Symposium Series, No. 406; American Chemical Society: Washington, DC, 1989, chapter 3.
19. Panagiotopoulos, A.Z. In *Supercritical Fluid Science and Technology*; Johnston, K.P.; Penninger, J.M.L., Eds., American Chemical Society Symposium Series, No. 406; American Chemical Society: Washington, DC, 1989, chapter 4.
20. Cochran, H.D.; Pfund, D.M.; Lee, L.L. *Sep. Sci. Tech.* **1988**, *23*, 2031.
21. Petsche, I.B.; Debenedetti, P.G. *J. Chem. Phys.* **1989**, *91*, 7075.
22. Debenedetti, P.G. *Chem. Eng. Sci.* **1987**, *42*, 2203.
23. Debenedetti, P.G.; Kumar, S.K. *AIChE J.* **1988**, *34*, 645.
24. Debenedetti, P.G.; Mohamed, R.S. *J. Chem. Phys.* **1989**, *90*, 4528.
25. Cochran, H.D.; Lee, L.L.; Pfund, D.M. *Fluid Ph. Equil.* **1988**, *34*, 161.
26. Cochran, H.D.; Lee, L.L.; Pfund, D.M. *Fluid Ph. Equil.* **1987**, *34*, 219.
27. Cochran, H.D.; Lee, L.L. *AIChE J.* **1987**, *33*, 1391.
28. Cochran, H.D.; Lee, L.L. *AIChE J.* **1988**, *34*, 170.
29. Betts, T.A.; Bright, F.V. *Appl. Spectrosc.* **1990**, *44*, 1196.
30. Betts, T.A.; Bright, F.V. *Appl. Spectrosc.* **1990**, *44*, 1203.
31. Weber, G.; Farris, F.J. *Biochemistry* **1979**, *18*, 3075.
32. Chapman, C.F.; Fee, R.S.; Maroncelli, M. *J. Phys. Chem.* **1990**, *94*, 4929.
33. Fedderson, B.A.; Piston, D.W.; Gratton, E. *Rev. Sci. Instrum.* **1989**, *60*, 2929.
34. Spencer, R.D.; Weber, G. *J. Chem. Phys.* **1970**, *52*, 1654.
35. Beechem, J.M.; Gratton, E. In *Time-Resolved Laser Spectroscopy in Biochemistry*; Lakowicz, J.R., Ed.; Proc. SPIE 909: Los Angeles, CA, 1988, pp 70.

36. Beechem, J.M.; Ameloot, M.; Brand, L. *Chem. Phys. Lett.* **1985**, *120*, 466.
37. Beechem, J.M.; Ameloot, M.; Brand, L. *Anal. Instrumn.* **1985**, *14*, 379.
38. Bright, F.V.; Betts, T.A.; Litwiler, K.S. *C.R.C. Crit. Rev. Anal. Chem.* **1990**, *21*, 389.
39. Lakowicz, J.R.; Lackzo, G.; Gryczynski, I.; Szmacinski, H.; Wiczk, W. *J. Photochem. Photobiol. B: Biol.* **1988**, *2*, 295.
40. Jameson, D.M.; Gratton, E.; Hall, R.D. *Appl. Spectrosc. Rev.* **1984**, *20*, 55.
41. Barrow, G.M. *Physical Chemistry* 4th Ed.; McGraw-Hill: New York, NY, 1979, chapter 16.
42. Brennecke, J.F., Ph.D. Thesis, University of Illinois **1989**.
43. Lakowicz, J.R. *Principles of Fluorescence Spectroscopy;* Plenum Press: New York, NY, 1983, chapter 7.
44. Cherkasov, A.S.; Dragneva, G.I. *Opt. Spectrosc.* **1961**, *10*, 283.
45. Piterskaya, I.V.; Bakhshiev, N.G. *Bull Acad. Sci. USSR, Phys. Ser.* **1963**, *27*, 625.
46. Demas, J.N. *Excited State Lifetime Measurements*; Academic Press: New York, NY, 1983.
47. Alcala, J.R.; Gratton, E.; Prendergast, F.G. *Biophys. J.* **1987**, *51*, 587.
48. James, D.J.; Ware, W.R. *Chem. Phys. Lett.* **1985**, *120*, 455.
49. James, D.J.; Ware, W.R. *Chem. Phys. Lett.* **1986**, *126*, 7.
50. Gryczynski, I.; Wiczk, W.; Johnson, M.L.; Lakowicz, J.R. *Biophys. Chem.* **1988**, *32*, 173.
51. Lakowicz, J.R.; Cherek, H.; Gryczynski, I.; Joshi, N.; Johnson, M.L. *Biophys. Chem.* **1987**, *28*, 35.
52. Eftink, M.; Ghiron, C.A. *Biophys. J.* **1987**, *52*, 467.
53. Bright, F.V.; Catena, G.C.; Huang, J. *J. Am. Chem. Soc.* **1990**, *112*, 1344.
54. Huang, J.; Bright, F.V. *J. Phys. Chem.* **1990**, *94*, 8457.
55. Catena, G.C.; Bright, F.V. *J. Fluor.* **1991**, *1*, 31.
56. Jarzeba, W.; Walker, G.C.; Johnson, A.E.; Kahlow, M.A.; Barbara, P.F. *J. Phys. Chem.* **1988**, *92*, 7039.

RECEIVED November 25, 1991

Chapter 5

Local Density Augmentation in Supercritical Solutions

A Comparison Between Fluorescence Spectroscopy and Molecular Dynamics Results

Barbara L. Knutson[1], David L. Tomasko[1], Charles A. Eckert[1],
Pablo G. Debenedetti[2], and Ariel A. Chialvo[2]

[1]School of Chemical Engineering, Georgia Institute of Technology,
Atlanta, GA 30332–0100
[2]Department of Chemical Engineering, Princeton University,
Princeton, NJ 08544–5263

Substantial evidence suggests that in highly asymmetric
supercritical mixtures the local and bulk environment of a solute
molecule differ appreciably. The concept of a local density
enhancement around a solute molecule is supported by
spectroscopic, theoretical, and computational investigations of
intermolecular interactions in supercritical solutions. Here we
make for the first time direct comparison between local density
enhancements determined for the system pyrene in CO_2 by two
very different methods--fluorescence spectroscopy and molecular
dynamics simulation. The qualitative agreement is quite
satisfactory, and the results show great promise for an improved
understanding at a molecular level of supercritical fluid solutions.

Experimental, theoretical, and computational investigations of molecular
interactions in supercritical mixtures (1-17) have led to a growing body of
evidence suggesting that in typical supercritical mixtures the local environment
surrounding solute molecules can be considerably different from the bulk. A
molecular-based understanding of these systems is essential to develop accurate
predictive models of their phase behavior.

Typically such systems are highly asymmetric, with a large solute molecule,
having a relatively large characteristic interaction energy, dissolved in a smaller,
more weakly interacting solvent. These have been termed attractive mixtures;
they are characterized by large, negative solute partial molar properties, and the
microstructure around the solute is characterized by a large augmentation of
solvent density with respect to its bulk value (12-13). The divergence of the
solute's infinite dilution partial molar properties is a critical phenomenon; it is
indicative of long-ranged density fluctuations. The local density augmentation, on
the other hand, is short-ranged and not necessarily restricted to the critical region.
Predictive criteria for attractive behavior in terms of solute-solvent differences in
size and interaction energy (for spherical molecules) as well as chain length, are
available (13).

0097–6156/92/0488–0060$06.00/0

In this work we deal only with such systems in the limit of infinite dilution, and there are both practical and theoretical reasons for this. First many studies are available in the limit of infinite dilution. Next, we avoid both the experimental and computational complications of solute-solute interactions. Many supercritical solutions of practical interest are in fact quite dilute. But perhaps most important, in the limit of infinite dilution we can deal directly with solute-solvent interactions without the complications of defining a composition dependence.

Experimental measurements of thermodynamic quantities such as solubility (*18*) and partial molar properties (*19-22*) have contributed significantly to our current understanding of supercritical mixtures. However, obtaining a quantitative relation between measurements of bulk thermodynamic properties and their underlying molecular causes is not always straightforward. Recovering macroscopic properties from a molecular-based model involves, inevitably, idealizations in the representation of intermolecular forces, and simplifications in the calculation of the partition function. These simplifications, though necessary, are not in general separable.

Spectroscopic techniques, such as ultra-violet (*9*), Infrared (*23*), Nuclear Magnetic Resonance (*24*), and Fluorescence spectroscopies (*5-8*), constitute direct probes of specific events occurring at the molecular scale. When a quantitative interpretation is possible, spectroscopy provides very detailed microscopic information. Unfortunately however, the interpretation of spectra in terms of molecular events is often complex. Yet another approach that probes events at the molecular scale involves the use of tracers, such as chromophores (*1-2,25*). Again, the complexity of the tracer imposes limitations on the extent to which the data can be interpreted quantitatively.

Finally, "data" can be obtained from computer simulations (*26*), whether deterministic (molecular dynamics) or stochastic (Monte Carlo). This approach provides a level of microscopic detail not available with any of the above experimental techniques. Results from computer simulations, furthermore, can be both qualitative (for example, observation of cavity dynamics in repulsive supercritical systems (*12*)) as well as quantitative. However, because true intermolecular potentials are not known exactly, simulation results must be interpreted with caution, especially if they are used to study the behavior of real systems. Through simulations, therefore, one obtains exact answers to ideal (as opposed to real) problems.

BACKGROUND

Among the above-mentioned experimental techniques, two designed specifically to study molecular-level events have been particularly useful in the study of supercritical solutions: solvatochromism and fluorescence spectroscopy. In solvatochromic experiments (*1-2,25*), the measured quantity is the wavelength of maximum absorption of an indicator dye (λ_{max}). The technique hinges on the fact that the presence of the solvent affects λ_{max} through short-ranged solute-solvent interactions. Thus, by measuring this quantity as a function of temperature and pressure, it has been possible to estimate local solvent densities around the solute dye under supercritical conditions. This was manifested in the solvatochromic experiments by a pronounced shift toward higher wavelengths in λ_{max} brought about by an increase in the solvent's polarizability per unit volume due to solvent compression.

In fluorescence spectroscopy (5-7), use is made of the fact that first-shell interactions between a solute and the surrounding solvent provide mechanisms for symmetry disruption. For the specific case of pyrene fluorescence, a ratio of the intensities of two pyrene spectrum peaks is a very sensitive measure solute-solvent interactions. The intensity ratio of the first peak (I_1), a symmetry-forbidden and solvent-sensitive transition, and the third peak (I_3), a strong, allowed, and solvent-insensitive transition, indicates the relative solvent strength of the local environment around a pyrene molecule. The sensitivity of this ratio to solvent effects has been documented by several studies (27-31) and is explained by the disruption of molecular symmetry through solute-solvent interactions. Dong and Winnik (32-33) used solvent effects on the vibronic fine structure of pyrene fluorescence to establish the Py polarity scale, correlating solvent strengths with observed I_1/I_3 values in 94 solvents. Using density as a measure of solvent strength, Eckert and coworkers extended the use of the I_1/I_3 ratio to supercritical fluid solutions and measured an apparent density increase in the proximity of a pyrene solute molecule.

The qualitative interpretation of the shifts in λ_{max} and intensity ratios in terms of local density enhancements around the solute molecule is straightforward. It is far less simple, however, to quantify these observations in molecular terms. Neither the magnitude of the local density augmentation, nor its characteristic length scale can be quantified unambiguously. In the solvatochromic experiments, for example, the reference (bulk) line is theoretical, not measured; in the fluorescence investigations, its exact location is somewhat arbitrary. In neither case, furthermore, can a length scale be deduced from the measurements. On the other hand, the density augmentation can be easily quantified (both in magnitude and length scale) in computer simulations. In the study of complex phenomena, no single approach is likely to provide a complete picture. Rather, fundamental understanding often results from looking at the phenomenon of interest from different viewpoints. This is the approach which we have taken in this paper, in which we compare fluorescence spectroscopy and molecular dynamics simulation results. The aim is not to "check" the accuracy of the quantitative interpretation of the experiments, but to look at the same phenomenon from two complementary perspectives: experimental reality, which yields trends that are easy to interpret but difficult to quantify unambiguously, and simulations, which can be easily quantified, but which in this work are based on highly idealized and oversimplified molecular models.

In what follows, we present new fluorescence spectra for pyrene in supercritical carbon dioxide. This is followed by molecular dynamics results on density augmentation in a mixture of Lennard-Jones atoms whose potential parameters were chosen so as to simulate pyrene and carbon dioxide. Finally, we compare the experimental and computational results, thereby obtaining information on the magnitude and extension of the density enhancements suggested by the experiments.

FLUORESCENCE EXPERIMENT

The steady-state fluorescence measurements of pyrene in supercritical CO_2 were made with a spectrometer assembly consisting mainly of Kratos optical parts. The custom built high pressure optical cell is equipped for detection at 90°. The emission was detected with a Hammamatsu 1P-28 photomultiplier tube. The spectrometer assembly, high pressure apparatus, and pressure and temperature control are described in detail elsewhere (5-6). SFC grade carbon dioxide from Scott Specialty Gases had a minimum purity of 99.99% and a maximum O_2 concentration of 2 ppm. The pyrene solute was Aldrich 99+% purity.

Fluorescence spectra of pyrene in supercritical CO_2 were recorded at two temperatures, 37.4 °C and 75.0 °C. These isotherms correspond to pure CO_2 reduced temperatures of 1.02 and 1.14. Data were collected over a density range of 4 - 20 gmole/l; the critical density of pure CO_2 is 10.65 gmole/l. A sample spectrum in which the I_1 and I_3 peaks are labeled is given in Figure 1. No excimer formation is observed at our pyrene concentration of $y_2=3.2 \times 10^{-8}$, suggesting that solute-solute interactions are negligible. At this low pyrene concentration, we assumed that the bulk solvent density was insensitive to the presence of pyrene since our conditions are relatively far removed from the critical scaling region (more than six degrees from the solvent critical point). Therefore, resulting I_1/I_3 ratios for both isotherms are plotted against the bulk density of the system, assuming pure CO_2 densities, in Figure 2.

The local density enhancement around a pyrene molecule in supercritical CO_2 is determined from the dependence of the I_1/I_3 ratio on solvent properties. Interpreting the solvent sensitivity of I_1/I_3 as a density effect, one can assume that the same value of I_1/I_3 corresponds to the same local density around a solute molecule. A high temperature reference state at which the bulk density and local density around the pyrene molecule are assumed to be equal must also be defined. At temperatures sufficiently removed from the critical point (reduced temperatures greater than 1.1, for example), the assumption that the bulk and local densities (averaged over a few solvation shells) are equal is a good approximation. Since we wish to study the system at constant density, substantially higher pressures are required at higher reduced temperatures. The pressure limitation of the fluorescence cell limited this work to a reduced temperature of 1.14 as an upper limit. Thus with $T_R=1.14$ as a reference state, the local density enhancement is the difference between the apparent density at $T_R=1.02$ and the bulk density. This is determined graphically as the horizontal distance from the lower temperature data to the linear fit of the higher temperature data, as shown in Figure 2. At the experimental pyrene concentrations, data interpretation was limited to bulk densities greater than approximately 4.8 gmol/l. Below this density, the fluorescence emission became too weak to measure accurately. Of course, as the density is decreased and the ideal gas limit is approached, the local and bulk densities must be equal due to the lack of molecular interactions. It will be shown that this low density convergence of local and bulk conditions is confirmed by molecular dynamics simulation at lower densities.

The resulting local and bulk densities for pyrene in CO_2 at $T_R=1.02$ are given in Table I. Local density enhancements around the pyrene solute, defined as the local density divided by the bulk density, are also included in Table I. These local density enhancements will be used later for direct comparison with simulation results.

COMPUTER SIMULATIONS

We studied the distribution of a supercritical solvent around an infinitely dilute solute molecule via molecular dynamics simulations. The Lennard-Jones potential

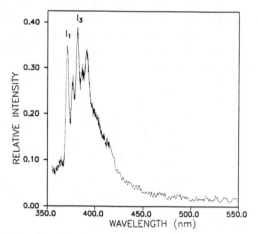

Figure 1: Pyrene spectrum at 37.4 °C and 83.0 bars.   The system concentration is $y_2 = 3.2 \times 10^{-8}$. The spectrum peaks I_1 and I_3 are labelled.

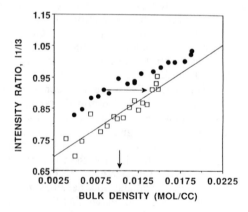

Figure 2: Intensity Ratio, I_1/I_3, for pyrene in supercritical carbon dioxide at 37.4 °C (●) and 75.0 °C (□). The vertical arrow denotes carbon dioxide's critical density; the horizontal arrow denotes the difference between local and bulk densities.

Table I. Local Densities for Pyrene in CO_2 at $T_R=1.02$

Bulk Density (gmole/l)	Local Density (gmole/l)	Density Enhancement (Local /Bulk)
18.8	21.5	1.14
18.7	20.9	1.12
18.0	19.7	1.09
16.8	19.5	1.09
16.0	19.5	1.16
15.1	18.6	1.22
14.2	17.8	1.23
12.8	17.5	1.36
12.0	16.1	1.34
12.0	15.8	1.32
11.2	15.7	1.39
10.1	16.5	1.64
8.99	14.0	1.55
8.37	14.5	1.73
7.71	13.0	1.73
6.89	13.0	1.89
5.82	11.0	1.90
4.87	10.0	2.05

was used for both solute and solvent. Potential parameters are listed in Table II; they are identical to those used by Wu et al. (*16*) to represent pyrene and carbon dioxide in their integral equation calculations. Technical details on the simulations are given in the Appendix.

The fluorescence spectra show clear evidence of an enhancement in the solvent's density around the solute at slightly supercritical temperatures. To investigate this observation quantitatively, we define an average local density,

Table II. Lennard-Jones Parameters used in Molecular Dynamics Simulations

Interaction	ε/k (K)	σ (A)
carbon dioxide-carbon dioxide (1-1)	225.3	3.794
carbon dioxide-pyrene (1-2)	386.4	5.467
pyrene-pyrene (2-2)	662.8	7.140

$$\rho(R) = \frac{<N_1(R)>}{\frac{4\pi}{3}\left[R^3 - \left(\frac{\sigma_2}{2}\right)^3\right]}$$

(1)

where angle brackets indicate average (thermodynamic) quantities, and $<N_1(R)>$ is the number of solvent molecules within a sphere of radius R centered around the solute. Note that the volume over which the density is being calculated excludes a "core" due to the solute. Except for the exact choice of the core's radius, which is somewhat arbitrary, this is precisely how one would compute a local density, should thus type of detailed microscopic information be available experimentally.

Figure 3 shows the relationship between the local and bulk density at two supercritical temperatures, and for three different values of R* (R* = R/s_1). Densities and temperatures are dimensionless ($\rho^* = \rho\sigma_1^3; T^* = kT/\varepsilon_1$). The best estimate of the critical point of the Lennard-Jones fluid is $\rho_c^* = 0.31; T_c^* = 1.31$ (34). Thus, the two temperatures correspond to T/T_c = 1.02 and 1.145. The three values of R* within which local densities were calculated are shown schematically in Figure 4. When comparing Figure 3 to experimental results, it is more meaningful to use the reduced temperature and density, because the potential parameters of Table II yield a critical density and temperature of 9.4 moles/liter and 295.1 K for carbon dioxide, whereas the actual values for this substance are 10.65 moles/liter and 304.2 K, respectively. The data of Figure 3 is listed in Table III, and replotted in Figure 5 in terms of density enhancement vs. bulk density. Local densities based on the total volume within a sphere of radius R, without excluding the solute core, are only slightly smaller than the reported values. The correction factors by which the reported densities must be multiplied to yield R^3-based densities are 0.886, 0.943, and 0.980 (R*= 1.94, 2.44, 3.44). To obtain densities in moles/l, the dimensionless densities given in Table III and in the figures should be multiplied by 30.4.

Table III: Local Densities Corresponding to Figures 3 and 5

Bulk Density	Local Density (T/T_c=1.02) R/σ_1			Local Density (T/T_c=1.145) R/σ_1		
	1.94	2.44	3.44	1.94	2.44	3.44
0.02	0.036	0.032	0.025	-	-	-
0.05	0.097	0.085	0.068	0.078	0.074	0.061
0.10	0.203	0.178	0.145	-	-	-
0.15	0.291	0.259	0.223	0.216	0.202	0.180
0.25	0.386	0.352	0.325	0.320	0.300	0.281
0.31	0.442	0.406	0.386	0.398	0.371	0.354
0.35	0.435	0.402	0.389	-	-	-
0.45	0.518	0.475	0.481	-	-	-

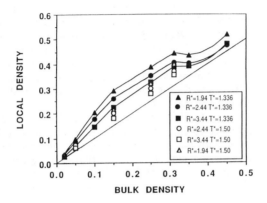

Figure 3: Relationship between local and bulk densities at two supercritical temperatures (T/T$_c$ = 1.02, 1.145) for an infinitely dilute mixture of Lennard-Jones atoms with potential parameters chosen so as to simulate pyrene in carbon dioxide (see Table II). Molecular dynamics simulation.

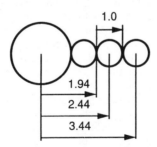

Figure 4: Location of the solvation shells within which local densities were calculated in the molecular dynamics simulations.

At a reduced temperature of 1.02 there exists a wide range of densities across which significant local density enhancements with respect to bulk conditions occur. This effect is more pronounced when $\rho^* = 0.1$ (bulk density at 32% of the critical density); the density enhancement is then more than 100% at $R^* = 1.94$, and stays as high as ca. 45% when $R^* = 3.44$ (third solvation shell). At lower bulk densities, differences between the local and bulk conditions disappear progressively due to the fact that the dilute solvent approaches ideal behavior. At high densities, the gradual diminution of local density enhancements is due to the decrease in the solvent's compressibility. Except at the critical density, the local density increases upon compression. When $\rho^* = 0.31$, however, ρ^* (R) actually decreases upon compression in the first two solvation shells, and only shows a very small increase (from 0.386 to 0.389 as the bulk density increases from 0.31 to 0.35) when three solvation shells are included. Local density enhancements become much less pronounced as the temperature is increased. Comparisons at constant bulk density (Figures 3, 5; Table III) show drastic decreases in the local density as the reduced temperature changes from 1.02 to 1.145.

Figure 6 shows a comparison between the density enhancements deduced from experiments and those calculated via simulation (the latter at $R^* = 1.94$). The reduced temperature in both cases is 1.02. This very good agreement suggests that the density augmentation measured in the fluorescence spectra corresponds to the first solvation shell.

DISCUSSION

Studies of local density enhancements around solute molecules in dilute supercritical mixtures by fluorescence spectroscopy and molecular dynamics yield very similar results, and suggest a common picture. For attractive mixtures of interest in supercritical extraction, the solvent's density in the vicinity of the solute is considerably higher than in the bulk. Even after averaging over three solvation shells, computer simulation results show local densities which can exceed the bulk value by as much as 50%. This local density enhancement disappears quite rapidly as the temperature is increased away from the critical point. The effect occurs over a wide density range which includes (but is not limited to) the solvent's critical density. At low bulk density (less than 33% of the critical density), enhancements become progressively less important as the solvent approaches an ideal gas. In the fluorescence experiment, bulk densities sufficiently low for direct comparison with simulation results were not reached. However, previous fluorescence spectroscopy data for pyrene in supercritical ethylene and supercritical fluoroform show a decrease in density enhancements as ideal gas densities are approached (7). At high bulk density, the solvent becomes gradually incompressible, which results in the corresponding disappearance of local density augmentations.

Local density enhancements, being by definition short-ranged, are not peculiar to the highly compressible near-critical region. Very close to the solute molecule, the local environment differs markedly from the bulk (for example, the local density in the first solvation shell at bulk near-critical conditions is $\rho^*(R) = 1.43 \, \rho_c$ when $\rho^* = 0.31$ and $T/T_c = 1.02$). However, even this region does not appear to have a liquid-like character, as suggested by other spectroscopic experiments (35-36).

The local, short-ranged phenomena discussed in this paper are of fundamental importance in determining the actual solubility of a solute in a given

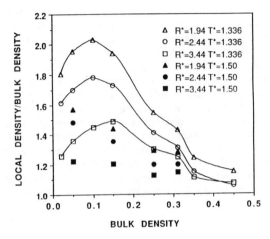

Figure 5: Relationship between local density augmentation (local/bulk density) and bulk density. Same system and conditions as in Figure 3. Molecular dynamics simulation.

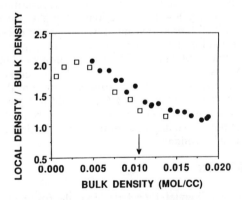

Figure 6: Comparison between local density augmentation deduced from fluorescence spectroscopy (●), and the corresponding molecular dynamics simulations at R* = 1.94 (□). Both curves are for a reduced temperature of 1.02. The arrow denotes the critical density of carbon dioxide.

solvent. On the other hand, by virtue of their finiteness, they contribute negligibly to the divergence in partial molar properties and in integrals of the correlation function. The latter are long-ranged phenomena, due to the proximity to the solvent's critical point. The insignificant contribution of short-ranged quantities to the magnitude of divergent properties should not be mistaken for their physical irrelevance. Quite the contrary is in fact true: the local microstructure around the solute is a distinguishing property of asymmetric, attractive mixtures. To adequately describe solubilities and phase behavior in supercritical mixtures, a rigorous, microscopically-based model must account for this local microstructure.

The picture of a solute molecule stabilized in solution by a local environment where the solvent's concentration differs considerably from the bulk value is consistent with experiments and simulation. The encouraging agreement between the basic trends found in experiments and simulations should not obscure the fact that Lennard-Jones atoms are a pedestrian representation of the actual molecules studied in the fluorescence experiments. Caution must therefore be exercised when comparing simulations and experiments. At the same time, the very fact that such a crude model is able to capture the essential physics of the phenomenon under investigation lends further support to the notion that local density augmentations are common to all attractive supercritical systems.

SUMMARY

The view of local density augmentations that has emerged from this comparison of experiments and simulations is satisfyingly consistent. Much remains to be done, however. First, the molecular models used in the present simulations must be made more realistic. In addition, experimental and theoretical work (*15-16,17*) seems to indicate that solute-solute interactions in supercritical solvents can be very important, even at very low mole fractions: they have been neglected in the present simulations. Finally, probably the most interesting question related to the microstructure surrounding the solute molecules in supercritical solvents is whether the extraordinary solubility enhancements brought about by the addition of small amounts of cosolvent (*37*) can be understood in terms of local cosolvent composition enhancements around the solute, a notion that appears to be supported by solvatochromic experiments (*1*). Should this prove to the case, and should it prove to be amenable to control via an appropriate choice of cosolvent, this opens up the exciting possibility of fine-tuning supercritical solutions for highly specific separations and reactions.

ACKNOWLEDGMENTS

Two of us (PGD and AAC) gratefully acknowledge the financial support of the U.S. Department of Energy, Office of Energy Sciences, Division of Chemical Sciences (Grant DE-FG02-87ER13714). Computer simulations were performed at the Florida State University Supercomputer Computations Research Institute, which is partially funded by the U.S. Department of Energy, through contract No. DE-FG02-85ER2500, and at the San Diego Supercomputer Center. PGD gratefully acknowledges the Camille and Henry Dreyfus Foundation, for a 1989 Teacher-Scholar Award and John Simon Guggenheim Memorial Foundation, for a 1991 Fellowship.

Three of us (BLK, DLT, and CAE) gratefully acknowledge financial support from the DuPont Co. and from the Department of Energy, through grant DE-FG22-88PC88922.

LITERATURE CITED

1. Kim, S.; Johnston, K.P. *AIChEJ.* **1987**, *33*, 1603.
2. Kim, S.; Johnston, K.P. *Ind. Eng. Chem. Res.* **1987**, *26*, 1206.
3. Johnston, K.P.; Kim, S.; Combes, J. In *Supercritical Fluid Science and Technology*; Johnston, K.P., Penninger, J.M.L., Eds.; ACS Symposium Series 406; American Chemical Society: Washington, DC, 1989; Chapter 5.
4. Brennecke, J.F.; Eckert, C.A. *AIChEJ.* **1989**, *35*, 1409.
5. Brennecke, J.F.; Eckert, C.A. *Proc. Int. Symp. Supercrit. Fluids (I) Nice, Fr.* **1988**, 263.
6. Brennecke, J.F.; Eckert, C.A. In *Supercritical Fluid Science and Technology*; Johnston, K.P., Penninger, J.M.L., Eds.; ACS Symposium Series 406; American Chemical Society: Washington, DC, 1989; Chapter 2.
7. Brennecke, J.F.; Tomasko, D.L.; Peshkin, J.; Eckert, C.A. *Ind. Eng. Chem. Res.* **1990**, *29*, 1682.
8. Brennecke, J.F.; Tomasko, D.L.; Eckert, C.A. *J. Phys. Chem,* **1990**, *94*, 7692.
9. Yonker, C.R.; Smith, R.D. *J. Phys. Chem.* **1988**, *92*, 2374.
10. Debenedetti, P.G. *Chem. Eng. Sci.* **1987**, *42*, 2203.
11. Petsche, I.B.; Debenedetti, P.G. *J. Chem. Phys.* **1989**, *91*, 7075.
12. Debenedetti, P.G.; Mohamed, R.S. *J. Chem. Phys.* **1989**, *90*, 4528.
13. Petsche, I.B.; Debenedetti, P.G. *J. Phys. Chem.* **1991**, *95*, 386.
14. Cochran, H.D.; Pfund, D.M.; Lee, L.L. *Proc. Int. Symp. Supercrit. Fluids (I) Nice, Fr.* **1988**, 245.
15. Cochran, H.D.; Lee, L.L. In *Supercritical Fluid Science and Technology*; Johnston, K.P., Penninger, J.M.L., Eds.; ACS Symposium Series 406; American Chemical Society: Washington, DC, 1989; Chapter 3.
16. Wu, R.S.; Lee, L.L.; Cochran, H.D. *Ind. Eng. Chem. Res.* **1990**, *29*, 977.
17. McGuigan, D.B.; Monson, P.A. *Fluid Phase Equilib.* **1990**, *57*, 227.
18. Kurnik, R.T.; Reid, R.C. *Fluid Phase Equilib.* **1982**, *8*, 93.
19. Ehrlich, P.; Fariss, R. *J. Phys. Chem.* **1969**, *73*, 1164.
20. Wu, P.C; Ehrlich, P. *AIChEJ.* **1973**, *19*, 541.
21. Eckert, C.A.; Ziger, D.H.; Johnston, K.P.; Ellison, T.K. *Fluid Phase Equilib.* **1983**, *14*, 167.
22. Eckert, C.A.; Ziger, D.H.; Johnston, K.P.; Kim, S. *J. Phys. Chem.* **1986**, *90*, 2738.
23. Yoshino, T. *J. Chem. Phys.* **1956**, *24*(1), 76.
24. Foster, R.; Fyfe, C.A. *Trans. Faraday Soc.* **1965**, *61*, 1626.
25. Yonker, C.R.; Frye, S.L.; Kalkwarf, D.R.; Smith, R.D. *J. Phys. Chem.* **1986**, *90*, 3022.
26. Allen, M.P.; Tildesley, D.J. Computer Simulation of Liquids; Clarendon Press: Oxford, 1987.
27. Nakajima, A. *Bull. Chem. Soc. Jpn.* **1971**, *44*, 3272.
28. Nakajima, A. *Bull. Chem. Soc. Jpn.* **1976**, *61*, 467.
29. Kalyanasundaram, K.; Thomas J.K. *J. Am. Chem. Soc.* **1977**, *99*, 2039.
30. Lianos, P.; Georghiou, S. *Photochem. Photobiol.* **1979**, *29*, 843.

31. Lianos, P.; Georghiou, S. *Photochem. Photobiol.* **1979**, *30*, 355.
32. Dong, D.C.; Winnik, M.A. *Photochem. Photobiol.* **1982**, *35*, 17.
33. Dong, D.C.; Winnik, M.A. *Can. J. Chem.* **1984**, *62*, 2560.
34. Smit, B; De Smedt, P.; Frenkel, D. *Mol. Phys.* **1989**, *68*, 931.
35. Combes, J. R.; Johnston, K. P.; O'Shea, K.; Fox, M. A., This Symposium.
36. Tomasko, D. L.; Knutson, B. L., Eckert, C. A.; Haubrich, J. E., Tolbert, L. M., This Symposium.
37. Johnston, K.P.; McFann, G.; Peck, G.; Lemert, D. *Fluid Phase Equilib.* **1989** b, *52*, 337.
38. Haile, J.M.; Gupta S. *J. Chem. Phys* **1983**, *79*, 3067.
39. Chialvo, A.A.; Debenedetti, P.G. *Comput. Phys. Comm.* **1990**, *60*, 215.
40. Chialvo, A.A.; Debenedetti, P.G. *Comput. Phys. Comm.* **1991**, *64*, 15.
41. Sengers, J.V.; Levelt Sengers, J.M. *Ann. Rev. Phys. Chem.* **1986**, *37*, 189.

APPENDIX

The computer simulations employed the molecular dynamics technique, in which particles are moved deterministically by integrating their equations of motion. The system size was 864 Lennard-Jones atoms, of which one was the solute (see Table II for potential parameters). There were no solute-solute interactions. Periodic boundary conditions and the minimum image criterion were used (*16*). The cutoff radius for binary interactions was 3.5 σ_1 (see Table II). Potentials were truncated beyond the cutoff.

Computations were done in the canonical ensemble (constant volume, number of particles and temperature), using the momentum scaling technique (*38*) to impose isothermality. An automated Verlet neighbor list algorithm (*39-40*) was implemented to increase the speed of the simulations. Every run was preceded by an equilibration period. This involved melting the initial face-centered cubic configuration at $T^*=1.9$, a condition which was judged to be attained when the root mean squared particle displacement reached ca. $4.5\sigma_1$. Thereafter, the system was quenched to the desired temperature and equilibrated for ca. 2000 time steps before beginning an actual simulation. The actual production runs varied in length from 80,000 to 200,000 time steps, the latter figure corresponding to the low-density simulations. The time step was $0.003\ \sigma_1(m_1/\varepsilon_1)^{1/2}$, where σ_1 and ε_1 are given in Table II, and m_1 is the mass of a carbon dioxide molecule. Statistics for the local density calculations were gathered at every step.

The correlation length (ξ) corresponding to simulations at the critical density and slightly supercritical temperature was calculated from the scaling equation $\xi = \xi_o\ (T_r -1)^{-0.63}$ where T_r is the reduced temperature, and ξ_o, a substance-specific amplitude (*41*). At $T_r = 1.02$, and using $\xi_o = 1.5$ A for carbon dioxide (*41*), the correlation length is 17.6 A. The length of the computational cell, $L = s_1\left(N/\rho^*\right)^{1/3}$, where N is the number of molecules in the simulation, is 77 A at the critical density. Thus, the system size was adequate to accommodate long-ranged correlations in the relative vicinity of the critical point.

RECEIVED January 21, 1992

Chapter 6

Investigation of Pyrene Excimer Formation in Supercritical CO_2

JoAnn Zagrobelny and Frank V. Bright

Department of Chemistry, Acheson Hall, State University of New York at Buffalo, Buffalo, NY 14214

We report on steady-state and time-resolved fluorescence of pyrene excimer emission in sub- and supercritical CO_2. Our experimental results show that, above a reduced density of 0.8, there is no evidence for ground-state (solute-solute) interactions. Below a reduced density of 0.8 there are pyrene solubility complications. The excimer formation process, analogous to normal liquids, only occurs for the excited-state pyrene. In addition, the excimer formation process is diffusion controlled. Thus, earlier reports on pyrene excimer emission at rather "dilute" pyrene levels in supercritical fluids are simply a result of the increased diffusivity in the supercritical fluid media. There is not any anomalous solute-solute interaction beyond the diffusion-controlled limit in CO_2.

For any pure chemical species, there exists a critical temperature (T_c) and pressure (P_c) immediately below which an equilibrium exists between the liquid and vapor phases (1). Above these critical points a two-phase system coalesces into a single phase referred to as a supercritical fluid. Supercritical fluids have received a great deal of attention in a number of important scientific fields. Interest is primarily a result of the ease with which the chemical potential of a supercritical fluid can be varied simply by adjustment of the system pressure. That is, one can cover an enormous range of, for example, diffusivities, viscosities, and dielectric constants while maintaining simultaneously the inherent chemical structure of the solvent (1-6). As a consequence of their unique solvating character, supercritical fluids have been used extensively for extractions, chromatographic separations, chemical reaction processes, and enhanced oil recovery (2-6).

Many of the present models used to describe fluid-solid phase equilibria require one to assume that the solute is at infinite dilution. That is, researchers have often assumed that solute-solute interactions are nonexistent. Recently, Brennecke et al. used the fluorescent probe pyrene to investigate the possibility of solute-solute interactions in CO_2, C_2H_4, and CF_3H (7-9). Pyrene is an interesting probe because it can form a characteristic excited-state dimer (excimer) during its excited-state

0097–6156/92/0488–0073$06.00/0

lifetime ($10,11$). In normal liquids, this bimolecular process is diffusion controlled ($10,11$) and is often described using the energy-level diagram shown in Figure 1. For all liquid systems studied to date, significant excimer is only observed at pyrene concentrations in the high millimolar range.

Brennecke et al. (7-9) observed pyrene excimer in pure supercritical fluids when the pyrene concentrations were in the high micromolar range and concluded that there were solute-solute interactions occurring on the submicrosecond time scale. However, the origin of the observed excimer emission was not completely determined in this earlier work. Specifically, it was not clear if the apparent enhancement in excimer (on a per mole pyrene basis) was a consequence simply of: 1) the increased diffusivity in the supercritical fluid (i.e., the excited- and ground-state pyrene molecules could diffuse faster, thus increasing the probability of interaction during the excited-state lifetime) or 2) actual static (ground-state) solute-solute interactions. To address these issues one must use steady-state and time-resolved fluorescence in concert ($10,11$).

In this paper, we present a preliminary analysis of the steady-state and time-resolved fluorescence of pyrene in supercritical CO_2. In addition, we employ steady-state absorbance spectroscopy to determine pyrene solubility and determine the ground-state interactions. Similarly, the steady-state excitation and emission spectra gives us qualitative insights into the excimer formation process. Finally, time-resolved fluorescence experiments yield the entire ensemble of rate coefficients associated with the observed pyrene emission (Figure 1). From these rates we can then determine if the excimer formation process is diffusion controlled in supercritical CO_2.

Experimental

Instrumentation. A schematic of the instrumentation used for the time-resolved fluorescence experiments is shown in Figure 2. A pulsed nitrogen laser (337 nm; Laser Science, model LS 337) is used for excitation, and a 340 nm band-pass filter (BPF) is used to eliminate extraneous plasma discharge. Typically, we operate at a repetition rate of 10-15 Hz. A portion (10%) of the laser is split off using a beam splitter (BS) and directed to a reference photodiode (PD) that serves to trigger the boxcar averager. Prior to entering the sample cell, the laser beam is conditioned with a 150 mm focal length lens.

The sample chamber consists of a light-tight aluminum housing and was constructed in house specifically for our supercritical fluid high pressure optical cells (12). These cells are constructed from 303 stainless steel with four fused-silica windows oriented at 90° to one another. The window seals are made using a set of in house designed lead and brass ring seals and the total cell volume is about 5 mL. Initially, viton o-ring seals were used, but a fluorescent impurity was continually extracted into the cell. Attempts to use Teflon seals alleviated the impurity problems; however, pyrene sorbed strongly to or into the Teflon seals. The metal-based seals alleviated all these problems.

Following excitation the resulting fluorescence is collected by a lens (50 mm focal length) and collimated, filtered using a bandpass filter (10 nm bandpass), and focussed with a second lens (50 mm focal length) onto the photocathode of a

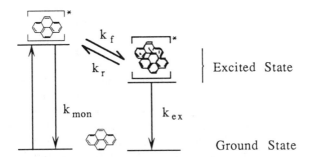

Figure 1. Energy-level diagram for excimer formation. Symbols represent: $h\nu$, absorbed photon; k_{mon} emissive rate from the monomer species; k_f, bimolecular rate coefficient for formation of the pyrene excimer; k_r, unimolecular rate coefficient for dissociation of the excimer; and k_{ex}, emissive rate from the excimer species. Note: no ground-state association is indicated.

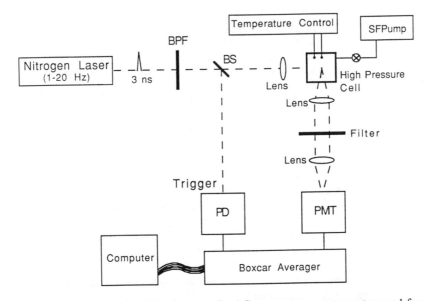

Figure 2. Schematic of the time-resolved fluorescence spectrometer used for the study of pyrene in supercritical fluids. Abbreviations represent: BPF, bandpass filter; BS, beam splitter; PD, photodiode; and PMT, photomultiplier tube.

photomultiplier tube (PMT; Hamamatsu model R372). The PMT was typically operated at a biasing potential of between - 600 to - 700 V DC using a high voltage power supply (SRS; model PS 350). The PMT dynode circuitry is similar to that described elsewhere (13,14).

The output from the PMT anode is directed first to a 5x amplifier (SRS; model SR445) then processed via a gated integrator boxcar averager (SRS; Model SR250). All data are acquired using a Epson Equity I+ personal computer and software developed in our laboratory. Data analysis was performed using a commercially available software package (Globals Unlimited; Urbana, IL).

All steady-state measurements are performed on a SLM 48000 modified to accommodate the optical cells (12). A Xe-arc lamp is used for excitation, and both excitation and emission monochromators are used for wavelength selection.

Pressures within the optical cells are adjusted using a microprocessor-controlled supercritical fluid syringe pump (Isco; model SFC-500). The temperature of the cylinder head is regulated using a VWR 1140 temperature bath. The output from the pump is directed through a 2 μm fritted filter and a series of valves into the optical high pressure cell which is temperature controlled ($\pm$ 0.1 °C) by a Lauda RLS-6 temperature bath. The local temperature of the optical cell is determined using a thermocouple (Cole Palmer) placed directly into the cell body.

Sample Preparation. SFC grade CO_2 (< 5 ppm O_2) was purchased from Scott and pyrene (99 %) was obtained from Aldrich. The pyrene purity was checked by reversed phase HPLC (C_{18}) and all reagents were used as received. Stock solutions of pyrene were prepared in absolute ethanol.

In order to prepare a sample for study, an appropriately sized aliquot of the stock pyrene solution (1 mM) is pipetted directly into the optical cell and the solvent is removed by placing the cell in a heated (60 °C) oven for several hrs. Following this, the cell valving system is assembled onto the cell and attached to the valving from the high pressure pumping system (12). To remove residual O_2, a 50 μm Hg vacuum is maintained for 10-15 min. The cell vacuum is monitored using a vacuum gauge (Huntigton, model TGC-201).

The pump head and optical cell are brought to the same temperature to minimize temperature gradients. The cell is charged to the desired pressure and allowed to equilibrate for about 30 minutes. When density studies are performed the optical cell remains attached to the high pressure pumping system and the pressure is adjusted as required (low → high). Densities are determined from literature compilations for CO_2 (15) and viscosities are calculated using the Lucas and Reichenberg formulations (16). For temperature-dependent studies, we found it easier to operate with the cell sealed off and disconnected from the pump. The temperature within the cell is adjusted to the desired value and allowed to equilibrate for 30 min. prior to the start of each experiment.

Results and Discussion

Steady-State Experiments. Figure 3 shows a series of density-dependent steady-state emission spectra for 100 μM pyrene in CO_2. The long wavelength, structureless excimer emission is clearly evidenced. (Again, at this concentration in a normal liquid like cyclohexane there is not any hint of excimer formation.) A plot of the excimer-to-monomer intensity ratio (I_{EX}/I_M) as a function of reduced density (ρ_r) is shown in Figure 4. These results provide insight into the efficiency of the excimer formation vs. density. Interestingly, at low density one sees very little excimer formation. As one progresses to higher densities the fraction of excimer emission increases; maximizing near a ρ_r of unity. At higher densities, beyond the critical value, the fraction of excimer steadily decreases. Because the observed trend tracks the isothermal compressibility of this system well, one might be inclined to attribute these results to some type of clustering or fluid condensation process. However, a detailed analysis of this system shows this to be incorrect.

On the low density side of these traces (below $\rho_r = 1$), the rise in I_{EX}/I_M (as one approaches a reduced density of unity) is a result of pyrene continually being solubilized by the fluid (Figure 5). This observation is consistent with Johnston's report of pyrene solubility in CO_2 (*17*). Again, recall that excimer formation is a generally believed (*10,11*) to be a bimolecular process, i.e., the rate is concentration dependent. Thus, because the actual analytical concentration of solubilized pyrene **in** the solution is lower due to solubility, the amount (fraction) of excimer is "artificially" lower. Thus, in the low density region ($\rho_r = 0.5$ -0.8) the observed trends are simply a result of solubility effects. However, solubility does not help to explain the results seen over the remainder ($\rho_r = 0.8$ - 1.8) of the density range investigated.

It is well documented that under certain conditions pyrene can form ground-state dimers (*18-20*). If this were the case here, we would need to be concerned with this process as it would be convolved with the simpler diffusion-controlled scenario. There are basically two methods for detecting the presence of ground-state processes: 1) studying the absorbance spectra and 2) investigation of the emission wavelength-dependent fluorescence excitation spectra. Figure 6 shows the density-dependent absorbance spectra for 100 μM pyrene in CO_2. At low density one sees fairly well resolved vibrational features much like those observed in gas-phase spectroscopy. This is not too surprising considering that we are in fact in the gas phase albeit at high pressure. Once we near the critical region the absorbance spectra broaden (pressure broadening and solubility), but they do not shift to any significant extent throughout the density range investigated. These results are inconsistent with the formation of ground-state pyrene dimers (*20*).

Figure 7 shows two representative sets of steady-state excitation spectra for 100 μM pyrene in sub- and supercritical CO_2. Results at all other densities were essentially identical to these and are omitted for clarity. For each density, the two corresponding excitation spectra were acquired while monitoring the emission at 380 (monomer region) and 460 nm (excimer region). If ground-state pyrene association were occurring the spectra obtained at these emission wavelengths would be significantly different because of the different ground-state subpopulations (cf. ref 19). Clearly, the spectra obtained in CO_2 are identical and again inconsistent with ground-state association. In summary, the two sets of experiments described above indicate that there is no apparent ground-state pyrene association. Thus, the simple model given in Figure 1 should suffice to model completely the kinetics for this system (*10,11*).

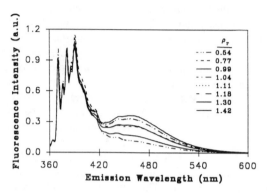

Figure 3. Steady-state fluorescence emission spectra for 100 μM pyrene in sub- and supercritical CO_2 as a function of density. $T_r = 1.02$. $\lambda_{ex} = 337$ nm. Spectral bandpass = 2 nm.

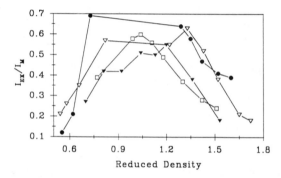

Figure 4. Density-dependent excimer-to-monomer intensity ratios (I_{EX}/I_M) for 100 μM pyrene in sub- and supercritical CO_2 at several temperatures. (symbol, T_r): ($\bullet$, 1.00); ($\triangledown$, 1.01); ($\blacktriangledown$, 1.02); ($\square$, 1.05).

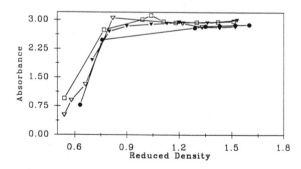

Figure 5. Steady-state absorbance (at 326 nm) for a sample that contains 100 μM pyrene as a function of CO_2 density. (symbol, T_r): ($\bullet$, 1.00); ($\triangledown$, 1.01); ($\blacktriangledown$, 1.02); ($\square$, 1.05).

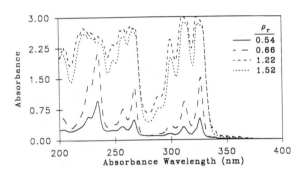

Figure 6. Absorbance spectra for 100 μM pyrene sample vs. CO_2 density. $T_r = 1.01$. Above $\rho_r = 0.8$, all pyrene is solubilized.

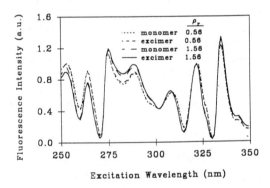

Figure 7. Collection of emission-wavelength-dependent steady-state excitation scans for 100 μM pyrene in sub- and supercritical CO_2. For the monomer and excimer scans, the emission wavelength (16 nm bandpass) is adjusted to 380 and 460 nm, respectively.

Time-Resolved Experiments. Clearly, once all the pyrene becomes solubilized in our system (near $\rho_r = 0.8$), the amount of excimer decreases as density increases. However, it is not clear from these steady-state experiments alone what causes the observed decrease in I_{EX}/I_M with density. In order to address this question one needs information about the rates of the various radiative and non-radiative processes occurring in this system. By using time-resolved fluorescence spectroscopy (10,11) we set about to determine the ensemble of kinetic parameters given in Figure 1.

In all cases investigated, the excited-state model used to fit the excimer formation in shown in Figure 1 (10,11). Traditionally, the rates were recovered by determining the apparent rates ($\lambda_{1,2}$) as a function of pyrene concentration (10,11). These apparent rate terms were plotted versus pyrene concentration to yield estimates of the actual rate coefficients. Presently, the kinetic terms are recovered by using a global analysis protocol where several experiments at multiple emission wavelengths and pyrene concentrations are analyzed simultaneously to obtain a self consistent set of kinetic parameters (21-26). The advantages of using this global approach include: 1) using multiple experiments to help improve accuracy and 2) recovering the rate terms directly from the experimental data. The goodness-of-fit is judged based on the reduced chi-squared, χ^2, random residuals, and random auto-correlation function. For all cases investigated here, the simple model shown in Figure 1 has sufficed to describe the observed photophysics.

Figure 8 shows a pair of typical time-resolved fluorescence decay traces for 100 μM pyrene in supercritical CO_2 ($T_r = 1.02$; $\rho_r = 1.17$). Note that the ordinate is logarithmic. The upper and lower panels show results for selective observation in the monomer (400 $\pm$ 10 nm) and excimer (460 $\pm$ 10 nm) regions of the pyrene emission spectrum. Several interesting features are apparent from these traces. First, both decay processes are not single exponential. Second, the excimer emission has a significant contribution from a species that "grows in" between 30 - 75 ns; this is a result of the excimer taking time to form (i.e., k_{MD} in Figure 1). Third, the fits between the experimental data and the model shown in Figure 1 are good. Detailed analysis of these decay traces (10,11,21-26) yields the entire ensemble of photophysical kinetic parameters for the pyrene excimer in supercritical CO_2.

Figure 9 shows the temperature dependence of the recovered kinetic rate coefficients for the formation (k_{MD}; bimolecular) and dissociation (k_{DM}; unimolecular) of pyrene excimers in supercritical CO_2 at a reduced density of 1.17. Also, shown is the bimolecular rate coefficient expected based on a simple diffusion-controlled argument (11). The value for the theoretical rate constant was obtained through use of the Smoluchowski equation (26). As previously mentioned, the viscosities utilized in the equation were calculated using the Lucas and Reichenberg formulations (16). From these experiments we obtain two key results. First, the reverse rate, k_{MD}, is very temperature sensitive and increases with temperature. Second, the forward rate, k_{DM}, is diffusion controlled. Further discussion will be deferred until further experiments are performed nearer the critical point where we will investigate the rate parameters as a function of density.

Conclusions

The formation of pyrene excimer in normal liquids has been well documented

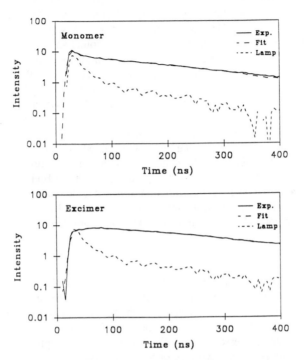

Figure 8. Time-resolved fluorescence decay traces for 100 μM pyrene in supercritical CO_2. $T_r = 1.02$; $\rho_r = 1.17$. Upper and lower panels represent monomer (400 nm) and excimer (460 nm) emission, respectively.

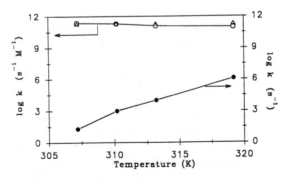

Figure 9. Recovered forward ($\triangle$) and reverse ($\bullet$) rates for the excimer system described in Figure 1 compared to that expected for a diffusion-controlled reaction ($\circ$).

and is accurately described by a simple diffusion controlled model ($10,11$). By studying this same system in supercritical fluids we can improve our understanding of solvation processes in this medium. The experiments reported here lead to several interesting conclusions. [1] From our steady-state absorbance and fluorescence experiments we conclude that there is not any solute-solute interactions between ground-state pyrene molecules (at 100 μM pyrene) above a reduced density of 0.8. Below $\rho_r = 0.8$ there are solubility problems to contend with that convolve the excimer results. [2] Pyrene excimer is observed in supercritical fluid media when the pyrene concentrations are far below those required for observing excimer emission in normal liquids. [3] The excimer formation process is completely diffusion controlled and only occurs in the excited state.

Presently, we are expanding our time-resolved experiments to move closer to the critical point to determine if there are any anomalous deviations from the diffusion control process reported here. These experiments include observing the behavior of the forward and reverse rates as a function of density. In addition, we are beginning to investigate other supercritical fluids (e.g., C_2H_4 and CF_3H) to determine if these conclusion are general enough to be ascribed to all supercritical fluids.

Acknowledgements

This work has been generously supported by the United States Department of Energy (DE-FGO2-90ER14143). Special thanks are also extended to Kevin S. Litwiler for helping with the development of the time-domain acquisition software. We also thank Gary Sagerman for his continued help with instrument construction. Finally, we thank Professors Brennecke and Eckert for helpful discussions and for providing preprints of their work prior to publication.

Literature Cited

1.   van Wasen, U.; Swaid, I.; Schneider, G.M., *Angew. Chem. Int. Ed. Engl.* **1980**, *19*, 575.

2.   *Supercritical Fluid Chromatography*; Smith, R.M., Ed.; Royal Society of Chemistry Chromatography Monograph; The Royal Society of Chemistry: London, 1988.

3.   McHugh, M.; Krukonis, V. *Supercritical Fluid Extraction: Principles and Practice*; Butterworths: Boston, MA, 1986.

4.   *Supercritical Fluid Science and Technology;* Johnston, K.P.; Penninger, J.M.L., Eds.; American Chemical Society Symposium Series, No. 406; American Chemical Society: Washington, DC, 1989.

5.   Reid, R.C.; Prausnitz, J.M.; Poling, B.E. *The Properties of Gases and Liquids* 4th Ed.; McGraw-Hill: New York, NY, 1987.

6.   Brennecke, J.F.; Eckert, C.A. *AIChE J.* **1989**, *35*, 1409.

7.   Brennecke, J.F.; Eckert, C.A. In *Supercritical Fluid Science and Technology* Johnston, K.P.; Penninger, J.M.L., Eds.; American Chemical Society Symposium Series, No. 406; American Chemical Society: Washington, DC, 1989, chapter 2.

8.   Brennecke, J.F., Ph.D. Thesis, University of Illinois, **1989**.

9.   Brennecke, J.F.; Tomasko, D.L.; Eckert, C.A. *J. Phys. Chem.* **1990**, *94*, 7692.

10.   Birks, J.B.; Dyson, D.J.; Munro, I.H. *Proc. Roy. Soc. A.* **1963**, *275*, 575.

11.   Birks, J.B., *Photophysics of Aromatic Molecules*; Wiley-Interscience: New York, NY, 1970.

12.   Betts, T.A.; Bright, F.V. *Appl. Spectrosc.* **1990**, *44*, 1196.

13.   Harris, J.M.; Lytle, F.E.; McCain, T.C. *Anal. Chem.* **1976**, *48*, 2095.

14.   Lytle, F.E. *Anal. Chem.* **1974**, *46*, 545A.

15.   Angus, A.; Armstrong, B.; de Reuck, K.M.; Altunin., V.V.; Gadetski, G.G.; Chapela, G.A.; Rowlinson, J.S. *International Thermodynamic Tables of the Fluid State Carbon Dioxide*; Pergamon Press: New York, NY, 1976.

16.   Reid, R.C.; Prausnitz, J.M.; Poling, B.E. *The Properties of Gases and Liquids* 4th Ed.; McGraw-Hill: New York, 1987, chapter 9.

17.   Johnston, K.P.; Ziger, D.H.; Eckert, C.A. *Ind. Eng. Chem. Fundam.* **1982**, *21*, 191.

18.   Yamazaki, I.; Tamai, N.; Yamazaki, T. *J. Phys. Chem.* **1987**, *91*, 3572.

19.   Yamanaka, T.; Takahashi, Y.; Kitamura, T.; Uchinda, K. *Chem. Phys. Lett.* **1990**, *172*, 29.

20.   Lochmuller, C.H.; Wenzel, T.J. *J. Phys. Chem.* **1990**, *94*, 4230.

21.   Beechem, J.M.; Gratton, E. In *Time-Resolved Laser Spectroscopy in Biochemistry*; J.R. Lakowicz, Ed.; Proc. SPIE 909: Los Angeles, CA, 1988, pp 70.

22.   Beechem, J.M.; Ameloot, M.; Brand, L. *Chem. Phys. Lett.* **1985**, *120*, 466.

23.   Beechem, J.M.; Ameloot, M.; Brand, L. *Anal. Instrumn.* **1985**, *14*, 379.

24.   Alcala, J.R.; Gratton, E.; Prendergast, F.G. *Biophys. J.* **1987**, *51*, 587.

25.   Huang, J.; Bright, F.V. *J. Phys. Chem.* **1990**, *94*, 8457.

26.   Lakowicz, J.R.; Cherek, H. *Chem. Phys. Lett.* **1985**, *122*, 380.

27.   Debye, P. *Trans. Electrochem. Soc.* **1942**, *82*, 205.

RECEIVED November 25, 1991

Chapter 7

Fluorescence Investigation of Cosolvent–Solute Interactions in Supercritical Fluid Solutions

David L. Tomasko[1], Barbara L. Knutson[1], Charles A. Eckert[1,3],
Jeanne E. Haubrich[2], and Laren M. Tolbert[2]

[1]School of Chemical Engineering, Georgia Institute of Technology,
Atlanta, GA 30332–0100
[2]School of Chemistry and Biochemistry, Georgia Institute of Technology,
Atlanta, GA 30332–0400

One of the challenges to the development of supercritical fluid (SCF) processes is understanding the solution thermodynamics on a molecular scale. Fluorescence spectroscopy has been shown to probe the local environment around chromophores. We now extend its use to SCF systems containing a cosolvent.

Often a cosolvent is chosen to interact specifically with a given solute through hydrogen bonding or chemical complexation. In this work we investigate such interactions by fluorescence spectroscopy. Probe molecules such as 2-naphthol and its 5-cyano-derivative are effective chromophores for studying acid/base interactions since both are relatively strong photo-acids. In addition, 2-naphthol is a common solute for which SCF solubility and physical property data exist. Ultimately, spectroscopic information will be used to develop a clearer picture of the specific interactions which induce large cosolvent effects on solubility in SCF solutions.

Supercritical fluids (SCF's) offer a unique alternative to traditional separation techniques, particularly for specialty chemical applications. In addition to tuning the solvent power of an SCF with pressure, the addition of a cosolvent can increase the solubility of solutes by several orders of magnitude via specific interactions (1-2). Unfortunately, current phase equilibria models, which use predominantly bulk phase properties, cannot reproduce the behavior of these highly asymmetric solutions, especially those containing strong cosolvent/solute interactions. Some measure of the strength and number of cosolvent interactions is therefore needed and one method of gaining insight into these intermolecular interactions is fluorescence spectroscopy.

Spectroscopy and computer simulation have given an abundance of information regarding solvent/solute clustering (3-8). In fact, it is reasonably

[3]Corresponding author.

0097–6156/92/0488–0084$06.00/0

well established that the local density of solvent molecules around a solute molecule is quite different from the bulk density. If the solvent clusters it seems reasonable to assume that the cosolvent will do so as well especially when the cosolvent is capable of hydrogen bonding or otherwise complexing with the solute.

Solvatochromism is most often invoked as experimental evidence for local composition enhancements in liquid solutions. The non-linear dependence of absorption or emission energy on solvent composition is used to fit the "index of preferential solvation" to describe the difference between local and bulkcompositions. Various models have been proposed to account for preferential solvation in mixed solvents; two recent examples are those of Suppan (9) and Chatterjee and Bagchi (10). These models rely on data taken over the entire range of composition but in SCF/cosolvent mixtures one is interested primarily in the range 0-10 mole percent cosolvent. An alternative approach for this application arises from charge-transfer complexes and excited state reactions which may be very sensitive to local compositions (11). These processes can easily be monitored spectroscopically which may give some insight into the local composition about a solute.

Background

As a probe of local compositions around a solute, proton transfer from 2-naphthol was chosen since it has been shown to be sensitive to solvent structure in liquids. Naphthols are useful probe molecules due to the large change in acidity of the proton upon excitation (12). The pK_a of 2-naphthol is approximately 9.5 in the ground state and 2.7 in the excited state making it a fairly strong photo-acid (13). The solvent effect on proton transfer is shown schematically in Figure 1, where the solvated neutral naphthol is the most stable species in the ground state. However, with the correct solvent structure in the excited state, the proton may be solvated and the naphtholate anion may become more stable. This stabilization of the anion species in solution is sensitive to the orientation of solvent around the solute as determined from studies in liquid solvents (14). The dissociation of 2-naphthol is specific for water and is apparently independent of the dielectric constant of the solvent (15). The fluorescence quenching curve (Stern-Volmer plot), is non-linear with water concentration (see Figure 3) which indicates combined dynamic (diffusional) and static (ground state complexation) quenching. Finally, a random walk solvation model which incorporates the probability of forming water clusters of specific sizes around the naphthol implies that a minimum cluster size of four water molecules is required to facilitate proton transfer (16).

In supercritical fluids, the possibility of local composition enhancements of cosolvent about a solute suggests that we should see enhancement of anion fluorescence if the water cosolvent clusters effectively about the 2-naphthol solute. Although in liquids the water concentration must be >30% to see anion emission, the higher diffusivity and density fluctuations in SCF's could allow stabilization of the anion at much lower water concentrations provided that the water molecules provide sufficient structure. Therefore the purpose of these experiments was to investigate 2-naphthol fluorescence in supercritical CO_2 with water cosolvent in the highly compressible region of the mixture to probe the local environment about the solute.

Experimental

The fluorescence apparatus used in this work has been described previously (7). The SCF solutions of solute/cosolvent/solvent were prepared by first coating a small filter paper with a desired amount of solute from liquid solution and allowing the solvent to evaporate. The filter paper was attached to the top plug of a 1.6 liter pressure vessel and the entire vessel evacuated for thirty minutes down to < 0.15 torr. The solutes used in this study have a significantly lower vapor pressure than 0.15 torr so it is unlikely that any sample was lost during this procedure. A predetermined amount of cosolvent was weighed into a smaller pressure vessel which was connected in series with the larger vessel. After sealing the smaller vessel and evacuating the larger, the two were opened to each other and gaseous solvent (CO_2 in this case) was flushed three times through both at about 100 psi in order to remove any air in the cosolvent vessel and connecting tubing. The cosolvent vessel was heated 20-40°C above the boiling point of the cosolvent to ensure vaporization of the cosolvent and complete transfer into the larger mixing vessel. The solvent was compressed to the desired density then the large vessel was isolated from the cosolvent vessel and compressor. Heating tape was placed on one side of the vessel to induce thermal currents and the solution was allowed to mix for a period of 24 hours.

2-Naphthol was obtained from Aldrich (99+%) and used as received. 5-Cyanonaphthol was prepared as described previously (17).

Results and Discussion

Experiments were carried out in supercritical CO_2 with 0.003 mole fraction water at 35 and 50°C over a range of pressures which corresponded to specifically chosen density values. Our objective was to sample the regions above and below the critical density through the region of high compressibility since this is the region where maximum cosolvent composition enhancement would be expected.

The effect of water cosolvent on the emission of 2-naphthol is shown in Figure 2. There is some loss of fine structure but no detectable peak in the 400-500 nm range indicative of the anion species. This lack of anion emission is likely due to the low concentration of water in the solution, and since the concentration is limited by the water/CO_2 phase equilibria (18), there is a need for a more sensitive probe.

The sensitivity of the naphthols for water is determined by their Stern-Volmer quenching curves, or the inverse of the normalized quantum yield of the neutral species as a function of water concentration. The quenching curves for 2-naphthol and 5-cyano-2-naphthol (5CN2N) in tetrahydrofuran at ambient temperature are shown in Figure 3, from which it is clear that the addition of a cyano group increases the sensitivity of the probe to water, mirroring the increased excited state acidity (17,19). Insufficient data are available to completely separate the effects of static and dynamic quenching for the cyanonaphthol, however, following the analysis outlined by Lackowicz (20), the product of K_D and K_S (dynamic and static quenching constants respectively) is approximately 0.03 for 2-naphthol and is approximately 1.0 for 5CN2N indicating a tremendous enhancement of the interaction with water.

The cyano-naphthol has an additional benefit as a proton transfer probe in that the anion species is present in pure alcohols; not only will the probe be more

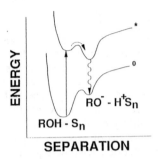

Figure 1. Schematic representation of solvent effect on proton transfer from naphthols.

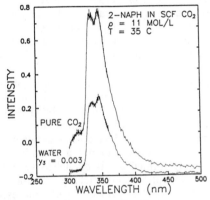

Figure 2. Effect of water cosolvent on the fluorescence emission of 2-naphthol.

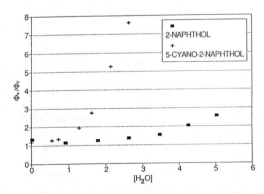

Figure 3. Stern-Volmer quenching curves for 2-naphthol and 5-cyano-2-naphthol.

sensitive to water but we may also observe proton transfer for CO_2 doped with methanol, which has a much higher solubility than water.

The effect of water (0.003 mole fraction) and methanol (0.02 mole fraction) cosolvents on the emission of 5CN2N is shown in Figure 4. For this probe there is complete loss of fine structure, a significant red shift in the neutral emission with methanol cosolvent, but again, no detectable anion emission.

The red shift of the 5CN2N emission due to methanol was determined relative to the second peak in the pure CO_2. The shift is shown in Figure 5 for two temperatures over the range of density studied. The reported mixture densities were found by multiplying the pure CO_2 density (interpolated from IUPAC data) by a calculated ratio of mixture to pure density. The ratio was calculated from the Soave-Redlich-Kwong equation of state using a binary interaction parameter (k_{12}) which had been fit to CO_2 - methanol VLE data at the temperature of interest. The parameters were 0.12 at 35 C and 0.14 at 50 C. Although the scatter in the data is slightly larger than the estimated uncertainty, there appears to be no clear dependence on either temperature or density which indicates that the thermodynamic properties of the solvent have little effect on the cosolvent/solute interaction in this case. The implication of this lack of dependence is that a measurement of the cosolvent/solute interaction in liquid solutions might be valid in the supercritical phase as well. A test of this supposition using equilibrium constants from liquid solutions in SCF phase equilibria models would be useful.

We suggest the following possible reasons for the absence of anion emission in these experiments: First, it is possible for water and methanol to form adducts with CO_2 which might inhibit their ability to solvate the proton. This possibility was checked using water as a cosolvent in supercritical ethylene, which will not interact with water; data do not indicate formation of an anion therefore solvent/cosolvent adducts are not a plausible explanation. Second, perhaps the anion is being formed but is quenched very efficiently. The anion species would be more susceptible to quenching than the neutral molecule. This possibility could be investigated in principle with fluorescence lifetime studies, but this has not been done. It is entirely possible that the anion emission is so weak as to be hidden underneath the tail of the neutral emission.

Finally, coupled with the phase equilibrium problem of concentrating the cosolvent, perhaps the composition of cosolvent around the probe is not sufficient to facilitate proton transfer. This implies that if the cosolvent is indeed "clustering" around the solute, there is insufficient structural integrity to solvate the proton.

Computer simulation visualizations (3) of local density augmentation show that the "clustering" is a dynamic process, with solvent molecules interchanging fairly rapidly in close proximity of the solute. Calculated viscosities are slightly higher in the region around the solute, and the diffusivity in SCF's is approximately two orders of magnitude greater than in liquids. Also, even with local density augmentation, the local density at the surface of the solute molecule is enough less than that in a liquid to create a very substantial increase in free volume, with a concomitant decrease in structure. If this is the case, a more sensitive probe molecule is needed to observe local composition enhancement.

There are a number of experiments using the proton transfer mechanism which may yet yield information regarding local compositions. A more polar solvent such as fluoroform may be able to support the proton in solution more

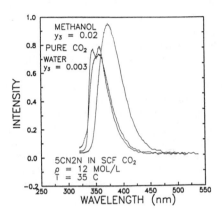

Figure 4. Effect of water and methanol cosolvents on the fluorescence emission of 5-cyano-2-naphthol.

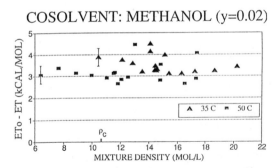

Figure 5. Methanol induced shift of 5-cyano-2-naphthol emission relative to the second peak in pure CO_2 as a function of density and temperature.

easily than the nonpolar solvents. Conversely, since information about the solute environment is desired, the real challenge is identifying probe molecules that are both sensitive to the small amounts of cosolvent and follow relatively simple photophysical kinetics.

Conclusions

Cosolvent effects on SCF solution behavior allow the tailoring of solvents for extractions and separations. The strong interactions in these systems currently defy prediction by popular computational methods. Only by understanding these interactions at a molecular level will we be able to guide the development of phase equilibria models successfully. One way of exploring the molecular level interactions is with spectroscopy of various kinds and we have demonstrated here an attempt to look at the cosolvent/solute interaction.

Proton transfer is sensitive to the local solute environment in liquid solutions as evidenced by the water quenching curves for 2-naphthol and its cyano- derivatives. We have used proton transfer as a mechanism to probe the cosolvent composition around a solute in supercritical fluids to discern any difference between local and bulk concentrations.   No proton transfer was observed from either 2-naphthol or 5-cyano-2-naphthol, presumably indicating insufficient structure in the SCF to solvate the proton. Although significant cosolvent effects on the fluorescence emission were observed, these appear to be independent of the thermodynamic variables.

Acknowledgments

LMT would like to acknowledge the support of the National Science Foundation through grant CHE-8805577. DLT, BLK, and CAE acknowledge financial support from the DuPont Co. and from the Department of Energy, through grant DE-FG22-88PC88922

Literature Cited

1.   Lemert, R.M., Johnston, K.P., "Chemical Complexing Agents for Enhanced Solubilities in Supercritical Fluid Carbon Dioxide," submitted to *I&EC Res.*, **1990**.
2.   Foster, N.R., University of New South Wales, personal communication, 1990.
3.   Kim, S., Johnston, K.P., "Molecular Interactions in Dilute Supercritical Fluid Solutions," *I&EC Res.*, **1987**, *26*, 1206.
4.   Kajimoto, O., Futakami, M., Kobayashi, T., and Yamasaki, K., "Charge-Transfer-State Formation in Supercritical Fluid: (N,N-Dimethylamino)benzonitrile in CF_3H," *J. Phys. Chem.*, **1988**, *92*, 1347.
5.   Petsche, I.B., Debenedetti, P.G., "Solute-solvent Interactions in Infinitely Dilute Supercritical Mixtures: A Molecular Dynamics Investigation," *J. Chem. Phys. ,* **1989**, *91*(11), 7075.
6.   Cochran, H.D., Lee, L.L., "Solvation Structure in Supercritical Fluid Mixtures based on Molecular Distribution Functions," *ACS Symp. Ser,* **1989**, *406*, 28.

7.   Brennecke, J.F., Tomasko, D.L., Peshkin, J., and Eckert, C.A.,
       "Fluorescence Spectroscopy Studies of Dilute Supercritical Solutions,"
       I&EC Res., **1990**, *29*, 1682.
8.   Knutson, B.L., Tomasko, D.L., Eckert, C.A., Petsche, I.B., and
       Debenedetti, P.G., this symposium.
9.   Suppan, P., "Time-resolved Luminescence Spectra of Dipolar Excited
       Molecules in Liquid and Solid Mixtures," *Farad. Disc. Chem. Soc.*, **1988**,
       85, 173.
10.   Chatterjee, P., Bagchi, S., "Study of Preferential Solvation in Mixed Binary
       Solvents by Ultraviolet-Visible Spectroscopy," *J. Chem. Soc. Farad.
       Trans.*, **1990**, *86*(10), 1785.
11.   Mataga, N., Kubota, T., Molecular Interactions and Electronic Spectra,
       Marcel Dekker Inc., New York, **1970**.
12.   Ireland, J.F., Wyatt, P.A.H., "Acid-Base Properties of Electronically
       Excited States of Organic Molecules," *Adv. Phys. Org. Chem.*, **1976**, *12*,
       131.
13.   Tsutsumi, K., Shizuka, H., "Proton Transfer and Acidity Constant in the
       Excited State of Naphthols by Dynamic Analyses," *Z. Phys. Chem.*, **1980**,
       122, 129.
14.   Lee, J., Griffin, R.D., and Robinson, G.W., "2-Naphthol: A Simple Example
       of Proton Transfer Effected by Water Structure," *J. Chem. Phys.*, **1985**,
       82(11), 4920.
15.   Huppert, D., Kolodney, E., Gutman, M., and Nachliel, E., "Effect of Water
       Activity on the Rate of Proton Dissociation," *J. Am. Chem. Soc.*, **1982**, *104*,
       6949.
16.   Lee, J., Robinson, G.W., Webb, S.P., Phillips, L.A., and Clark, J.H.,
       "Hydration Dynamics of Protons from Photon Initiated Acids," *J. Am.
       Chem. Soc.*, **1986**, *108*, 6538.
17.   Tolbert, L.M., Haubrich, J.E., "Enhanced Photoacidities of
       Cyanonaphthols," *J. Am. Chem. Soc.*, **1990**, *112*, 8163.
18.   Wiebe, R., "The Binary System Carbon Dioxide-Water Under Pressure,"
       Chem. Rev., **1941**, *29*, 475.
19.   Weller, A., "Study of Fast Reactions of Excited Molecules by a
       Fluorescence Technique," *Z. Elektrochem.*, **1960**, *64*, 55.
20.   Lackowicz, J.R., Principles of Fluorescence Spectroscopy, Plenum Press,
       New York, **1983**, pp 266-271.

RECEIVED November 25, 1991

Chapter 8

Investigations of Solute—Cosolvent Interactions in Supercritical Fluid Media

A Frequency-Domain Fluorescence Study

Thomas A. Betts and Frank V. Bright

Department of Chemistry, Acheson Hall, State University of New York at Buffalo, Buffalo, NY 14214

Static and dynamic fluorescence spectroscopy are used to investigate the local compositional changes in binary supercritical fluids on a picosecond time scale. A fluorescent solute molecule whose emission characteristics are sensitive to solvent polarity is used to probe the composition of the local solvent environment. The systems investigated were supercritical CO_2 with the addition of small amounts of the polar cosolvents, CH_3OH and CH_3CN. These systems exhibit a reorganization of the local solvent shell(s) about the fluorescent probe following optical excitation, a process known as solvent relaxation. Average rates for this dynamic solvation process were determined, and an Arrhenius analysis performed. It is shown that the binary supercritical fluid composed of CO_2 and CH_3OH can reorganize more rapidly than the CO_2-CH_3CN system.

Solvation in supercritical fluids depends on the interactions between the solute molecules and the supercritical fluid medium. For example, in pure supercritical fluids, solute solubility depends upon density (1-3). Moreover, because the density of supercritical fluids may be increased significantly by small pressure increases, one may employ pressure to control solubility. Thus, this density-dependent solubility enhancement may be used to effect separations based on differences in solute volatilities (4,5). Enhancements in both solute solubility **and** separation selectivity have also been realized by addition of cosolvents (sometimes called entrainers or modifiers) (6-9). From these studies, it is thought that the solubility enhancements are due to the increased local density of the solvent mixtures, as well as specific interactions (e.g., hydrogen bonding) between the solute and the cosolvent (10).

Cosolvent-modified supercritical fluids are also used routinely in supercritical fluid chromatography (SFC) to modify solute retention times (11-20). In these reports, cosolvents are used to alter the mobile and stationary phase chemistries (16,17,20). However, distinguishing between such effects in a chromatography

0097–6156/92/0488–0092$06.00/0

experiment is difficult at best unless one knows the specific effects and their magnitude.

Kim and Johnston (*21*), and Yonker and Smith (*22*) have used solute solvatochroism to determine the composition of the local solvent environment in binary supercritical fluids.   In our laboratory we investigate solute-cosolvent interactions by using a fluorescent solute molecule (a probe) whose emission characteristics are sensitive to its local solvent environment.   In this way, it is possible to monitor changes in the local solvent composition using the probe fluorescence.   Moreover, by using picosecond time-resolved techniques, one can determine the kinetics of fluid compositional fluctuation in the cybotactic region.

In this paper we focus on: 1) the kinetics of cosolvent solvation in supercritical media, and 2) determine how the nature of the cosolvent affects the solvation process.

Theory

To investigate changes in solvent composition about a fluorescent probe with time, one studies the time evolution of the probe emission spectrum.   The key segments of data in these experiments are the so called time-resolved emission spectra.   These are obtained by acquiring the time-resolved intensity decays at a series of emission wavelengths (λ_{em}) spanning the entire fluorescence spectrum.   In the simplest cases, if the system is accurately described by a two-state, excited-state model, the intensity decay at each wavelength will be bi-exponential.   In addition, the apparent decay times (τ_i) will remain constant across the emission spectrum and the pre-exponential amplitude factors ($\alpha_i(\lambda)$) will be wavelength dependent (*23*).   Thus, the wavelength-dependent time course of the fluorescence intensity decay is described by:

$$I(\lambda, t) = \sum_{i=1}^{n} \alpha_i(\lambda) \exp(-t/\tau_i) \tag{1}$$

where $\alpha_i(\lambda)$ and τ_i are recovered using frequency-domain fluorescence techniques (*24*).

In the frequency-domain, the experimentally measured quantities are the frequency- (ω) and wavelength- (λ) dependent phase shift ($\Theta_m(\lambda,\omega)$) and demodulation factor ($M_m(\lambda,\omega)$).   For any assumed decay model (equation 1), these values are calculated from the sine ($S(\lambda,\omega)$) and cosine ($C(\lambda,\omega)$) Fourier transforms.   If we assume the decay kinetics are described by a simple sum of exponential decay times we have (*24*):

$$S(\lambda,\omega) = \left(\sum_{i=1}^{n} \frac{\alpha_i(\lambda)\omega\tau_i^2}{(1 + \omega^2\tau_i^2)} \right) \div \left(\sum_{i=1}^{n} \alpha_i(\lambda)\tau_i \right) \tag{2}$$

$$C(\lambda,\omega) = \left(\sum_{i=1}^{n} \frac{\alpha_i(\lambda)\tau_i^2}{(1 + \omega^2\tau_i^2)} \right) \div \left(\sum_{i=1}^{n} \alpha_i(\lambda)\tau_i \right) \tag{3}$$

and for any set of $\alpha_i(\lambda)$ and τ_i, the calculated phase shift and demodulation factor are given by:

$$\theta_c(\lambda,\omega) = \arctan[S(\lambda,\omega)/C(\lambda,\omega)] \tag{4}$$

$$M_c(\lambda,\omega) = [S(\lambda,\omega)^2 + C(\lambda,\omega)^2]^{1/2}. \tag{5}$$

The decay parameters $[\alpha_i(\lambda)$ and $\tau_i]$ are recovered from the experimentally measured phase shift and demodulation factor by the method of non-linear least squares (24,25). The goodness-of-fit between the assumed model (c subscript) and the experimentally measured (m subscript) data is determined by the chi-squared (χ^2) function:

$$\chi^2 = \frac{1}{D} \left\{ \sum_{\omega,\lambda} \left(\frac{\theta_m(\omega,\lambda) - \theta_c(\omega,\lambda)}{\sigma_\theta} \right)^2 + \sum_{\omega,\lambda} \left(\frac{M_m(\omega,\lambda) - M_c(\omega,\lambda)}{\sigma_M} \right)^2 \right\} \tag{6}$$

where D is the number of degrees of freedom, and σ_θ and σ_M are the uncertainties in the measured phase angle and demodulation factor, respectively.

Time-resolved emission spectra are reconstructed from the normalized impulse response functions (26):

$$I'(\lambda,t) = N(\lambda)I(\lambda,t) \tag{7}$$

where $N(\lambda)$ is the wavelength-dependent normalization factor (26):

$$N(\lambda) = F(\lambda)\left\{ \int_0^\infty I(\lambda,t) \, dt \right\}^{-1} \tag{8}$$

and $F(\lambda)$ is the normalized steady-state fluorescence intensity at wavelength λ.

Once the time-resolved emission spectra are so generated, it is informative to monitor the time evolution of the emission spectra. To this end, it is convenient to focus on the time course of the emission center of gravity ($\nu(t)$) (27):

$$\nu(t) = \left(\int_0^\infty I'(\lambda,t)\nu \, d\nu \right) \left(\int_0^\infty I'(\lambda,t) \, d\nu \right)^{-1}. \tag{9}$$

Under our experimental conditions, multifrequency data are collected at six equally spaced emission wavelengths spanning the entire emission spectrum. For this particular case, the time-dependent emission center of gravity is then given by (27):

$$v(t) = 10,000 \left(\frac{\sum_\lambda I'(\lambda,t) \; \lambda^{-1}}{\sum_\lambda I'(\lambda,t)} \right). \tag{10}$$

In studies of normal liquids, time-dependent spectral shifts are often quantified by a spectral shift correlation function (S(t)) (*28-34*):

$$S(t) = \frac{v(t) - v(\infty)}{v(0) - v(\infty)} \tag{11}$$

where $v(0)$, $v(t)$, and $v(\infty)$ are the spectral centers of gravity at time zero (immediately following optical excitation), t (during the spectral relaxation process), and infinity (after the solute/solvent system has reached equilibrium), respectively. This S(t) function serves to normalize the observed spectral shift to the total shift and allows one to compare systems in which the absolute spectral shifts differ. In this paper we use the S(t) function to document the solvent reorganization in several binary supercritical fluid systems.

Experimental

SFC grade CO_2 was purchased from Scott Specialty Gases. 6-propionyl-2-dimethylaminonaphthalene (PRODAN) was obtained from Molecular Probes. HPLC grade CH_3OH and CH_3CN were from Aldrich, and all reagents were used as received.

The high pressure apparatus has been described previously (*35*). For the present studies it has been modified slightly to allow for injection of liquid cosolvents under pressure. In these experiments, the high pressure apparatus consisted of an Isco Model SFC-500 syringe pump which delivered CO_2 through a Rheodyne 7010 LC injector into a high pressure optical cell (*35*). The optical cell was constructed from 303 stainless steel. Four quartz windows (sealed with Teflon o-rings) were used to optically access the high pressure sample. A water jacket within the cell body was used to control temperature ($\pm$ 0.1 °C). The cell temperature was monitored with a thermocouple placed directly into the cell body.

Samples were prepared by pipetting an aliquot of PRODAN (dissolved in CH_3OH) into an optical cell. The cell was placed in a heated (65 °C) oven to evaporate the solvent. The cell was then sealed and connected to the remainder of the high pressure apparatus. A vacuum of 60 μm Hg was drawn on the entire system for about 1 hour to remove residual oxygen which would otherwise quench fluorescence. For this same reason, all cosolvents were deaerated by bubbling with N_2 for 20 minutes prior to injection. The optical cell was charged with CO_2 using the syringe pump and cosolvent was added under CO_2 flow by redirecting the CO_2 through the LC injector. This step also served to thoroughly mix the cosolvent with the CO_2. All experiments were conducted in the single phase, supercritical region.

All fluorescence measurements were performed with an SLM 48000 MHF multifrequency cross-correlation phase and modulation fluorometer. The excitation source was the 351.1 nm line from an argon-ion laser (Coherent model 90-6). Extraneous plasma discharge was removed from the excitation with a 340 nm band-pass filter. Either a monochromator or a series of band-pass filters (Oriel) were used to select the observed emission. All lifetime measurements were performed using magic angle polarization. POPOP (p-bis[2-(5-phenyloxazolyl)]benzene) in ethanol was used as the reference standard lifetime (1.35 ns) (*36*).

Time-resolved emission spectra were reconstructed from a set of multifrequency phase and modulation traces acquired across the emission spectrum (*37*). The multifrequency phase and modulation data were modeled with the help of a commercially available global analysis software package (Globals Unlimited). The model which offered the best fits to the data with the least number of fitting parameters was a series of bi-exponential decays in which the individual fluorescence lifetimes were linked across the emission spectrum and the pre-exponential terms were allowed to vary.

Results and Discussion

Steady-State Fluorescence. The fluorescence characteristics of PRODAN are extremely sensitive to the physicochemical properties of the solvent (*38*). As benchmarks, the steady-state emission spectra for PRODAN in several liquid solvents are presented in Figure 1. It is evident that the PRODAN emission spectrum red shifts with increasing solvent polarity. This red shift is a result of the dielectric properties of the surrounding solvent and the large excited-state dipole moment (ca. 20 Debye units) of PRODAN (*38*). It is the sensitivity of the PRODAN fluorescence that will be used here to investigate the local solvent composition in binary supercritical fluids.

Steady-state emission spectra of PRODAN in pure supercritical CO_2, and binary supercritical mixtures of CO_2 with CH_3OH or CH_3CN are illustrated in Figure 2. The addition of the polar cosolvents results in a red shift in the fluorescence emission of PRODAN relative to that in pure supercritical CO_2. The shift is not, however, due to a simple enrichment of the bulk dielectric properties of the supercritical solvent. With the addition of only 1.57 mol% CH_3OH, the emission maximum is already red shifted 38% of the way toward the emission maximum in pure CH_3OH. For 1.44 mol% CH_3CN, the shift is not nearly as great compared to CH_3OH, but it is still 11% of the way toward pure CH_3CN. These disproportionate shifts in the fluorescence emission are characteristic of specific solvent-solute interactions.

The pronounced shoulder in the emission spectrum of PRODAN in the binary supercritical fluid composed of CO_2 and CH_3OH indicates that at least two distinct species are emitting. That is, because PRODAN is the only fluorescent species in this system, emission must result from PRODAN in at least two different environments. Further, one environment must be more polar than the other(s). Figure 1 shows that a blue-shifted spectrum is indicative of a less polar environment. Thus, the shoulder on the blue edge of the emission spectrum of PRODAN in the CO_2/CH_3OH system is due to PRODAN in a nonpolar environment. The red edge is then due to PRODAN in a more polar environment. Therefore, at some point during the emission process, PRODAN emits from more than one environment. Time-resolved

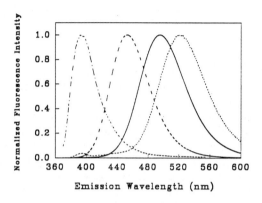

Figure 1. Steady-state emission spectra for 10 μM PRODAN in cyclohexane (— · —), acetonitrile (– – –), methanol (——) and water (------).

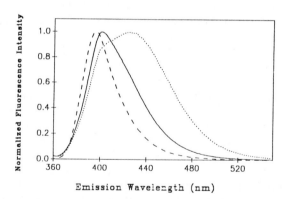

Figure 2. Steady-state emission spectra for 10 μM PRODAN in pure CO_2 (– – –), CO_2 + 1.44 mol% CH_3CN (——), and CO_2 + 1.57 mol% CH_3OH (------). All spectra were obtained at 45 °C and 81.4 bar.

emission spectra will show that the environment surrounding PRODAN becomes more polar with time after excitation.

Additional evidence for PRODAN emitting from two environments is given by the temperature dependence of emission spectra in the CO_2/CH_3OH binary supercritical fluid (Figure 3). As the temperature is increased (at constant density) the contribution from the red-shifted peak decreases while the relative intensity of the peak at 400 nm increases. This indicates that the less polar environment surrounding PRODAN is preferred at higher temperatures. In other words, the specific cosolvent-solute interactions responsible for the red shift in emission are disrupted at elevated temperatures.

Dynamic Fluorescence. Steady-state fluorescence can only provide insight into the average solvation process. In order to probe the solvation process in a detailed manner, one must utilize techniques which have greater time resolution (e.g., time-resolved fluorescence spectroscopy).

A typical set of recovered fluorescence decay times for PRODAN in binary supercritical fluids composed of CO_2 with CH_3OH or CH_3CN are compiled in Table I. Detailed global analysis of our data sets with a battery of far more complex models could not improve on the fits obtained with the bi-exponential decay law. The fact that our dynamic data are best fit by a bi-exponential decay model at each wavelength and that the decay times do not vary across the emission contour also supports the assertion that PRODAN exists in two different environments. Several interesting trends are apparent in these data (37), but the key point is that several of the pre-exponential factors (e.g., those at the red edge of CH_3OH) become negative. This is indicative of a process occurring in the excited state (23). Because of the complications induced by the excited-state process (23), it is not possible to assign any physical significance to the actual recovered lifetimes or pre-exponential factors (Table 1); these terms are convoluted with the excited-state reaction kinetics (37). Our goal here is to use the recovered fluorescence lifetimes and pre-exponential factors to accurately reconstruct the time-resolved emission spectra. As long as the model accurately describes the fluorescence decay, the fact that lifetimes and pre-exponential factors contain excited-state kinetic information will not influence the results presented here (27,37).

The time-resolved emission spectra were reconstructed from the fluorescence decay kinetics at a series of emission wavelengths, and the steady-state emission spectrum as described in the Theory section (37). Figure 4 shows a typical set of time-resolved emission spectra for PRODAN in a binary supercritical fluid composed of CO_2 and 1.57 mol% CH_3OH (T = 45 °C; P = 81.4 bar). Clearly, the emission spectrum red shifts following excitation; indicating that the local solvent environment is becoming more polar during the excited-state lifetime. We attribute this red shift to the reorientation of cosolvent molecules about excited-state PRODAN.

When PRODAN is excited to its lowest singlet state, it undergoes a change in dipole moment of approximately 20 Debye units (38). According to the Franck-Condon principle, absorption of a photon occurs faster than nuclei can move (31,40). Therefore, immediately following optical excitation, there will be an excited-state dipole moment produced, but this dipole is essentially surrounded by a ground-state solvent cage. That is, the system is not in equilibrium. With time, the solvent

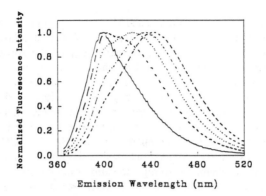

Emission Wavelength (nm)

Figure 3. Steady-state emission spectra for 10 μM PRODAN in CO_2 + 1.57 mol% CH_3OH at 36 (------), 40 (– · – ·), 45 (·····), 50 (– – –) and 60 (——) °C and constant CO_2 density (0.251 g/mL).

Table I. Recovered Fluorescence Lifetime Values for PRODAN in Several Binary Supercritical Solvent Mixtures[a]

Em λ (nm)	CO_2/1.57 mol% Methanol				CO_2/1.44 mol% Acetonitrile			
	α_1[b]	τ_1[c] (ns)	α_2[b]	τ_2[c] (ns)	α_1[b]	τ_1[c] (ns)	α_2[b]	τ_2[c] (ns)
400	0.902	0.242	0.098	1.67	0.821	0.563	0.179	1.83
420	0.688	0.242	0.312	1.67	0.697	0.563	0.303	1.83
440	0.441	0.242	0.559	1.67	0.675	0.563	0.325	1.83
460	0.130	0.242	0.870	1.67	0.655	0.563	0.345	1.83
480	-0.244	0.242	0.756	1.67	0.542	0.563	0.458	1.83
500	-0.428	0.242	0.572	1.67	0.579	0.563	0.421	1.83

a) Operating conditions were 45 °C and 81.4 bar.
b) $|\alpha_1 + \alpha_2| = 1$.
c) Linked parameter in global fit across the emission spectrum.

molecules will attempt to reorganize to minimize the unfavorable interactions created between themselves and the excited-state solute dipole. Minimization of these unfavorable interactions will lower the energy of the singlet excited state (28-37), producing a shift in the fluorescence to longer wavelengths (Figure 5). Thus, the red shift is a direct result of a solvent reorientation process, i.e., solvent relaxation (28,31).

The rate of shift from the unrelaxed (U state in Figure 5) to the relaxed (R state in Figure 5) state is a measure of the rate of solvent relaxation. To help visualize the time course of this shift, we have adopted the solvation correlation function (S(t); equation 11) which normalizes the shift at any time to the total shift in the emission spectrum (28-34). In this manner, one can easily compare systems which shift the emission spectra different magnitudes. Figure 6 shows typical S(t) functions obtained for PRODAN in binary supercritical fluids composed of CH_3OH or CH_3CN with CO_2. The binary supercritical mixture of CH_3OH and CO_2 clearly shift the emission spectrum faster compared to the CH_3CN + CO_2 system. This solvation process is extremely complex, but may be simplified by integrating the area under the S(t) curve to obtain an average solvation time constant ($<\tau_s>$) (28-34). Under these particular experimental conditions, $<\tau_s>$ for CO_2/CH_3OH and CO_2/CH_3CN are 128 ps and 432 ps, respectively. Of course, these two samples have slightly different cosolvent concentrations because we inject fixed volumes into the high pressure optical cells. Thus, it is necessary to compare the time constant per unit concentration of cosolvent. For the binary supercritical fluids composed of CO_2 with CH_3OH or CH_3CN (T = 45 °C; P = 81.4 bar), the recovered rate constants are $(8.07 \pm 0.31) \times 10^{10}$ $M^{-1}s^{-1}$ and $(3.09 \pm 0.14) \times 10^{10}$ $M^{-1}s^{-1}$, respectively. Clearly, even when the differences in concentration are accounted for, it is apparent that the CH_3OH-CO_2 binary supercritical fluid interacts much faster compared to CH_3CN-CO_2.

The difference between the steady-state emission spectra of PRODAN in the binary supercritical fluid composed of CO_2 and CH_3OH or CH_3CN (Figure 2) may now be partially explained. The excited-state lifetime of PRODAN in pure supercritical CO_2 at 40 °C is fairly short (150 ps). When PRODAN absorbs a photon, the CH_3CN in the supercritical solvent mixture does not interact fast or efficiently enough to populate the relaxed excited state before the unrelaxed state decays. Thus, the bulk of the fluorescence is emitted around 400 nm from the unrelaxed excited state. In the binary supercritical fluid composed of CO_2 and CH_3OH, the CH_3OH is able to effectively relax the initially excited state, populating the relaxed excited state, and thereby producing a red shift in the emission. In this case, there is still a contribution from the unrelaxed species (cf. shoulder in Figure 2).

Temperature Dependence. In order to understand why CH_3OH "solvates" faster compared to CH_3CN, the dependence of the solvation rate with temperature was investigated. Figure 7 shows the recovered temperature-dependent S(t) traces for PRODAN in the binary supercritical fluid composed of CO_2 and 1.57 mol% CH_3OH. The same experiment was carried out for the supercritical mixture of CO_2 and CH_3CN, and the trend is similar. In all cases, as temperature is increased, the solvent relaxation process becomes, as expected, faster. By determining the rate constants of the relaxation process as described above, we construct Arrhenius plots

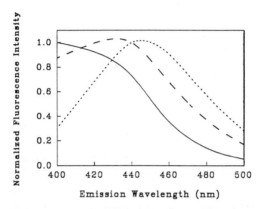

Figure 4. Time-resolved fluorescence emission spectra for PRODAN in CO_2 + 1.57 mol% CH_3OH at 100 ps (– – –), 400 ps (———), and 2.0 ns (------) after excitation. Operating conditions were 45 °C and 81.4 bar.

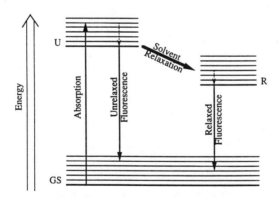

Figure 5. Energy level diagram for the excited-state solvent relaxation process. GS (ground state), U (unrelaxed singlet excited state), R (relaxed singlet excited state).

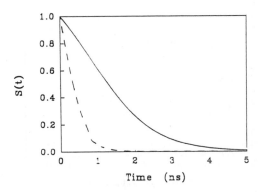

Figure 6. Recovered solvation correlation functions (S(t)) for PRODAN in CO_2 + 1.57 mol% CH_3OH (– – –), and CO_2 + 1.44 mol% CH_3CN (——). Operating conditions were 45 °C and 81.4 bar.

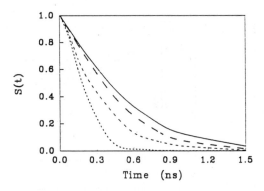

Figure 7. Recovered solvation correlation functions (S(t)) for 10 μM PRODAN in CO_2 + 1.57 mol% CH_3OH at 36 (——), 40 (– – –), 45 (-----), and 50 °C (⋯⋯) with constant bulk density (0.251 g/mL).

where the slope and intercept reveal the energy of activation and the collisional frequency, respectively.

The results from the Arrhenius analyses indicate that the CO_2/CH_3OH system has a higher activation barrier (74.2 $\pm$ 7.7 kJ/mol) for the solvation process compared to CO_2/CH_3CN (43.3 $\pm$ 18 kJ/mol). However, CO_2/CH_3OH has a faster solvation rate. The reason for this fast solvation rate lies in the collisional or Arrhenius frequency factor. The collisional frequency factor for CO_2/CH_3OH and CO_2/CH_3CN are (1.6 $\pm$ 0.3) x 10^{14} $\pm$ 10 M^{-1} s^{-1} and (4.2 $\pm$ 0.8) x 10^8 M^{-1} s^{-1}, respectively.

Preliminary results from absorbance analyses indicate that CH_3OH interacts more strongly with PRODAN in the ground state than does CH_3CN (results not shown). That is, the CH_3OH molecules are in closer proximity (on average) to the PRODAN compared to CH_3CN. This is not too unexpected considering the Lewis acid and base characteristics of the CH_3OH and PRODAN, respectively. Thus, when PRODAN absorbs a photon, the CH_3OH is already close, maybe already in contact, and has an opportunity to interact with the probe more frequently during the PRODAN excited-state lifetime compared to CH_3CN. In fact, if we look at the Arrhenius factors we see that the CH_3OH results are significantly faster than expected based solely on diffusion control (*41*) and we propose that there is some degree of pre-association between the CH_3OH and PRODAN.

It is also well known (*42,43*) that CH_3OH and CH_3CN self associate in the liquid state (*41,42*). When dissolved in a nonpolar supercritical fluid such as CO_2, it seems reasonable to believe that aggregates of CH_3OH and CH_3CN are still present (albeit to a lesser degree). However, because CH_3OH can hydrogen bond effectively with other CH_3OH (in addition to the solute) molecules it would take more energy to disrupt these aggregates compared to CH_3CN. Hence, one would expect the activation energy to be larger for CH_3OH compared to CH_3CN. Again, this is exactly what is seen experimentally. Therefore, although CH_3OH is in closer proximity to PRODAN it must also break hydrogen bonds with its neighboring cosolvents in order to interact with (i.e., relax) the PRODAN.

Conclusions

We have utilized the static and dynamic fluorescence characteristics of an environmentally-sensitive solute molecule, PRODAN, to investigate the local solvent composition in binary supercritical fluids. In the two solvent systems studied (CO_2/1.57 mol% CH_3OH and CO_2/1.44 mol% CH_3CN), specific cosolvent-solute interactions are clearly evident. Time-resolved fluorescence emission spectra indicate that the cosolvent-solute interactions become more pronounced with time after excitation. Hence, the local composition of cosolvent around the excited-state solute becomes greater than that surrounding the ground-state solute. That is, the photon-induced increase in excited-state dipole moment drives picosecond cosolvent augmentation about PRODAN.

The average rate of this solvent reorganization was faster for the CO_2/CH_3OH compared to the CO_2/CH_3CN system. The energy of activation for the CO_2/CH_3OH excited-state solvation process was greater compared to CO_2/CH_3CN and is probably a result of the strong hydrogen-bonding nature of the CH_3OH relative to CH_3CN.

The reason for the faster rate of solvation for CO_2/CH_3OH lies in the greater collisional frequency. This is a result of the closer proximity between PRODAN and CH_3OH compared to CH_3CN in the ground state.

Acknowledgments

This work has been generously supported by the United States Department of Energy (DE-FG02-90ER14143), an ACS Analytical Division Fellowship (TAB) sponsored by the Tennessee Eastman Company, and the SUNY-Buffalo Mark Diamond Research Fund.

Literature Cited

1.   Diepen, G.A.; Scheffer, F.E.C. *J. Phys. Chem.* **1953**, *57*, 575.
2.   Johnston, K.P.; Eckert, C.A. *AIChE J.* **1981**, *27*, 773.
3.   Kurnik, R.T.; Holla, S.J.; Reid, R.C. *Chem. Eng. Data.* **1981**, *26*, 47.
4.   Paulaitis, M.E.; Krukonis, V.J.; Kurnik, R.T.; Reid, R.C. *Rev. Chem. Eng.* **1983**, *1*, 179.
5.   McHugh, M.A. in *Recent Developments in Separation Science*; Li, N.N.; Carlos, J.M., Eds.; CRC Press: Boca Raton, FL, 1984; Vol. IX.
6.   Schmitt, W.J.; Reid, R.C. *Fluid Phase Equilib.* **1986**, *32*, 77.
7.   Dobbs, J.M.; Johnston, K.P. *Ind. Eng. Chem. Res.* **1987**, *26*, 1476.
8.   Dobbs, J.M.; Wong, J.M.; Lahiere, R.J.; Johnston, K.P. *Ind. Eng. Chem. Res.* **1987**, *26*, 56.
9.   Brunner, G. *Fluid Phase Equilib.* **1983**, *10*, 289.
10.   Walsh, J.M.; Ikonomou, G.D.; Donohue, M.D. *Fluid Phase Equilib.* **1987**, *33*, 295.
11.   Yonker, C.R.; Smith, R.D. *J. Chromatogr.* **1980**, *201*, 241.
12.   Hirata, Y.; Nakata, F. *J. Chromatogr.* **1984**, *315*, 295.
13.   Hirata, Y. *J. Chromatogr.* **1984**, *315*, 31.
14.   Blilie, A.L.; Greibrokk, T. *Anal. Chem.* **1985**, *57*, 2239.
15.   Mourier, P.A.; Eliot, E.; Caude, M.H.; Rosset, R.H.; Tambute, A.G. *Anal. Chem.* **1985**, 57, 2819.
16.   Leyendecker, D.; Schmitz, F.P.; Klesper, E. *Chromatographia* **1987**, *23*, 171.
17.   Levy, J.M.; Ritchey, W.M. *J. Chromatogr. Sci.* **1986**, *24*, 242.
18.   Leyendecker, D.; Schmitz, F.P.; Leyendecker, D.; Klesper, E. *J. Chromatogr.* **1987**, 393, 155.
19.   Schoenmakers, P.J.; Verhoeven, F.C.C.J.G.; Van Den Bogaert, H.M. *J. Chromatogr.* **1986**, 371, 121.
20.   Yonker, C.R.; McMinn, D.G.; Wright, B.W.; Smith, R.D. *J. Chromatogr.* **1987**, *393*, 19.
21.   Kim, S.; Johnston, K.P. *AIChE J.* **1987**, *33*, 1603.
22.   Yonker, C; Smith, R.D. *J. Phys. Chem.* **1988**, *92*, 2374.

23.   Lakowicz, J.R. *Principles of Fluorescence Spectroscopy*; Plenum: New York, 1983, Chapter 12.

24.   Gratton, E.; Jameson, D.M.; Hall, R.B. *Ann. Rev. Biophys. Bioengr.* **1984**, *13*, 105.

25.   Bevington, P.B. *Data Reduction and Error Analysis for Physical Sciences*; McGraw-Hill, New York, 1969.

26.   Easter, J.H.; DeToma, R.P.; Brand, L. *Biophys. J.* **1976**, *15*, 571.

27.   Lakowicz, J.R.; Cherek, H. *Chem. Phys. Lett.* **1985**, *122*, 380.

28.   Maroncelli, M.; Fleming, G.R. *J. Chem. Phys.* **1987**, *86*, 6221.

29.   Nagarajan, V.; Brearly, A.M.; Kang, T.-J.; Barbara, P.F. *J. Chem. Phys.* **1987**, *86*, 3183.

30.   Maroncelli, M.; Castner, E.W.,Jr.; Bagchi, B.; Fleming, G.R. *Faraday Discuss. Chem. Soc.* **1988**, 85, 199.

31.   Simon, J.D. *Acc. Chem. Res.* **1988**, *21*, 128.

32.   Maroncelli, M.; MacInnis, J.; Fleming, G.R. *Science* **1989**, *243*, 1674.

33.   Rips, I.; Klafter, J.; Jortner, J. *J. Chem. Phys.* **1988**, *89*, 4288.

34.   Kahlow, M.A.; Kang, T.; Barbara, P.F. *J. Chem. Phys.* **1988**, *88*, 2372.

35.   Betts, T.A.; Bright, F.V. *Appl. Spectrosc.* **1990**, *44*, 1196.

36.   Lakowicz, J.R.; Cherek, H.; Balter, A. *J. Biochem. Biophys. Methods* **1981**, *5*, 131.

37.   Betts, T.A.; Bright, F.V. *Appl. Spectrosc.* **1990**, *44*, 1203.

38.   Weber, G.; Farris, F.J. *Biochemistry* **1979**, *18*, 3075.

39.   Lakowicz, J.R.; Balter, A. *Biophys. Chem.* **1982**, *16*, 99.

40.   Ware, W.R.; Lee, S.K.; Brant, G.J.; Chow, P.P. *J. Chem. Phys.* **1971**, *54*, 4729.

41.   Birks, J.B. *Photophysics of Aromatic Molecules*; Wiley-Interscience, New York, 1987.

42.   Wenzel, H.; Krop, E. *Fluid Phase Equilib.* **1990**, *59*, 147.

43.   Uosaki, Y.; Matsumura, H.; Ogiyama, H. *J. Chem. Thermo.* **1990**, *22*, 797.

RECEIVED January 15, 1992

Chapter 9

Laser Flash Photolysis Studies of Benzophenone in Supercritical CO_2

John E. Chateauneuf[1], Christopher B. Roberts[2], and Joan F. Brennecke[2]

[1]Radiation Laboratory and [2]Department of Chemical Engineering,
University of Notre Dame, Notre Dame, IN 46556

Supercritical fluids (SCF's) exhibit much potential as solvents for reactions, in addition to their traditional use as extraction solvents. However, very little is known from a molecular standpoint of the effect of the supercritical solvent on reaction rates. With the objectives of understanding these molecular level interactions in SCF's, and the possibility of "tuning" chemical reactivity in mind, we have initiated a broad range of photochemical investigations in SCF's. Foremost is the use of time-resolved spectroscopies in combination with well-characterized photochemical probes to elucidate solvent/solute interactions in these unique fluids. Herein, we present our initial laser flash photolysis measurements of observed and bimolecular reaction rate constants as a function of pressure for the reaction of benzophenone triplet with isopropanol in supercritical carbon dioxide. The triplet is produced by laser flash photolysis and its disappearance measured directly with time-resolved UV-VIS absorption. The pressure effect on the bimolecular reaction rate constant is negative; i.e., the reaction rate decreases as the pressure is increased and the conditions moved further from the critical point. This is a definitive example where the reaction rate can be increased by operation near the critical point. While the results of this and subsequent studies indicate the influence of changing local compositions near the critical point, one must also consider the observed trends in terms of transition state theory and possible cage effects. The results of this study are important because they open the possibility of enhancing reaction rates by operation in a supercritical fluid near the critical point and address the important effects of the intermolecular interactions between the solvent and the reactants and transition state. A better understanding of these interactions may lead to the prediction of reactions best suited for operation in supercritical fluids.

Supercritical fluids (SCF's) have received much attention as potential solvents for extractions (Paulaitis et al., 1982; Rizvi et al., 1986; McHugh and Krukonis, 1986; Brennecke and Eckert, 1989). However, they may also be well suited as solvents for

0097–6156/92/0488–0106$06.00/0

reactions. High diffusion coefficients and low viscosities may cut down on mass transfer resistance in diffusion controlled reactions, the products may be separated easily downstream with a simple change in temperature or pressure, and the solvent effect on the reaction rate constant and the reaction rate may be significant.

While studies of reactions in supercritical fluids abound, only a few researchers have addressed the fundamental molecular effects that the supercritical fluid solvent has on the reactants and products that can enhance or depress reaction rates. A few measurements of reaction rate constants as a function of pressure do exist. For instance, Paulaitis and Alexander (1987) studied the Diels Alder cycloaddition reaction between maleic anhydride and isoprene in SCF CO$_2$. They observed bimolecular rate constants that increased with increasing pressure above the critical point and finally at high pressures approached the rates observed in high pressure liquid solutions. Johnston and Haynes (1987) found the same trends in the unimolecular decomposition of α-chlorobenzyl methyl ether in supercritical difluoroethane. In the naphthalene/triethylamine exciplex formation (Brennecke et al., 1990a) absolute bimolecular rate constants were not reported; however, the pressure effect on those rate constants showed the same trend as above - increasing rate with increasing pressure. Penninger and coworkers (1989) studied the chemistry of methoxynaphthalene in supercritical water. While the simultaneous hydrolysis and pyrolysis reaction made the results somewhat more difficult to interpret, they were able to calculate a bimolecular rate constant, which decreased with increasing pressure. While this trend is opposite of those mentioned above, the actual <u>rate</u> of the reaction did increase with increasing pressure because the reaction is acid-catalyzed and the concentration of H$_3$O$^+$ is a strong function of pressure and temperature in supercritical water. Similarly, Narayan and Antal (1990) found a decrease in the rate constant with increasing pressure for the acid catalyzed dehydration of n-propanol in supercritical water. Finally, Gehrke et al. (1990) reported rates for the unimolecular photoisomerization of diphenylbutadiene, studied with picosecond absorption spectroscopy. This is the one other study reported in the literature using time-resolved absorption to examine a reaction in supercritical fluids. An acceleration of the isomerization was attributed to a lowering of the potential energy barrier induced by solvent clustering in several SCF's.

A very useful way to study systems on a molecular scale is with spectroscopic techniques. In fact a number of investigators have gained insight about the equilibrium molecular structure in supercritical fluids from steady state UV absorption and fluorescence spectroscopies (Yonker et al., 1986; Yonker and Smith, 1988; Kim and Johnston, 1987a,b; Brennecke et al., 1990a,b). In this paper we use laser flash photolysis to study the reaction of benzophenone triplet with isopropanol in supercritical carbon dioxide. When benzophenone is irradiated with laser light the molecules are excited to their first excited singlet state. Almost instantaneously these molecules experience intersystem crossing to the ground triplet state. This triplet state is very reactive towards hydrogen atom donors, such as an alcohol. The triplet and the isopropanol react to form a ketyl radical. The triplet has an absorption spectrum in the UV-VIS range so its depletion by reaction can be monitored with time-resolved absorption spectroscopy. The reaction proceeds as follows:

benzophenone   excited singlet   triplet

triplet   isopropanol   ketyl radical

Given sufficient time the triplet benzophenone also will decay back to the ground state singlet in the absence of a hydrogen atom donor in a unimolecular process as follows:

benzophenone triplet   ground state singlet

These are well understood reactions that have been studied extensively in liquid solutions (Turro, 1978; Small and Scaiano, 1978), micelles (Scaiano et al., 1982; Jacques et al., 1986), and high pressure liquids (Okamoto and Teranishi, 1986; Okamoto, 1990).

The reaction mixture contains a dilute solution of benzophenone solute, a few mole percent isopropanol and the remainder supercritical carbon dioxide solvent. This is analogous to a solute dissolved in a supercritical fluid/cosolvent mixture. These types of systems are important because in many applications researchers have found that the addition of a small amount of cosolvent (such as acetone or an alcohol) of volatility intermediate between that of the solute and the SCF can greatly enhance the solubility of the solute (Van Alsten, 1986).

The purpose in studying this reaction is to examine the solvent effect on the reaction rate and how that can be controlled by operation in a supercritical fluid. We seek to gain a fundamental understanding of the interactions with the solvent and how those interactions affect the rate of the reaction between benzophenone triplet and isopropanol.

Experimental

Apparatus. The aforementioned hydrogen abstraction reaction was studied with laser flash photolysis. A diagram of this system is shown in Figure 1. An 8 ns pulse from a Quanta Ray DCR-1 Nd:YAG laser provided ~8 mJ excitation at 355 nm. The laser beam entered a high pressure optical cell at 90° from the line of the transient absorption measurement. A detailed design of the optical cell can be found elsewhere (Brennecke, 1989). The only modification is the use of silicone o-rings, that have been rinsed repeatedly with cyclohexane to remove any possible contaminants. A 1000 W pulsed xenon lamp serves as the absorption light source. The light transmitted through the cell of path length 1.3 cm is directed through a 420 nm cut off filter to eliminate any scattered light from the laser and overtones of the ketyl radical absorption.

The desired wavelength, primarily 580 nm, was monitored with a monochromater and a photomultiplier tube. The absorption signal is then digitized by

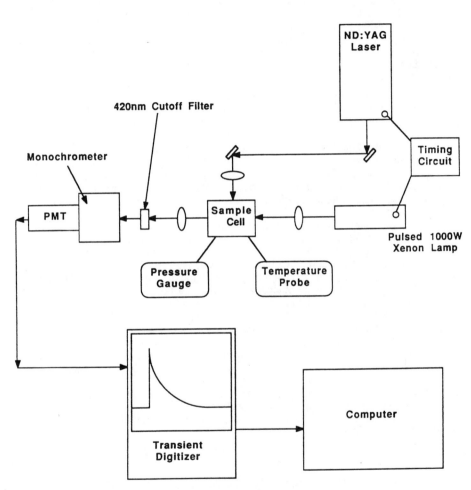

Figure 1   Laser flash photolysis apparatus.

a transient digitizer which is interfaced with a VAX-11/780 computer system for data collection and manipulation. This equipment has been described in detail previously (Fessenden and Nagarajan, 1985).

Recrystallized Gold Label benzophenone (99+%) was obtained from Aldrich Chemical. SCF grade carbon dioxide was received from Scott Specialty Gases and 2-propanol (certified ACS grade) was obtained from Fisher Chemical. Both the CO_2 and the 2-propanol were used as received.

Absolute rate constants were measured for alcohol concentrations of 0.25 to 7.2 mole % (0.01 to 1.3 Molar) while benzophenone concentrations studied were in the range of .0002 to .0006 mole fraction (0.001 to 0.01 Molar). Pressures were studied from 71 bar, which is slightly below the critical pressure, up to 110 bar. A model 901A Heise pressure gauge was used to monitor the pressure of the sample cell and the temperature of the cell, tightly fitted with insulation tape, was maintained at $33.0^oC \pm 0.1^oC$ with an Omega (model CN-6070A) temperature controller equipped with a Watlow Firerod cartridge heater. The critical pressure and temperature of CO_2 are 73.98 bar and 31^oC, respectively. In all cases, care was taken to insure that the samples were in the one phase region.

Sample Preparation. The sample preparation apparatus consists of the SCF grade CO_2 cylinder fitted with a dip tube to deliver liquid CO_2, an Isco high pressure syringe pump (model 1200), a reservoir containing 2-propanol, and the high pressure optical cell. Each component was connected with 1/16" stainless steel high pressure tubing and compartmentalized with high pressure (HIP) line valves. All valves were fitted with teflon o-rings. During sample preparation great care was taken to minimize molecular oxygen since O_2 is an efficient quencher of triplet benzophenone. Prior to sample preparation the entire assemble was extensively purged with CO_2 at low pressure. The 2-propanol was de-airated with high purity nitrogen and kept under a nitrogen atmosphere. A given amount of N_2-saturated alcohol was then drawn into the syringe pump and the remaining volume of the pump filled with liquid CO_2. With this known amount of alcohol and liquid carbon dioxide contained in the pump, the mole fractions of the system were determined with the IUPAC CO_2 density data (Angus et al., 1976). The pump was heated, which increased the pressure such that the solution was one phase at all times (Radosz, 1986). Part of the contents of the Isco pump was then loaded into the sample cell, which contained an appropriate amount of crystalline benzophenone, and the cell heated to facilitate mixing and raise the contents above the critical point. Benzophenone concentrations were determined with a Cary 1 UV-VIS spectrophotometer, and were typically 0.8 absorbance at the laser excitation wavelength. Laser experiments were run from high to low pressures by allowing the homogeneous solution to escape via an external valve. Ample time was allotted for equilibration at the individual pressures, as indicated by a stabilization in temperature.

Results

A spectrophotometric investigation of the ground-state absorption of benzophenone in pure SC CO_2 and in the presence of 2-propanol was performed in order to determine both benzophenone solubility and the potential for solute/solute or co-solvent/solute aggregations. Linear Beer's law plots of absorbance versus concentration of benzophenone over a concentration range of 1-15 mM in SC CO_2 insured that the ketone was fully disolved and in a non-associated form under the conditions of the laser experiments (~ 10 mM). Analysis of the absorption spectrum of benzophenone in SC CO_2 ranging from high to low pressure, resulted in absorption bands nearly identical to those observed in liquid cyclohexane. A similar analysis in SC CO$_2$ with 3.5 mole % 2-propanol resulted in only ± 2nm displacement of the spectral bands of benzophenone, which is within the resolution of the spectrophotometer.

Laser flash photolysis of benzophenone in pure SC CO$_2$ at 33°C and 110 bar resulted in generation of triplet benzophenone within the laser excitation pulse (Figure 2, insert) and a triplet lifetime (1/k) of ~ 2.5 μs, which is similar to that observed in N$_2$-saturated acetonitrile. Monitoring the absorption signal 200 ns after laser excitation and construction of a point by point absorption spectrum gave the characteristic spectrum of triplet benzophenone (Figure 2), which has maxima at 320 and 525 nm and, for comparison, is nearly identical to that observed in liquid acetonitrile (Figure 2). Reaction kinetics were examined by monitoring the changes in the rate of decay of the absorption signal in the presence of added substrate. Throughout the experiments, 580 nm was chosen as the monitoring wavelength in order to eliminate interference from ketyl radical, which has a similar visible absorption band, however, having negligible intensity at 580 nm. The absorption decay traces of the benzophenone triplet were measured at various pressures on a single isotherm. The experiments were done at the following alcohol mole fractions: x = 0.0025; x = 0.0087; and x = 0.0118; x = 0.0207; x = 0.0364; x = 0.0514 and x = 0.0719. Since these alcohol concentrations greatly exceed the concentration of the triplet benzophenone generated, the kinetics observed may be treated as psuedo-first-order. Indeed, kinetic analysis of a typical time-resolved decay trace (Figure 2, insert) fits a single exponential, yielding an observed psuedo-first-order rate constant, k_{obs}. For each of the alcohol concentrations studied, the observed rate constants were found to increase with a decrease in pressure. A plot of observed rate constants versus pressure, as in Figure 3, illustrates a sharp increase in k_{obs} as the critical point is approached.

The rate of benzophenone triplet decay is a combination of the unimolecular decay and the bimolecular reaction with the alcohol and can be described by:

$$\frac{-d[BP^3]}{dt} = k_{obs} [BP^3] = k_0 [BP^3] + k_{bm} [BP^3] [RH] \qquad (1)$$

where: k_{obs} = observed rate constant

k_0 = rate constant in absence of alcohol

k_{bm} = bimolecular rate constant [M^{-1}s^{-1}]

$[BP^3]$ = benzophenone triplet concentration

$[RH]$ = alcohol concentration [M]

Dividing by $[BP^3]$, the bimolecular rate constant is the slope of k_{obs} vs. alcohol concentration at constant pressure and k_0 is the intercept. k_0 represents all modes of triplet decay in the absence of alcohol, including any reaction with impurities. Examples of bimolecular quenching plots at various pressures are shown in Figure 4. The pressure dependence of the bimolecular rate constant is obtained by plotting $\ln(k_{bm})$ against pressure as given in Figure 5. Again the bimolecular rate constant increases as the critical pressure is approached from higher pressures.

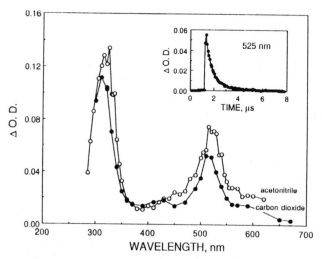

Figure 2   Transient absorption spectrum observed 200 ns after 355 nm laser
            excitation of benzophenone in supercritical CO_2 at 117 bar and
            33°C compared with that obtained in N_2-saturated CH_3CN. Insert:
            Typical benzophenone triplet decay trace observed at 525 nm in
            supercritical CO_2.

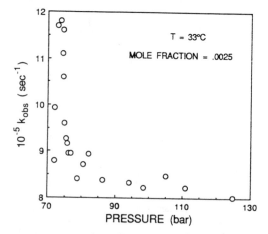

Figure 3   The observed pseudo-first order rate constants as a function of
            pressure for benzophenone triplet in supercritical CO_2 at 33°C and
            isopropanol mole fraction of 0.0025.

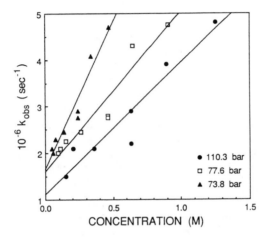

Figure 4   The determination of the bimolecular rate constant for the hydrogen abstraction from isopropanol by benzophenone triplet at 110.3, 77.6, 73.8 bar and 33°C in supercritical CO$_2$.

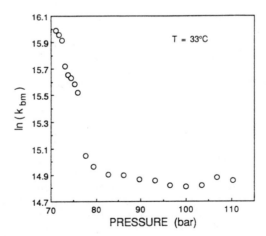

Figure 5   The bimolecular rate constant for isopropanol + benzophenone triplet at 33°C as a function of pressure in supercritical CO$_2$.

Discussion

Supercritical fluid solvents can act in a variety of ways to affect reaction rates. Since the reaction rate is the product of the rate constant and the concentrations of the reactants, one must consider the solvent effect on the rate constant itself (discussed below), as well as changes in concentrations. It is this second possibility that has not been addressed until this study; i.e., the possible influence of changes in the local concentrations of the reactants in the compressible region near the critical point.

Looking at the wavelength of maximum absorption of a solvatochromic dye in supercritical fluid mixtures, Kim and Johnston (1987a, b) showed that the local concentration of a cosolvent around a solute molecule increased to as high as seven times that in the bulk as the pressure of the sample was lowered and the critical point approached. Since the reaction investigated in this study mimics the systems investigated by Kim and Johnston (a dilute solute in a supercritical fluid solvent with 0.5 - 5 mole % cosolvent present), the increase in the apparent bimolecular rate constant, which is based on bulk concentrations, may actually reflect an increase in the local concentrations of the isopropanol. In fact, the measured rate constants increased by a factor of 2 over approximately the same pressure range that resulted in a 4 fold increase in the local compositions measured by Kim and Johnston. It is interesting to note that in the work of Kim and Johnston the local compositions at a given pressure increase by a constant factor regardless of the bulk concentration of cosolvent. For instance, at 85 bar the local composition is approximately 2 times the bulk for acetone concentrations from 1 to 5.25 mole %. Clearly, as the bulk concentration is increased further this can no longer be true and for the reaction presented here one might expect some curvature at higher concentrations of the plots shown on Figure 4. However, our experiments are within the approximate concentration range investigated by Kim and Johnston.

Also, it is interesting to note that the bimolecular rate constants obtained in supercritical CO_2 near the critical pressure are greater than can be obtained in liquid isopropanol at any pressure, as shown in Figure 6. Moreover, as the pressure increases in the supercritical fluid, the rate constants appear to approach those in liquid solutions. The liquid data are those of Okamoto and coauthors (Okamoto and Teranishi, 1986; Okamoto, 1991) for the reaction in neat liquid isopropanol. Unlike most of the examples cited in the introduction where the rate constant was very low near the critical point and increased to nearly the liquid value at high pressures, in this study we actually obtain rate constants five times those that can be obtained in liquids. This is an example of a type of reaction that may benefit from operation in a supercritical fluid solvent.

At this point we believe that changing local compositions is the dominant effect in the apparent rate increase for the reaction of benzophenone triplet and isopropanol in supercritical CO_2 and this conclusion is supported by subsequent studies of the reaction of the triplet with other quenchers (Roberts et al., 1991). However, one must also consider the thermodynamic pressure effect on the rate constant in terms of transition state theory and possible cage effects.

The thermodynamic pressure effect on the reaction rate constant can be explained in terms of transition state theory (Evans and Polanyi, 1935), when the reactants are in thermodynamic equilibrium with a transition state. Once the transition state complex is formed it proceeds directly to products. With this analysis the pressure effect on the reaction rate constant can be given as follows:

$$A + B \rightleftharpoons [\text{transition state}]^{\neq} \longrightarrow \text{products}$$

$$RT \frac{\partial \ln k_{bm}}{\partial P} = -\Delta v^{\neq} - RT\, k_T \tag{2}$$

where k_{bm} = bimolecular rate constant $(M^{-1}s^{-1})$

$$\Delta v^{\neq} = \bar{v}_{ts} - \bar{v}_A - \bar{v}_B$$

$\bar{v}_i$ = partial molar volume

k_T = isothermal compressibility

The last term in the equation is a result of the reactant concentrations changing with pressure. If the rate expressions are written in pressure independent units (i.e., mole fraction) that last term drops out of the equation. While partial molar volumes in liquids are usually only a few cc/gmol, they can be very large negative values in supercritical fluids, as shown in Figure 7 (Ziger et al., 1986). As a result, the pressure effect on the reaction rate constant can be very significant. This emphasizes that in supercritical fluids the pressure effect on the reaction rate constant reflects the relative strengths of the intermolecular interactions (both the repulsive size effects and the attractive interactions) between the reactants and the transition state with the SCF solvent, as born out in the experimentally determined partial molar volumes. This is very different than the situation in liquid solvents where partial molar volumes are on the order of 5-10 cc/gmol and the pressure effect on rate constants can be used to discern information on the size change that occurs in going from the reactants to the transition state. For further discussion see Johnston and Haynes (1987).

While we believe the local composition effect discussed above to be an important influence on the measured rates, we could not come to this conclusion without examining the pressure dependence of the bimolecular rate constant for the benzophenone triplet/isopropanol reaction. From Figure 8, where the experimental $RT \{\partial \ln k_{bm}/\partial P\}$ is plotted as a function of pressure, it is clear that it is negative and experiences a sharp dip in the region of the critical pressure. This means that the rate constant decreases with increasing pressure and that the rate is higher near the critical point. This is in sharp contrast to most of the other examples of measured rate constants in the literature and opens the possibility of increasing reaction rates by operation in the region near the critical point. The right side of equation 2 is as follows:

$$- (\bar{v}_{ts} - \bar{v}_{benzophenone\ triplet} - \bar{v}_{alcohol}) - RT\, k_T$$

Partial molar volumes and the isothermal compressibility can be calculated from an equation of state. Unfortunately, these equations require properties of the components, such as critical temperature, critical pressure and the acentric factor. These properties are not known for the benzophenone triplet and the transition state. However, they can be estimated very roughly using standard techniques such as Joback's modification of Lyderson's method for T_c and P_c and the standard method for the acentric factor (Reid et al., 1987). We calculated the values for the benzophenone triplet assuming a structure similar to ground state benzophenone. The transition state was considered to be a benzophenone/isopropanol complex. The values used are shown in Table 1.

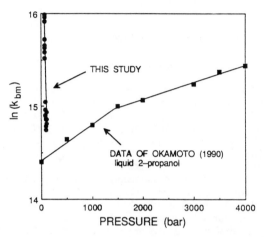

Figure 6   Comparison of the rate constants in supercritical carbon dioxide to those in liquid isopropanol at high pressures (Okamoto, 1990).

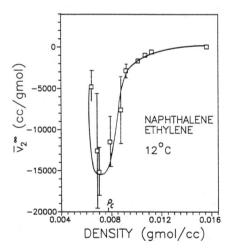

Figure 7   Measured infinite dilution partial molar volume of naphthalene in supercritical ethylene at 12°C (Ziger et al., 1986).

Table 1. Physical Properties Used in Estimations of Activation Volumes

Component	T$_c$(K)	P$_c$ (bar)	ω
Carbon Dioxide[*]	304.2	73.8	.225
isopropanol[**]	508.3	47.6	.248
Triplet Benzophenone[***]	821.0	33.0	.557
Transition State[***]	994.0	24.58	.993

[*]Angus et al., 1976
[**]Reid et al., 1987
[***]Estimated, see text.

For simplicity, the Peng-Robinson equation of state should suffice in providing reasonable estimates of the partial molar volumes of the benzophenone triplet, isopropanol and transition state, as well as the isothermal compressibility. Since no better information was available, the binary interaction parameters for all pairs were set to zero. The predicted values of RT ∂lnk$_{bm}$/∂P are shown in Figure 9.

These calculations predict the pressure effect on the rate constant to be positive; i.e. based on transition state theory and the partial molar volumes the rate constants should increase with increasing pressure. This is the exact opposite trend from the measurements reported here, as can be seen by comparing the experimental results in Figure 8 and the calculated numbers in Figure 9. Even though there is significant uncertainty in the estimation of the critical properties and acentric factors, as well as the ability of the Peng-Robinson equation to predict partial molar volumes, the sign of the activation volume is undoubtedly predicted correctly. As a result, the measured rate constants can not be explained by the thermodynamic pressure effect predicted from transtion state theory.

Since the observed results cannot be explained by partial molar volumes, the possibility of cage effects is another situation that must be considered. A cage effect is any situation in which the solvent orientation around the reactants unduly speeds or hinders the reaction. Arguments could be made that the solvent ordering around a solute, as born out in the large negative partial molar volumes, could be considered a cage. Moreover, there is some precedence to support the possibility of cage effects hindering the benzophenone triplet reaction, especially at very high pressures, even though the reaction is not diffusion controlled. The data of Okamoto and coworkers (Okamoto and Teranishi, 1986; Okamoto, 1990) was shown earlier in Figure 6 for benzophenone in liquid 2-propanol at very high pressures. Below 2000 bar they observe an increase in the rate constant with pressure. However, at higher pressures the slope changes and the rate constant actually decreases for some alcohols. Their explanation is that the solvent forms a cage around the intermediates, effectively preventing the product from escaping. For such a mechanism to dominate in the present observations the solvent cage near the critical region would require considerable integrity. It has, however, recently been suggested by O'Shea et al. (1991) that no such solvent organization occurs in SC CO$_2$ since no cage effect was observed in the decarbonylation of dibenzyl ketones in SC CO$_2$ near the critical point.

In summary, the measured rate constants (based on bulk concentrations) increase as the pressure is decreased near the critical point. This cannot be explained solely on the basis of the pressure effect on the rate constant predicted from transition state theory or cage effects. As a result, we believe that local composition increases near the critical point play an important role in the rate increase.

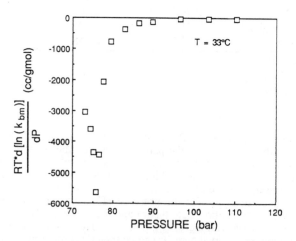

Figure 8   Plot of the pressure effect on the bimolecular rate constant as a
            function of pressure for benzophenone triplet + isopropanol in
            carbon dioxide at 33°C.

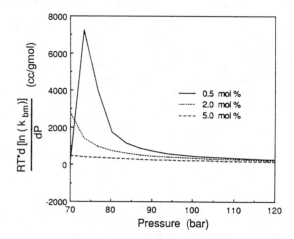

Figure 9   Values of the pressure effect on the reaction rate constant of
            benzophenone triplet + isopropanol in SCF CO_2 predicted from the
            Peng-Robinson equation of state at various alcohol concentrations.

Conclusions

In this, our initial investigation into absolute reactivity in SCF's, the pressure effect on the reaction rate constant for the hydrogen abstraction reaction between benzophenone triplet and isopropanol in supercritical CO$_2$ has been investigated by laser flash photolysis. The bimolecular rate constant increased as the pressure was decreased in approaching the critical point. This study is unique in presenting experimental evidence of the pressure effect on a reaction that is in the opposite direction of that observed in liquid solutions and presenting a definitive example of an increased rate near the critical point. This emphasizes the possibility of enhancing reactivity under relatively mild conditions by operation in a supercritical fluid. We present three possibilities to explain the experimentally observed increase in the reaction rate near the critical point, including the pressure effect on the reaction rate constant, local composition effects and cage effects. Considering the present data, the local composition effect appears to be dominant in this reaction.

Acknowledgments

This work was supported by the National Science Foundation under grant NSF-CTS 90 09562, the American Chemical Society Petroleum Research Fund grant 22947-G5 and the Office of Basic Energy Sciences of the Department of Energy (Notre Dame Radiation Laboratory Contribution No. NDRL-3374).

Literature Cited

Angus, S.; Armstrong, B.; de Reuck, K. M.; eds., International Thermodynamic Tables of the Fluid State: Carbon Dioxide, Pergamon Press, Oxford, 1976.

Brennecke, J. F.; Intermolecular Interactions in Supercritical Fluid Solutions from Fluorescence Spectroscopy, PhD Thesis, University of Illinois, Urbana, 1989.

Brennecke, J. F.; Eckert, C. A.; AIChE J., 1989, 35(9), 1409-1427.

Brennecke, J. F.; Tomasko, D. L.; Eckert, C.A.; J. Phys. Chem., 1990a, 94, 7692-7700.

Brennecke, J. F.; Tomasko, D. L.; Peshkin, J.; Eckert, C. A.; Ind. Eng. Chem. Res., 1990b, 29(8), 1682-1690.

Evans, M.G.; Polanyi, M.; Trans. Faraday Soc., 1935, 31, 875.

Fessenden, R. W.; Nagarajan, V.; J. Phys. Chem., 1985, 89, 2330-2335.

Gehrke, Ch.; Schroeder, J.; Schwarzer, D.; Troe, J.; Voβ, F.; J. Chem. Phys., 1990, 92(8), 4805-4816.

Jacques, P.; Lougnot, D. J.; Fouassier, J. P.; Scaiano, J. C.; Chem. Phys. Lett., 1986, 127(5), 469-474.

Johnston, K. P.; Haynes, C.; AIChE J., 1987, 33(12), 2017-2026.

Kim, S.; Johnston, K. P.; AIChE J., 1987a, 33(10), 1603.

Kim, S.; Johnston, K. P.; Ind. Eng. Chem. Res., 1987b, 26, 1206.

McHugh, M.A.; Krukonis, V. J.; Supercritical Fluid Extraction: Principles and Practice, Butterworths, Boston, 1986.

Narayan, R; Antal, J. M.; J. Am. Chem. Soc., 1990, 112, 1927.

Okamoto, M.; J.Phys. Chem., 1990, 94, 8182-8186.

Okamoto,M.; Teranishi, H.; J. Am. Chem. Soc., 1986, 108, 6378-6380.

O'Shea K. E.; Combes, J. R.; Fox, M. A.; Johnston, K. P.; Photochem. and Photobiol., 1991, 54, 571-576.

Paulaitis, M. E.; Alexander, G. C.; Pure & Appl. Chem., 1987, 59(1), 61-68.

Paulaitis, M. E.; Krukonis, V. J.; Kurnik, R. T.; Reid, R.C.; Rev. Chem. Eng.,
1982, 1(2), 179.
Peng, D.-Y.; D.B. Robinson; Ind. Eng. Chem. Fund., 1976, 15(1), 59.
Penninger, J. M. L.; Kolmschate, J. M. M.; In Supercritical Fluid Science and
Technology; Johnston, K.P.; Penninger, M. L., Eds.; ACS Symposium Series No.
406; American Chemical Society: Washington , DC, 1989; 242-258.
Radosz, M.; J. Chem. Eng. Data., 1986, 31, 43-45.
Reid, R. C.; Prausnitz, J. M.; Poling, B. E.; The Properties of Gases and Liquids,
4th ed., McGraw-Hill, New York, 1987.
Rizvi, S.S.H.; Benado, A. L.; Zollweg, J. A.; Daniels J. A.; Food Technol., 1986,
40(6), 55.
Roberts, C. B.; Chateauneuf, J. E.; Brennecke, J. F.; J. Am. Chem. Soc.,
manuscript in preparation.
Small, R. D., Jr.; Scaiano, J.C.; J. Phys. Chem., 1978, 82(19), 2064-2066.
Scaiano, J. C.; Abuin, E. B.; Stewart, L. C.; J. Am. Chem. Soc., 1982, 104, 5673-
5679.
Turro, N. J.; Modern Molecular Photochemistry, Benjamin/Cummings, Menlo Park,
CA, 1978, p. 362.
Van Alsten, J. G.; Structural and Functional Effects in Solutions with Pure and
Entrainer-Doped Supercritical Solvents, PhD Thesis, Univ. of Illinois, Urbana,
1986.
Yonker, C.R.; Smith, R.D.; J. Phys. Chem., 1988, 92, 2374.
Yonker, C.R.; Frye, S.L.; Kalkwarf, D. R.; Smith, R.D.; J. Phys. Chem., 1986,
90, 3022.
Ziger, D. H; Johnston, K. P.; Kim, S; Eckert, C. A.; J. Phys. Chem., 1986, 86,
2738.

RECEIVED November 25, 1991

Chapter 10

Spectroscopic Investigations of Organometallic Photochemistry in Supercritical Fluids

Steven M. Howdle, Margaret Jobling, and Martyn Poliakoff

Department of Chemistry, University of Nottingham, University Park, Nottingham, United Kingdom NG7 2RD

IR spectroscopy is used to monitor photochemical reactions of organometallic molecules in supercritical fluids at near ambient temperatures eg scXe, scCO$_2$ & scC$_2$H$_4$. The ability to create high concentrations of reactant gases in supercritical fluids has been exploited to generate so-called "non-classical" dihydrogen complexes such as W(CO)$_5$(H$_2$) and CpMn(CO)$_2$(H$_2$). The unique spectroscopic transparency of scXe allows these species to be identified using FTIR spectroscopy. Photochemical reactions in scC$_2$H$_4$ are also discussed, and a method for recovering compounds from our reactors is outlined.

Abbreviation Throughout this paper, we use the prefix "sc" to denote a material in its supercritical state (ie at temperatures and pressure above, but close to, their critical values.)

Our research group at Nottingham has been one of the first to apply supercritical fluids to the study of organometallic chemical reactions (*1-6*). Organometallic species are important in fields as diverse as industrial catalytic reactions and biological processes (*7*), and the application of supercritical fluids to such reactions offers possibilities which would be difficult to achieve in conventional solvents. Most of the organometallic reactions which we have studied involve metal carbonyl species. These are chemical complexes which in general contain a number of carbon monoxide groups bonded through the carbon atom to a transition metal center in the form M(CO)$_x$. In addition, metal carbonyl complexes may also contain simple organic ligands such as benzene, cyclopentadienyl (C$_5$H$_5$) and ethylene. Such metal carbonyl complexes are particularly amenable to spectroscopic studies in supercritical fluids because in general, they are reasonably volatile and thus have excellent solubilities in supercritical fluids at near-ambient temperatures; many reactions can be induced cleanly by use of ultra violet (UV) light; the infra red (IR) bands associated with the C-O stretching vibrations, v(C-O), of metal carbonyl groups are very intense and the number and relative intensities of v(C-O) bands observed in the IR spectrum allow one to

0097-6156/92/0488-0121$06.00/0
© 1992 American Chemical Society

distinguish very easily between reactants and products, and indeed to deduce the nature of the products (8,9).

We have already described the design of our miniature reactors, the overall concept and the details of the window seals (1,10). Briefly, the cell has a small volume, ca. 2ml, on grounds of both safety and cost (when using scXe!). It is constructed from stainless steel and can be filled from an external syringe pump or by condensing gaseous material into the cold finger. Although our original cell was fitted with only one port, the current cell has two ports to allow fluids to be flowed through. In addition, we now use chromatography valves (SSI) which have a negligible dead volume compared to the cell itself. The contents of the cell are stirred by a steel ball bearing which is moved magnetically up and down the cold finger by a solenoid.

Our initial experiments in supercritical fluids were designed to exploit the unique spectroscopic transparency of supercritical xenon (scXe). Because of its monatomic nature, Xe possesses no vibrational characteristics and hence is completely transparent to IR radiation (11). The IR spectrum of $CpMn(CO)_3$ (Cp = η^5-C_5H_5) dissolved in scXe (1100 psi, ambient temperature) is shown in Figure 1. Within the region 4500 - 1000 cm^{-1}, one can observe at least 50 different IR bands which may be assigned to the organometallic species. The important features to note in this spectrum are: the ν(C-H) vibrations in the region 3400 - 2900 cm^{-1} which allow one to characterise the organic ligand; the ν(C-O) vibrations in the region 2100 - 1800 cm^{-1}, which in this case are totally absorbing; the 2 x ν(C-O) overtone and combination vibrations in the region 4200 - 3900cm^{-1}. These latter bands are approximately an order of magnitude weaker in intensity than the ν(C-O) vibrational bands. Later in this paper, we show how these 2 x ν(C-O) bands are extremely useful for monitoring photochemical reactions of organometallic species in more conventional supercritical fluid media.

Supercritical Xenon is also transparent throughout the UV and near-IR thus allowing excellent characterization by UV spectroscopy. Figure 2 shows the UV visible spectrum of the same supercritical solution of $CpMn(CO)_3$. Because of the relatively high concentration of $CpMn(CO)_3$ in this experiment, ca. $10^{-3}M$, the spectrum is totally absorbing in the far UV. This means that it is important that our supercritical solution be stirred in order to avoid "inner filter effects" during the course of a photochemical reaction. In the near UV and visible regions, the absorption maxima may be measured accurately permitting the progress of a photochemical reaction of the organometallic species to be monitored quantitatively (12).

Photochemical Reactions of Metal Carbonyls with Dihydrogen

Over the past few years, there has been great interest in the reactions of organometallic species with dihydrogen, H_2. The product of such a reaction may take one of two forms. Usually, irradiation by UV light causes loss of a CO ligand from the metal to be replaced by two individual M-H bonds. Thus, dihydrogen is added oxidatively to the metal, the H-H bond is broken, and a classical dihydride species results.

$$M(CO)_x + H_2 \longrightarrow M(CO)_{x-1} (H)_2 + CO$$

classical dihydride complex

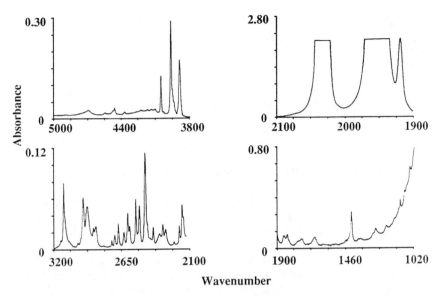

Figure 1. IR spectrum of $CpMn(CO)_3$ (ca. $10^{-3}M$) dissolved in scXe (1100 psi) at ambient temperature, obtained with an optical pathlength of 1.8mm. The unique spectroscopic transparency of scXe reveals approximately 50 bands associated with $CpMn(CO)_3$ in the region 4500 - 1000 cm^{-1}. The region below 1000 cm^{-1} is cut off by the CaF_2 window material employed in our cell.

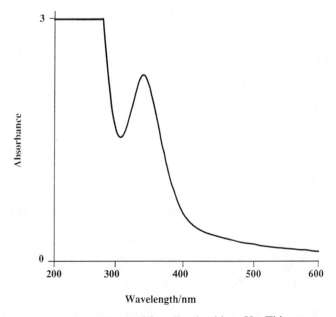

Figure 2. UV spectrum of $CpMn(CO)_3$ dissolved in scXe. This spectrum was recorded immediately after, and from the same sample, as that described in Figure 1.

However, in 1984, a second stable bonding mode was discovered where the dihydrogen bonds sideways to the metal centre as a single ligand with the H-H bond remaining intact and the oxidation state of the metal unchanged (13). This is the so called "non-classical" dihydrogen group (14,15,16).

$$M(CO)_x + H_2 \longrightarrow M(CO)_{x-1} (H_2) + CO$$

$$\text{M} \overset{\text{H}}{\underset{\text{H}}{\rule[-0.5em]{0.4pt}{1.5em}}}$$

non - classical
dihydrogen complex

These two types of metal/hydrogen bonding can be distinguished using IR spectroscopy by monitoring not only the changes which take place in the $v(C-O)$ bands on reaction but also the appearance of other features in the IR spectrum which indicate the nature of bonding of dihydrogen more directly (13,14). For example, the vibration of the H-H bond in gaseous dihydrogen causes no change in the dipole moment of the molecule, and hence no IR band is observed. When dihydrogen is bound to a metal centre however, $v(H-H)$ becomes weakly IR active because the dipole moment along the $M-(H_2)$ axis changes as the H-H bond stretches. Hence, observation of this weak IR band provides unequivocal proof of the presence of a non-classical dihydrogen group (4,17).

Our attempts to study such reactions with dihydrogen gas in scXe led us to uncover a previously unrecognized advantage of supercritical fluids. In a conventional hydrocarbon solvent, the reaction of an organometallic complex with a reactant gas usually proceeds in the liquid phase with the dissolved gas in equilibrium with the gas phase reservoir. In such a system, an increase in the pressure of the reactant gas will lead to an increase in the concentration of dissolved gas as predicted by Henry's Law. The solubility of H_2 in a hydrocarbon solvent is quite low; an overpressure of 15psi (1 atm.) will dissolve in cyclohexane to give a solution of 3.8×10^{-3} M at 25°C (18). If a supercritical fluid is used, however, one has a single phase system and the hydrogen is completely miscible with the "solvent". In our supercritical cell, a pressure of 1 atm. of H_2 produces a solution of 4×10^{-2} M. This concentration of H_2 is an order of magnitude greater than that for a conventional solvent at atmospheric pressure. Under very much higher pressures of H_2 (typically we use 1500psi), one might expect this difference in H_2 concentration to be even more pronounced because it is frequently difficult to equilibrate completely liquids and gases in conventional high pressure cells.

The high concentration of dihydrogen gas in our experiments has allowed us to stabilize complexes which previously had only a fleeting existence at ambient temperatures. For example, the species $W(CO)_5(H_2)$ has a lifetime in conventional solvents of somewhat less than one second at room temperature. Under our supercritical conditions, however, the lifetime may be extended to more than three minutes. Furthermore, the stability conferred upon this molecule by our supercritical system, and the unique spectroscopic transparency of scXe have allowed us to detect the very weak $v(H-H)$ band of coordinated dihydrogen using only a conventional FTIR spectrometer and a powerful UV lamp (4) as shown in Figure 3.

We have also applied these unique experimental conditions to the study of new organometallic complexes of dihydrogen, for example the photochemical reactions of two quite similar species $CpM(CO)_3$ (M=Mn & Re) with dihydrogen (2). This example illustrates the ease with which the $v(C-O)$ vibrations of photoproducts may be compared with those of the parent molecules to indicate the nature of the product.

The spectra in Figure 4 show the $v(C-O)$ regions of $CpMn(CO)_3$ and $CpRe(CO)_3$ from two separate experiments after photolysis with UV light. The complexes were initially dissolved in scXe in the presence 1500psi H_2 giving a total pressure ca. 3000psi at 300K. The IR spectra for these two $CpM(CO)_3$ complexes are almost identical. On photolysis, new bands (colored black) appear in the $v(C-O)$ region of both complexes. The $v(C-O)$ bands of these photoproducts are quite different for the two metals. The bands of the Re species appear **between** those of $CpRe(CO)_3$, while the bands of the Mn photoproduct are shifted **down** in wavenumber relative to $CpMn(CO)_3$

The Re photoproduct can be identified from its $v(C-O)$ spectrum as the known (19) classical dihydride, $CpRe(CO)_2H_2$, formed by oxidative addition of H_2 to the Re centre, thus changing the oxidation state from +I to +III. The $v(C-O)$ bands of the Mn photoproduct are actually very close in position to those of the known (20) species $CpMn(CO)_2(N_2)$, thus indicating that the oxidation state in this photoproduct (+I) has remained the same as the starting material (2). The photoproduct almost certainly contains H_2 bonded to the metal because there are small but definite shifts in the separation of the $v(C-O)$ bands when D_2 is used instead of H_2. Despite the IR transparency of scXe, no bands directly associated with coordinated η^2-H_2 were observed for the photoproduct in the case of $CpMn(CO)_3$ probably because there are relatively strong $v(C-H)$ bands due to the cyclopentadienyl group in the region ca. 3100 - 2700 cm^{-1} where $v(H-H)$ bands might be expected to occur (17) as shown in Figure 1. Thus, these observations indicate that photolysis of $CpMn(CO)_3$ in the presence of H_2 leads to the formation of a new non-classical dihydrogen complex $CpMn(CO)_2(H_2)$ rather than a dihydride, equation 1.

$$CpMn(CO)_3 + H_2 \xrightarrow{\text{UV, scXe}} CpMn(CO)_2(H_2) + CO \qquad (1)$$

$CpMn(CO)_2(H_2)$ is surprisingly stable in scXe at room temperature under a pressure of H_2, the intensities of the IR bands decay by around 50% overnight, and the parent bands are regenerated. If however, the H_2 is vented and the cell is refilled with N_2, the compound reacts smoothly with N_2 to form the known compound $CpMn(CO)_2(N_2)$ (Figure 5). This dinitrogen complex may also be generated photochemically directly from $CpMn(CO)_3$. The change of the reactant gases is achieved by simply cooling the cold finger of the reaction cell in liquid nitrogen, thus freezing the Xe as a solid. The H_2 may then be removed by pumping on the cell through the vacuum frame followed by addition of a high pressure of N_2 gas. Of course, a small amount of the H_2 does remain trapped in the solid Xe, but the added N_2 is far in excess of this. Under the conditions used, this exchange reaction was complete in less than 70 min, see equation 2

$$CpMn(CO)_3 + N_2 \xrightarrow{\Delta, \text{scXe}} CpMn(CO)_2(N_2) + H_2 \qquad (2)$$

In contrast to this behaviour, the dihydride, $CpRe(CO)_2H_2$, does not react thermally with N_2, even over periods of many hours in the presence of high pressures

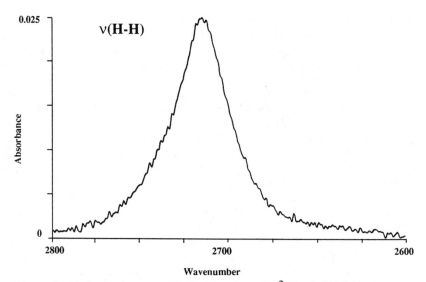

Figure 3. IR band due to ν(H-H) vibration of η^2-H$_2$ in W(CO)$_5$(H$_2$). The spectrum was recorded at ambient temperature with a total pressure (H$_2$ + scXe) of 3000psi and a 1.8 mm optical pathlength. Under these conditions, W(CO)$_5$(H$_2$) has a lifetime of approximately 3 min. The spectrum was recorded whilst the sample was being irradiated with UV light.

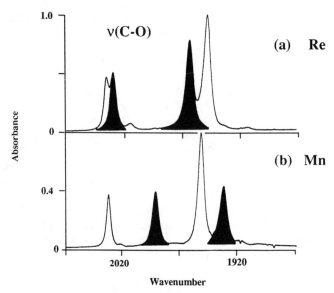

Figure 4. IR spectra in the ν(C-O) region, obtained after (a) 10 min u.v. photolysis of CpRe(CO)$_3$ and (b) 5 sec u.v. photolysis of CpMn(CO)$_3$ in scXe and H$_2$ at 25°C; the colored bands are assigned respectively to the photoproducts, CpRe(CO)$_2$H$_2$ and cpMn(CO)$_2$(H$_2$); the unlabelled bands are due to residual CpM(CO)$_3$.

of N_2. These experiments outline our use of the unique spectral transparency of scXe and the high concentration of added gas which allow us to generate new organometallic species and to characterize them spectroscopically. Such high concentrations of reactant gases in supercritical fluids have recently been exploited by other workers *(21)*

Photochemical Synthesis of Ethylene Complexes

The study of species in which ethylene is coordinated to transition metal centres holds great interest in areas of catalytic and polymerization chemistry *(7)*. The bonding of the ethylene ligand to the metal centre in such species has been compared to that of the dihydrogen complexes described above *(14,15,22)*. Photolysis of chromium hexacarbonyl, $Cr(CO)_6$, in conventional solvents in the presence of dissolved ethylene gas is known to lead initially to a highly labile species in which one CO ligand is replaced by ethylene. Further photolysis leads to a more stable compound which contains two ethylene ligands trans to each other across the metal centre *(23)*, equation 3. The conventional synthesis is experimentally difficult; the two photochemical

$$Cr(CO)_6 \underset{\Delta}{\overset{UV,\ C_2H_4}{\rightleftharpoons}} Cr(CO)_5(C_2H_4) \underset{\Delta}{\overset{UV,\ C_2H_4}{\rightleftharpoons}} trans\text{-}\ Cr(CO)_4(C_2H_4)_2 \quad (3)$$

reactions must be carried out at low temperatures ca. -50 °C in order to prevent the highly labile intermediate complex, $Cr(CO)_5(C_2H_4)$, decaying thermally. However, the success of our spectroscopic experiments involving dihydrogen complexes suggested that such ethylene complexes might be prepared, and stabilized, at ambient temperatures, in supercritical fluids containing high concentrations of ethylene. Fortuitously, ethylene itself becomes supercritical at ambient temperature and modest pressures (T_c 9°C, P_c 50.4 bar). Hence, very high concentrations of ethylene may be achieved simply by using scC_2H_4 both as solvent and reactant.

Figure 6 shows the v(C-O) region of the IR spectrum of a solution of $Cr(CO)_6$ dissolved in scC_2H_4 at 25°C and 1900psi. Plainly, there are IR bands in this spectrum which arise from the ethylene solvent itself. In order to prepare a synthetically significant quantity of product, it is necessary to use a high concentration of the starting species, $Cr(CO)_6$. As a consequence of this, the IR bands of both $Cr(CO)_6$ and ethylene mask any IR evidence for formation of $Cr(CO)_4(C_2H_4)_2$. Hence, one might imagine that monitoring of any photochemical processes using IR spectroscopy will be extremely difficult.

After 30 sec. UV photolysis of $Cr(CO)_6$ in scC_2H_4, a new band appears in the only part of the v(C-O) spectrum which is not obscured by the absorptions of scC_2H_4. This v(C-O) band can be assigned to the first photoproduct, $Cr(CO)_5(C_2H_4)$, which rapidly reaches a steady state concentration. Although the parent compound $Cr(CO)_6$ continues to be consumed, no other new IR bands can be detected in the v(C-O) region of the spectrum. However, as described earlier for $CpMn(CO)_3$ dissolved in scXe, (Figure 1) the overtone/combination bands (2x v(C-O)) of the carbonyl vibrations can provide excellent characterisation of the organometallic species present in solution. Fortunately, this region of the IR spectrum is unobscured in the scC_2H_4 solution. Hence in the region, 4200 - 3900 cm^{-1} (Figure 7), it is relatively easy to monitor the depletion of the parent species and subsequent growth of the stable product, $Cr(CO)_4(C_2H_4)_2$. Because the steady state concentration

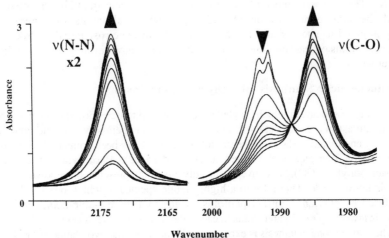

Figure 5. IR spectra in the ν(N-N) and part of the ν(C-O) regions showing the thermal reaction of CpMn(CO)$_2$(H$_2$) with N$_2$ in scXe at 25°C. CpMn(CO)$_2$(H$_2$) was generated photochemically before the addition of N$_2$. The first spectrum shown in the Figure was recorded as soon as the cold finger had warmed to room temperature and subsequent spectra were taken at 10 min intervals. The band labelled ▼ is due to CpMn(CO)$_2$(H$_2$) and those labelled ▲ are due to the known compound CpMn(CO)$_2$(N$_2$). Note that the ν(N-N) band has been plotted with a x2 expansion in the ordinate scale.

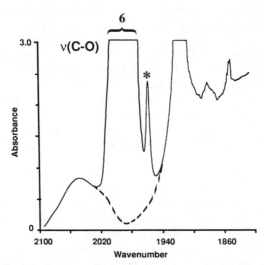

Figure 6. IR spectrum in the ν(C-O) region of a solution of Cr(CO)$_6$ dissolved in scC$_2$H$_4$ (1900psi) at ambient temperature. The IR absorptions of the ethylene solvent effectively mask large regions of this spectrum and make monitoring of photochemical processes extremely difficult. The dashed line indicates the appearance of the spectrum in the absence of Cr(CO)$_6$. The band labelled 6 indicates natural abundance Cr(CO)$_6$ and the band labelled* is natural abundance Cr(CO)$_5$(^{13}CO).

of the intermediate photoproduct, $Cr(CO)_5(C_2H_4)$, remains comparatively low, the overtone/combination vibrations of this species are not observed.

The maximum yield of $Cr(CO)_4(C_2H_4)_2$ is achieved after ca. 30 minutes photolysis and the compound is sufficiently stabilized in scC_2H_4 at ambient temperature for significant quantities to remain even 12 hours after turning off the photolysis lamp. (In the conventional synthesis, trace impurities appear to catalyse the decomposition of $Cr(CO)_4(C_2H_4)_2$ in solution until it is purified.)

Manipulation of Reaction Products

Many organometallic species are significantly more stable in the solid state than they are in solution. Thus it may well be possible to prepare new organometallic compounds in supercritical solution and then to isolate them by rapid precipitation into the solid state. Although solubility in a supercritical fluid decreases with applied pressure, the recovery of reaction products requires more subtlety than simply releasing the pressure at the end of the reaction. This procedure would merely coat the solid product over the entire inner surface of our reactor, making it extremely difficult to remove. Figure 8 shows a method, currently under development in our laboratory, which overcomes this limitation. The photochemical reaction is carried out in our cell exactly as described above. The cell is then attached to a computer controlled syringe pump filled with $scCO_2$, and the reaction mixture is flushed out of the cell through an expansion valve. As the supercritical fluid passes through the expansion valve, the pressure is released and the products are deposited as solids into a suitable collection vessel (eg. NMR tube).

This method has a number of positive features; it may be applied to most supercritical fluids with critical temperatures close to ambient; deposition of the solid product occurs in a controlled manner, if necessary under an inert atmosphere; and the high pressure "stabilizing" conditions are maintained right up to the point of precipitation. The precipitated solid product may then be analysed and characterised by other off-line spectroscopic techniques. In our example, the ^{13}C-NMR spectrum of the solid material, redissolved in d^8-toluene, shows the same resonances as those observed with a genuine sample of $Cr(CO)_4(C_2H_4)_2$.

Serendipitously this "flushing" procedure has also proved to be the ideal way of cleaning our supercritical cells! Supercritical CO_2 is used to flush out the contents of the cell, thus removing all detectable traces of soluble material and preparing the cell for the next supercritical synthetic reaction.

Conclusions

Supercritical fluids have already provided routes to a considerable variety of new and unusual organometallic species. In each case, the reactions have been monitored and the products identified solely by non-destructive spectroscopic techniques. Although our reactions have only involved milligram quantities of organometallic species, supercritical fluids give us the freedom to manipulate such quantities with unusual delicacy and precision.

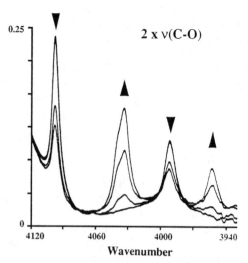

Figure 7. IR spectra in the combination/overtone region showing the photochemical reaction of $Cr(CO)_6$ in scC_2H_4 at ambient temperature. The first spectrum shown in the Figure was recorded just before the UV lamp was switched on, and subsequent spectra were taken at ca. 10 min intervals. The bands labelled ▼ are those of $Cr(CO)_6$ and those labelled ▲ show the growth of the final photoproduct $Cr(CO)_4(C_2H_4)_2$. The steady state concentration of the intermediate species $Cr(CO)_5(C_2H_4)$ is relatively low, hence the overtone/combination bands of this species are not observed in this experiment.

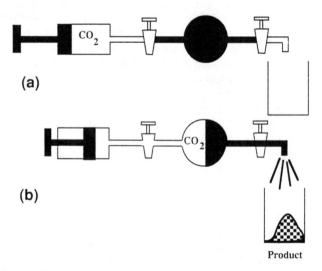

Figure 8: Schematic representation of our technique for recovering solid products from reactions in supercritical fluids (a) shows the supercritical reactor connected to a computer controlled syringe pump (Lee Scientific Model 501) filled with $scCO_2$. (b) shows how the $scCO_2$ is used to drive the supercritical reaction mixture (colored) through an expansion valve (Jasco 880/81 Back-pressure Regulator) where the products dissolved in the fluid are precipitated.

Acknowledgements

We thank the BP Venture Research Unit, SERC, the donors of the Petroleum Research Fund administered by the ACS and the E.C. Science Program (Contract ST0007) for their support. We are grateful to Professor F-W. Grevels for an authentic sample of $Cr(CO)_4(C_2H_4)_2$, Professor J.J. Turner, Dr M.A. Healy, Dr G. Davidson, Dr D.J.M. Ray, Dr T.J. Jenkins, Mr J.M. Whalley, Mr D.R. Dye and Mr J.G. Gamble for their help and advice.

Literature Cited

1. Poliakoff, M.; Howdle, S. M.; Healy, M. A.; Whalley, J. M.
In *Proc. Internat. Symp. on Supercritical Fluids, Nice*;
Perrut, M. Ed; Soc. Franc. Chim, 1988, 967;
2. Howdle, S. M.; Poliakoff, M.
J. Chem. Soc. Chem. Commun. **1989**, 1099.
3. Howdle, S. M.; Grebenik, P.; Perutz, R. N.; Poliakoff, M.
J. Chem. Soc. Chem. Commun. **1989**, 1517.
4. Howdle, S. M.; Healy, M. A., Poliakoff, M.
J. Am. Chem. Soc. **1990**, *112*, 4804.
5. Jobling, M.; Howdle, S. M.; Healy, M. A.; Poliakoff, M.
J. Chem. Soc. Chem. Commun. **1990**, 1287.
6. Jobling, M.; Howdle, S. M.; Poliakoff, M.
J. Chem. Soc. Chem. Commun. **1990**, 1762.
7. Collman, J. P.; Hegedus, L. S.; Norton, J. R.; Finke, R. G.
Principles and Applications of Organotransition Metal Chemistry,
University Science Books, 1987.
8. Braterman P.S.; Metal Carbonyl Spectra, Academic Press, London, 1975
9. Timney, J. A.; *Inorganic Chemistry*, **1979**, *18*, 2502.
10. Poliakoff, M; Howdle, S. M.; Jobling, M.; George, M. W.
Proc. 2nd Internat. Conf. on Supercritical Fluids, Boston, 1991, 189.
11. Healy, M. A.; Jenkins, T. J.; Poliakoff, M.
Trends in Analytical Chemistry **1991**, *10*, 92.
12. Geoffroy, G. L.; Wrighton, M. S. *Organometallic Photochemistry*, Academic Press, New York, (1979)
13. Kubas, G. J.; Ryan, R. R.; Swanson, B. I.;Vergamini, P.; Wasserman, H.
J. Am. Chem. Soc. **1984**, *106*, 451.
14. Kubas, G. J. *Acc. Chem. Res.* **1988**, *21*, 120.
15. Crabtree, R. H.; Hamilton, D. G. *Adv. in Organomet. Chem.* **1988**,*28*, 299.
16. Ginsburg, A. G.; Bagaturyants, A. A.
Metalloorganicheskaya. Khim. **1989**, *2*, 249.
17. Upmacis, R. K.; Poliakoff, M.; Turner, J. J.
J. Am. Chem. Soc., **1986**, *108*, 3645;
18. Makranczy, J.; Megyery-Balog, K.; Rusz, L.; Patyi, L.
Hung. J. Ind. Chem., **1976**, *4*, 26.
19. Hoyano, J. K.; Graham, W. A. G. *Organometallics*, **1982**, *1*, 783.
20. Sellmann, D. *Angew. Chem. Int. Ed. (Eng.)* **1971**, *10*, 919.
21. Rathke, J.W.; Klingler, R.J.; Krause, T.R. *Organometallics*, **1991**, *10*, 1350
22. Burdett, J.K.; Phillips, J.R.; Pourian, M.A.; Poliakoff, M.; Turner, J. J.;
Upmacis, R. K. *Inorganic Chemistry*, **1987**, *26*, 3054.
23. Grevels, F-W.; Jacke, J.; Ozkar, S. *J. Am. Chem. Soc.*, **1987**, *109*, 7536.

RECEIVED November 25, 1991

Chapter 11

Use of Solvatochromic Dyes To Correlate Mobile Phase Solvent Strength to Chromatographic Retention in Supercritical Fluid Chromatography

Terry A. Berger and Jerome F. Deye

Hewlett–Packard Company, P.O. Box 900, Route 41 and Starr Road, Avondale, PA 19311–0900

The bulk solvent strength of binary and ternary supercritical mixtures, measured with the solvatochromic dye Nile Red (E_{NR}), is a non-linear function of composition. Both acids and bases produced non-linear plots of log k` vs % methanol in carbon dioxide but linear plots of log k` vs E_{NR}. Four different stationary phases, including sulfonic acid and aminopropyl, produced linear plots of log k` vs E_{NR} for benzylamine. Thus, the non-linearity in retention is related to bulk mobile phase behavior independent of surface phenomena. This indicates that the retention mechanism does not involve competitive adsorption but instead fluid/fluid partition with non-linear mobile phase solvent strength. The effective stationary phase appears to consist primarily of an adsorbed film of mobile phase components relatively invariant in thickness or composition above approximately 2 % methanol.

The most widely used supercritical fluids are not very polar and cannot elute polar compounds in reasonable times. The addition of a more polar modifier to such fluids increases the mobile phase solvent strength allowing the elution of more polar solutes. Methanol appears to be the most polar modifier that is completely miscible with carbon dioxide. On an absolute scale, however, methanol is not very polar (1) and most polar solutes cannot be eluted from standard stationary phases with methanol-carbon dioxide mixtures.

We have had considerable success (2-5) eluting polar solutes by including very polar additives in the mobile phase. These additives are strong acids or bases that are present as trace components. The additive is usually dissolved in a modifier and then the binary modifier is dynamically mixed with the main fluid. Additives

0097–6156/92/0488–0132$06.00/0

include such compounds as trifluoroacetic acid (TFA), citric acid, t-butylammonium hydroxide (TBAH), and isopropylamine (IPAm). Concentrations range from 0.01 to 2 % in the modifier (typically 0.005 % in the complete mobile phase). Hydroxy-benzoic acids (2), polycarboxylic acids (3), benzylamines (4), PTH-amino acids (5) and many other families of polar solutes only elute when an appropriate additive is present in the mobile phase.

With binary and ternary supercritical mixtures as chromatographic mobile phases, solute retention mechanisms are unclear. Polar modifiers produce a non-linear relationship between the log of solute partition ratios (k`) and the percentage of modifier in the mobile phase. The only form of liquid chromatography (LC) that produces non-linear retention is liquid-solid adsorption chromatography (LSC) where the retention of solutes follows the adsorption isotherm of the polar modifier (6). Recent measurements confirm that extensive adsorption of both carbon dioxide (7,8) and methanol (8,9) occurs from supercritical methanol/carbon dioxide mixtures. Although extensive adsorption of mobile phase components clearly occurs, a classic adsorption mechanism does not appear to describe chromatographic behavior of polar solutes in packed column SFC.

In this work correlations between mobile phase solvent strength and chromatographic retention of a number of different solute families will be presented. The first solvent strength measurements on ternary mobile phases will also be presented. Finally, a retention mechanism for packed column SFC is proposed.

Mobile Phase Solvent Strength

Retention mechanisms in LC (including LSC) all contain an implicit assumption that mixtures of liquids form ideal solutions (1,6,10)(simple statistical mixing). Although generally ignored by chemists, the chemical engineering literature contains numerous studies (10-14) of the non-ideality of binary supercritical mixtures where solvent strength cannot be found by linear extrapolation between the solvent strengths of the components. Most solvent strength measurements have used solvatochromic dyes whose spectra shift depending on the strength of the solvent they are dissolved in.

Nile Red was recently introduced as a solvatochromic dye for studying supercritical fluids (10). Although not ideal, Nile Red does dissolve in both non-polar and polar fluids and does not lose its color in the presence of acids, like some previously used dyes. Major criticisms of Nile Red include the fact that it measures several different aspects of "polarity" simultaneously (polarizability and acidity (15)) yet it is insensitive to bases (10). However, in chromatography other single dimension polarity scales, like P`, are routinely used. Measurements with Nile Red and other dyes indicate that the solvent strength of binary supercritical fluids is often a non-linear function of composition (10-14). For example, small additions of methanol to carbon dioxide produce large increases in solvent strength.

Solvent strength measurements using two different solvatochromic dyes (see structures), including Nile Red, are plotted against each other in Figure 1. The experimental set-up was described previously (10). Most of the data in Figure 1 indicates the relative solvent strengths of pure liquids. Note that Nile Red is much less sensitive to acidic solvents than is E_{t30}. The solvent strengths of several methanol/carbon dioxide mixtures is also presented in Figure 1 and indicates the dramatic change accompanying small additions of methanol. Even on the Nile Red scale, 10 % methanol raises the apparent solvent strength of the mixture to nearly half way between pure carbon dioxide and pure methanol. On the E_{t30} scale, 5 % methanol in carbon dioxide appears to be more polar than pure methylene chloride or acetonitrile and approaches the polarity of pure 1-propanol.

Correlation Between Retention and Solvent Strength

Binary systems. A typical plot of the log of partition ratios (k`) vs. % modifier is presented in Figure 2. Like solvent strength, chromatographic retention is a non-linear function of mobile phase composition. However, plots of log k` vs solvent strength (i.e., E_{NR}) are linear, for at least some solutes, as shown for phenols (16) in Figure 3. The concentrations of methanol producing the E_{NR} values are shown at the top of the Figure. Similar but preliminary results had previously been obtained with a PTH-amino acid (17). Since E_{NR} is a mobile phase measurement, linear plots as in Figure 3 suggest that, at modifier concentration above a few percent, the non-linearities in log k` vs % modifier are due to mobile phase effects independent of surface phenomena (i.e., not due to adsorption).

Tertiary systems. With methanol/carbon dioxide mixtures the addition of even the most polar additives has only a small impact on the mobile phase solvent strength as measured with Nile Red. With TFA concentrations below 1 to 2 % in methanol, ternary mixtures of TFA/methanol/carbon dioxide produce the same apparent solvent strength as binary methanol/carbon dioxide mixtures. As much as 5 or 10 % TFA in methanol is required to noticeably increase the solvent strength of TFA/methanol/carbon dioxide mixtures above those for binary methanol/carbon dioxide mixtures, as shown in Figure 4.

Measurements with Nile Red were ineffective in measuring the solvent strength of even pure bases (10). Measurements with tertiary systems containing less than 1 % t-butylammonium hydroxide in methanol showed no difference from methanol/carbon dioxide mixtures.

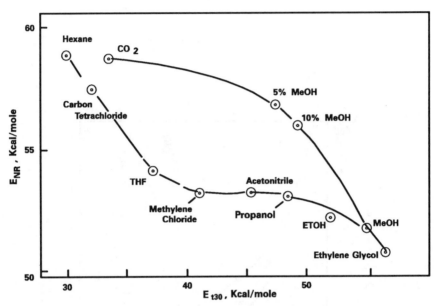

Figure 1. Solvent strength measured with two different solvatochromic dyes. On both scales solvent strength is a non-linear function of composition. Note that E_{t30} is more sensitive to hydrogen bonding than E_{NR}.

Nile Red

Et(30)

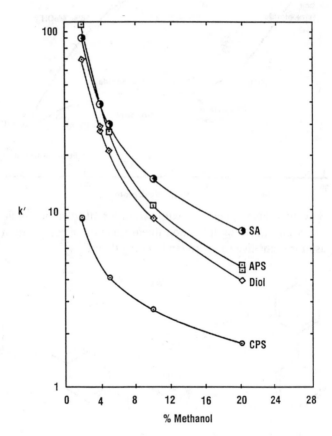

Figure 2. Plot of log k` vs % modifier for benzylamine on 4 different stationary phases. 40°C, 182 bar (outlet).

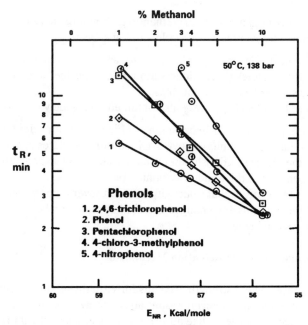

Figure 3. Plot of the log of the retention time vs. solvent strength measured with Nile Red for 5 phenols at 50° and 138 bar (outlet). The percent methanol in carbon dioxide is indicated at top of figure. Column: 4.6 x 200 mm, 5 μm Lichrosorb Diol. Flow: 2.5 ml/min.

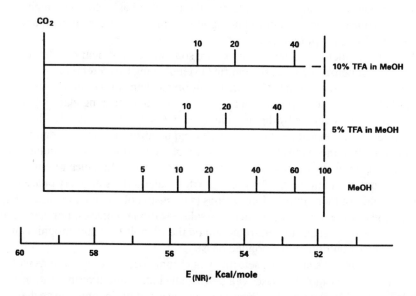

Figure 4. Solvent strength (E_{NR}) of ternary mixtures of trifluoroacetic acid (TFA)/methanol/carbon dioxide indicating that the additive creates only small changes in solvent strength with relatively large concentrations.

Using ternary mixtures of additive/methanol/carbon dioxide as the chromatographic mobile phase, linear plots of log k` vs E_{NR} were obtained for hydroxybenzoic acids (2), and benzylamines (4), as shown in Figures 5 and 6, respectively. Other solutes (i.e. amides, sulfonamides, purines and pyrimidines, etc.) produced similar linear relationships.

Linear plots of log k` vs E_{NR} were also obtained on a wide variety of bonded stationary phases, as shown for benzylamine in Figure 7. Both acidic (i.e., sulfonic acid, Diol) and basic (i.e., aminopropyl, SAX) stationary phases produced similar results. Thus, the linear relationship between mobile phase solvent strength and log k` is independent of the bonded stationary phase.

Other results (4) indicate that selectivity between several benzylamines was similar on all the columns and changed in a similar manner when the modifier concentration was changed.

Proposed Chromatographic Retention Mechanism

Bulk mobile phase solvent strength accounts for virtually all non-linearities in solute retention at modifier concentrations above a few percent. This means that non-linear surface phenomena, like adsorption, play an insignificant role in the solute retention mechanism in this concentration region. Yet, it is clear that there is extensive adsorption of both methanol and carbon dioxide (and additives) on chromatographic stationary phases. Others have measured monolayer coverage by methanol (8,9) and multi-layer coverage by carbon dioxide (7,8) when the methanol concentration was only 1-2 % in the mobile phase. However, higher methanol concentrations produced no significant further increase in methanol or carbon dioxide adsorption (8,9).

Below 1-2 % methanol, plots of log k` are somewhat non-linear (see left of Figure 6) and peaks become asymmetric (18) indicating two retention mechanisms are operational. Multiple retention mechanisms create poorly shaped chromatographic peaks. On the chromatographic packing material, less than monolayer coverage by the modifier (i.e., below 1-2 % methanol) allows some interaction between solute molecules and uncovered silanols, creating a second retention mechanism. If the stationary phase is noticeably less polar than the silanols, tailing by polar solutes becomes more severe (18) since retention on the more polar silanols should be much greater than on the stationary phase (a small number of more intense interactions (solute-silanols) competing against a large number of less intense interactions (solute-stationary phase). For stationary phases more polar than silanols, retention on the silanols becomes insignificant compared to retention on the bonded phase.

In at least some instances the selectivity of chromatographic separations with binary and tertiary mobile phases can be virtually independent of the stationary phase identity (4). Both acidic (Diol, sulfonic acid) and basic (aminopropyl,

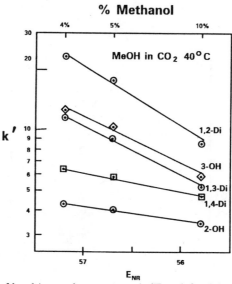

Figure 5. Plot of log k` vs solvent strength (E_{NR}) for 2 hydroxybenzoic acids, and 3 benzenedicarboxylic acids at 40°C, 150 bar. The mobile phase was methanol (containing 0.5 % trifluoroacetic acid) in carbon dioxide. Column: 2 x 100 mm, 7μm Nucleosil Diol (300 A).

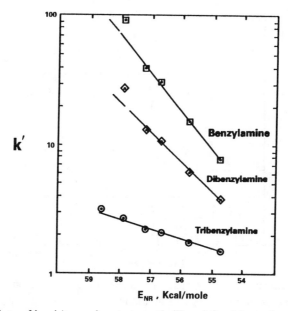

Figure 6. Plots of log k` vs solvent strength (E_{NR}) for 3 benzylamines at 40°C and 182 bar (outlet). Column: 2 x 100 mm, 5 μm Nucleosil SA (sulfonic acid). The mobile phase was methanol (containing 0.8 % isopropylamine) in carbon dioxide.

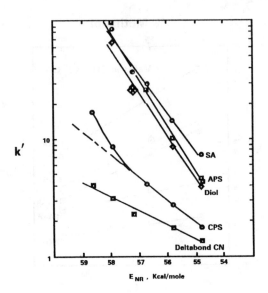

Figure 7. Plot of log k' vs solvent strength (E_{NR}) for benzylamine on 5 different stationary phases at 40°C, 182 bar. The mobile phase was methanol (containing 0.8 % isopropylamine) in carbon dioxide.

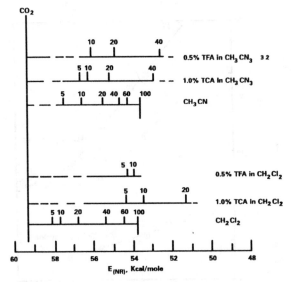

Figure 8. Solvent strength measurements as in Figure 7, except methylene chloride and acetonitrile replace methanol. With these lower polarity modifiers, the additive appears to be more polar.

strong anion exchanger) columns produced nearly the same selectivity. This strongly implies that the effective stationary phase on all the columns was similar.

All these results indicate that the adsorbed film of mobile phase components is the primary stationary phase for the separation and the bonded phase acts mostly to regulate the extent of such adsorption. Above a few percent modifier, a single fluid/fluid partition mechanism operates. At lower concentrations, a second mechanism, competitive adsorption, also becomes important and peak shapes degrade.

Future work. With modifiers less polar than methanol, low concentrations of some additives dramatically increase the apparent solvent strength of ternary mixtures, as shown in Figure 8. Thus, 20 % methylene chloride containing 1.0 % trichloroacetic acid (TCA) produces an apparent solvent strength greater than pure methanol (compare Figures 4 & 8).

Ternary mixtures of additives, less polar modifiers (like methylene chloride), and non-polar supercritical fluids have not been extensively used as chromatographic mobile phases. A few measurements suggest that they do not produce the dramatic decrease in solute retention (compared to the additive in methanol) implied by the increases in solvent strength in Figure 8.

However, additives adsorb extensively onto stationary phases even from methanol/carbon dioxide mixtures. Surface coverages as high as 20 % from 10^{-5} M solutions of additive have been measured (19). Adsorption of additive from mixtures containing a less polar modifier should be even more extensive than from mixtures containing methanol. With such high coverages it is likely that additives further increase the polarity of the stationary phase and increase retention. This would tend to counter the effect of a more polar mobile phase. Since chromatography is the sum of stationary phase and mobile phase interactions, more study will be required before such systems can be understood.

Literature cited.

1.) L.R. Snyder, and J.J. Kirkland, "Introduction to Modern Liquid Chromatography", 2nd Edition, John Wiley and Sons, New York, 1973, Chapter 7.

2.) T.A. Berger and J.F. Deye, *J.Chromatogr. Sci.,***29** (1991) 26-30.

3.) T.A. Berger and J.F. Deye, *J.Chromatogr. Sci.,***29** (1991) 141-146.

4.) 2.) T.A. Berger and J.F. Deye, "Effect of Basic Additives...." accepted by *J.Chromatogr. Sci.*

5.) T.A. Berger, J.F. Deye, M. Ashraf-Khorassani, and L.T.Taylor, *J.Chromatogr. Sci.,***27** (1991)105-110.

6.) L.R. Snyder, "Principles of Adsorption Chromatography", Marcel Dekker, New York, 1968, Chapter 8.

7.)J.R.Strubinger, H.Song, and J. F. Parcher, *Anal. Chem.*, **63**(1991)98-103.

8.) J.R.Strubinger, H.Song, and J. F. Parcher, *Anal. Chem.*, **63**(1991)104-108.

9.) C.H. Lochmuller and L.P. Mink, *J. Chromatogr.*,**471**(1989) 357-366.

10.) J.F. Deye, T.A. Berger and A.G. Anderson, *Anal. Chem.*, **62**(1990)615-622.

11.) J. Figueras, *J.Am.Chem.Soc.*,**93**(1971)3255-3263.

12.) K.-S. Nitsche, P. Suppan, *Chimia*, **36(9)**(1982)346-348.

13.) S. Kim, K.P. Johnston, *AIChE*, **33(10)**(1987)1603-1611.

14.) S.L. Frye, C.R. Yonker, D.R. Kalkwarf, and R.D. Smith, in "Supercritical Fluids:Chemical and Engineering Applications" T.G.Squires and M.E. Paulaitis, Eds.

15,) Susan Olesik, Ohio State University, private communication.

16.) T.A. Berger and J.F. Deye, *J.Chromatogr. Sci.*,**29**(1991)54-59.

17.) T.A. Berger and J.F. Deye, *Anal. Chem.*, **62**(1990)1181-1185.

18.) T.A. Berger, and J.F. Deye, "Effect of Mobile Phase Polarity...." accepted by *J.Chromatogr. Sci.*

19.) T.A. Berger, and J.F. Deye, accepted by *J.Chromatogr.*

RECEIVED November 25, 1991

SEPARATION SCIENCE

Chapter 12

Supercritical Fluid Extraction
New Directions and Understandings

Mary Ellen P. McNally, Connie M. Deardorff, and Tarek M. Fahmy[1]

Agricultural Products, Experimental Station, E402/3328B, E. I. du Pont de Nemours and Company, Wilmington, DE 19880–0402

Supercritical fluid extraction (SFE) has been demonstrated as a technique that has eliminated some of the tedious steps of current liquid-liquid and solid-liquid extraction procedures. SFE also offers cleaner extracts, less sample handling and equivalent or better recoveries to conventional technologies. As a technique, it is cost effective, time efficient and low in solvent waste generation.

The areas which need to be addressed to bring SFE into the routine laboratory are precision/reproducibility and multi-sample analysis. Without development in these directions, competition with conventional methodology does not exist. As demonstrated by applications to Agricultural Products, experiments will be presented which show the feasibility of using SFE as a routine analytical tool in sample preparation.

The adaptation of supercritical fluid extraction (SFE) in routine residue and metabolism analysis as well as other extraction/separation laboratories and applications has been slow. This is despite the demonstrated feasibility of using SFE for the removal of sulfonylureas, phenylmethylureas and their metabolites from soil and plant materials (1-2), as well as widespread demonstrated use of supercritical fluid extraction for other applications (3-6). The reason for this is simple. Although automated, SFE extraction apparatus typically only analyzes a single sample at a time. The technique could not compete effectively with the productivity of an experienced technician performing many sample extractions simultaneously. In essence, with a one vessel automated supercritical fluid extractor, operator attendance is high and throughput is about the same or even less than current conventional liquid-liquid and solid-liquid extraction techniques.

[1]Current address: Dupont Chemicals, Fluorochemicals Research and Development, Chamber Works, K37, E. I. du Pont de Nemours and Company, Deepwater, NJ 08023

The introduction of commercial instrumentation in this automated area has been too slow and too disappointing to meet the need for routine analysis of numerous samples. The options have been the extraction of one sample at a time or individual samples in parallel. Either of these options make the repetitive analysis of the same sample or the sequential analysis of different samples exceptionally time consuming. Parallel analysis, proposed by one manufacturer, is susceptible to cross-contamination and across the board sample loss with clogging of one extraction vessel. In order to move supercritical fluid extraction into the realm of routine operations for residue analysis, rapid analysis of multiple samples needed to be addressed.

We have developed a multi-vessel extractor in which six or twelve samples can be analyzed in a segmented parallel fashion in three hours or less. Timing has been set up so that periods of static and dynamic extraction overlap within the individual vessels. This time overlap eliminates deadtime which would occur if the samples were extracted serially.

In addition, our design has eliminated the use of a restrictor. Restrictors are the most common means of controlling the pressure or density of a supercritical process. With no restriction, flow is dead-ended (i.e. restricted) via a switching valve in our invention. Supercritical fluid extractions are then conducted in a static mode (no flow).

Once the extraction is complete, the dead-ended valve is repositioned to allow flow. Subsequently, pressure and density are rapidly reduced to prevent significant losses of the supercritical fluid and the extraction effluent is transferred for collection. With a non-restricted transfer, the flow of supercritical fluid effluent is rapid. This rapid depressurization was made possible by the invention of a delivery nozzle which would ensure collection of the extracted solutes without losses. It consists of a small inverted polyethylene delivery funnel, a few common stainless steel fittings and a spring. No loss of the extracted solutes and modifier has been observed with the use of this nozzle.

Details of the instrumental design as well as results obtained using it are included in this report. The challenge is for instrument manufacturers to examine the features and produce equivalent instrumentation on a commercial basis. Only with this rapid, multiple sample analysis will the technique of supercritical fluid extraction be exploited to full advantage.

Instrumental Design

The Model 50 Supercritical Fluid Microextractor from the Suprex Co. (Pittsburgh, PA) was adapted for this design. A schematic of our multi-vessel extractor design can be found in Figures 1 and 2, for six and twelve multi-vessel systems, respectively. The following is a detailed description of the main components. The design, explained here for the extraction from six and twelve vessels is in principle applicable to any number of vessels, provided that other components of the system are scaled up.

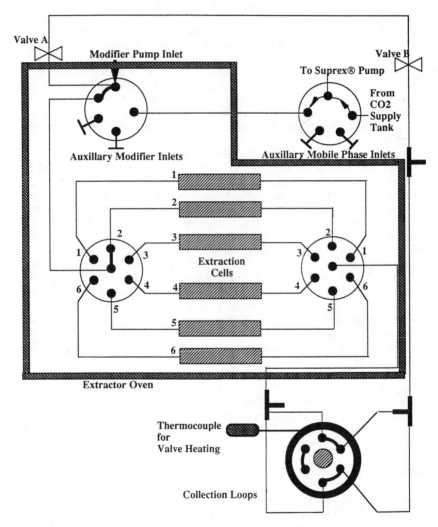

Figure 1. Schematic of Six Vessel Multi-vessel Extractor.

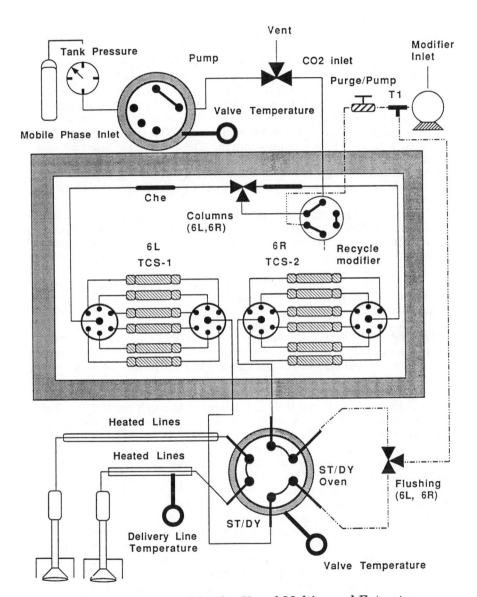

Figure 2. Schematic of Twelve Vessel Multi-vessel Extractor.

The description of the design changes to a conventional supercritical fluid extraction unit are presented here in order of introduction to the flow pattern of the fluid.

Modifier Pump. The first feature in our adapted design is the introduction of a liquid pump via an instrument controlled VALCO (Model E04, Valco Instruments, Houston, TX), four position selection valve. We have used an LKB Model 2150, dual piston pump for pumping modifier and entrainer fluids (LKB-Produkter AB, Bromma, Sweden). However, any suitable liquid pump could be substituted. Only pure fluids such as carbon dioxide have been introduced with the Suprex system syringe pump. With the addition of this second pump to deliver liquids, modifier is introduced directly into the extraction vessel. A wide range of alternative fluids and fluid mixtures can be rapidly selected with this dual pumping option. The criteria for selection of a modifier pump include: the ability of the pump heads to withstand pressures in the range of 100 to 300 atm and interfacing capabilities, i.e. the ability to be turned on and off by the Suprex contact closure controls.

In our configuration, this pump serves two purposes. First, it introduces modifier into the columns during extractions in modifier enhanced procedures. Second, the pump is used to introduce solvent for system cleanout after extraction (extraction vessels are not in position during this cleanout).

Introduction of the modifier is done via two ports. The first and most convenient, since it is controlled by a Suprex system method program, is introduced through position 2 on the 4-port selector valve in the oven unit. See Figure 3. The start/stop functions on the pump are controlled via LED sensing on the actuator of this 4-port valve. The cleanout path for this flow inlet is illustrated in Figure 4a. For the exit lines from the extraction vessels and the static/dynamic selector valve, cleanout is done by concurrently opening valves A and B. This counter flow cleanout is illustrated in Figure 4b. This second portion of the cleanout where the liquid fluid is introduced through ports 2 and 3 of the 4-port selector valve is not necessarily controlled by the method program of the instrument.

When the pump is being used for its primary purpose of introduction of modifier for extractions, the actual amount of modifier that is delivered is a function of four variables and these variables can be controlled. The variables include, the length and diameter of line between pump and inlet port to the column, the allowable pressure limit of the pump, the pump flow rate and the amount of material including air in the columns. The first factor can be accounted for by measuring the volume of the lines between the inlet port and the pump head. This volume can then be subtracted from the calculated amount delivered to the first extraction vessel filled in a series. The second and third variables are parameters that can be set on the pump. The fourth variable was rarely a problem, but can be controlled by placing a vacuum pump in position 3 or 4 to evacuate the column and the connected lines.

Autofill Valve. The VALCO autofill valve and actuator is an option available on the Suprex Model 50 extractor that presented problems during routine operation even without the multi-vessel adaptations. With the multi-vessel option, rapid refill of the syringe pump is a necessity for routine unattended operation. This is accomplished via a VALCO 1/16" five port valve (Model E04) mounted on an electric actuator. The valve has a pressure rating of 11,000 PSI. In the tank mode, the autofill valve directs flow from the mobile phase tank to the pump. In the column mode, flow is directed from the pump to the column.

For efficient operation, the Autofill valve needs to be heated. The autofill valve accommodates 1/16" tubing. As carbon dioxide flow is directed to the extraction oven from the pump, flow is hindered. The hindrance arises from the formation of a solid phase at the entrance to the rotor seal within the valve. For carbon dioxide, this solid phase can form at room temperature and at pressures above 75 atm. In addition, the presence of a sudden contraction--from 1/8" tubing to 1/16" tubing i.e. pump to valve or pump to filter connections--serves to introduce a pressure gain in the lines leading into and out of the pump. Thus, the pressures that form are sufficient for the formation of this condensed phase. When the valve is heated to temperatures near $31°C$, the carbon dioxide or other fluid passing through the valve is either a liquid or near its supercritical state. As a result, flow is not impeded.

4-Port Selector Valve. The inside of the Model 50 Extractor oven was originally equipped with a VALCO 4-port selector valve (Model No. E04). This valve is important in operations regularly utilizing additional co-solvents, inerts or vacuum, as in the introduction of modifiers in our case. When the valve is mounted in its normal fashion, one inlet port controls the flow to *four* outlet ports which are user selected by position number. When the valve is mounted in a reverse fashion, four inlet ports control the flow to *one* outlet port. In this manner, the inlet ports can be used to independently deliver various solvents. This reverse configuration is used in our system design. With this configuration, multiple fluids can be introduced to the extraction vessels with ease.

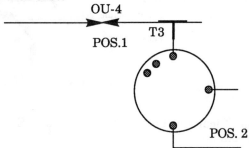

Figure 3. Four Port Selector Valve in Oven Unit Used in Reverse Configuration.

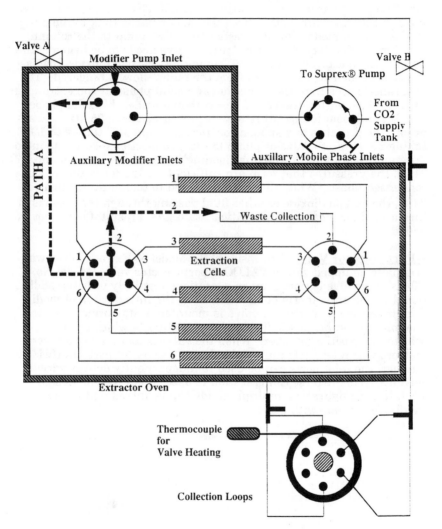

Figure 4a. Clean out Path A for Multi-vessel Extractor-Forward Flow Direction.

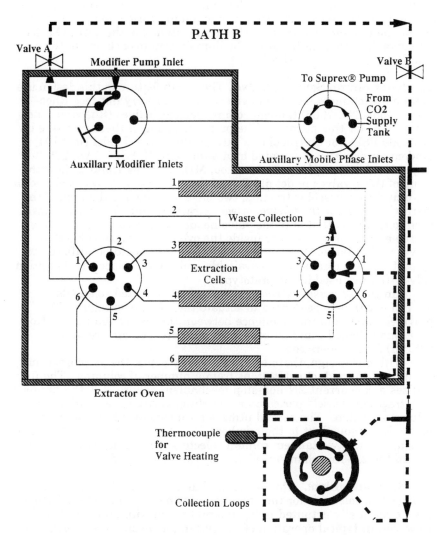

Figure 4b. Clean out Path B for Multi-vessel Extractor-Reverse Counter Flow Direction.

Briefly the positions of the valve are used in a manner illustrated in Figure 3. Position 1 is fitted to a 1/16" stainless steel tee (T3) fitting for delivery of fluid from the system syringe pump to the two 6-port tandem selectors for 12 column operation. In the 6 column system, this tee is eliminated. If six column operation is desired on the twelve column system, an on/off valve placed in line with one of the exit tubes (OU-4) from the stainless steel tee is manually closed and single delivery to one tandem selector is achieved. Position 2 of the VALCO selector valve delivers modifier from the liquid pump to each one of the extraction vessels automatically.

Tandem Column Selector Valve. The fourth feature in this design is a column switching valve. We have used 6-port rotary flat face Rheodyne (Model No. 7066) tandem column selector valves (Rheodyne Inc., Cotati, CA) for this purpose. The pressure rating of these valves is up to 10,000 PSI. For control of this valve, a Rheodyne (Model No. 5704) tandem air actuator is actuated by a Mini-Myte 10 PSI 7.0 W/120V 3-way solenoid (Humphrey, Kalamazoo, MI). Figures 1 and 2 show the setup of these valves in detail. One of these selection valves is placed in the unit for six-vessel extraction, two have been mounted on top of the extraction oven for the twelve-vessel extraction unit. The valves have been mounted so that only the valve heads and the connecting tubing for the extraction vessels are inside the oven. It is conceivable that valves with a larger number of sample positions could be utilized, an 18 port valve is commercially available and could replace the 6 position valves we have used in the prototypes.

In the 12-vessel extractor unit, two air actuated solenoids (60 PSI each) have been used to actuate the column selectors. This provides maximum independence between the two column sets. However, one solenoid could be used with an increased air pressure (80 PSI) for the operation.

With our design, there are two ways by which the selectors could be utilized to conduct extractions from a number of columns. The first utilizes a concurrent 2N non-stop mechanism, where N is the number of extractions vessels per column switching valve, for actuation and column selection. The second utilizes a concurrent 2N-1 stop mechanism. N is equivalent to the number of vessels. Thus, in the 2N mechanism, and for two tandem column selectors, 12 vessels can be selected. In the 2N-1 mechanism, 11 vessels can be selected.

2N Mechanism. This is the mechanism currently in use. It utilizes two solenoids for the actuation of two selectors. The selectors can be manually actuated independently for maximum flexibility. However, in typical operation of the system, an event-end or method-end pulse is sent to the contact closures from the control unit, the pulse actuates both solenoids concurrently, and both selectors are automatically actuated at the same time. When the selectors are actuated, a pair of extraction vessels is selected for the next extraction. Both of these extraction vessels are subjected to the same extraction conditions. Identical or different samples can be analyzed as members of the pair, i.e. a control and spiked or aged sample can be vessel pairs. However,the extraction conditions will be identical.

With this system, two extract delivery lines are required, one for each of the vessels as well as two delivery nozzles. The advantages of this mechanism are its simplicity, ease of operation, and its capacity. The mechanism is also compatible with the method-chaining program in the SUPREX. Since the extraction vessels are treated as pairs,when the program asks for a count number on the columns, an entry of six, which is the maximum allowable, will result in twelve extractions, i.e. 2N. An entry of four will result in eight extractions, etc. However, because at each actuation of the solenoids, a pair of columns is selected, this mechanism does not allow room for independent method use between the extraction vessel pairs.

Generally, this mechanism could be extended to accommodate any number of vessels with ease. For example, if 60 vessels were desired, 5 column-selector valves and 5 solenoids would be needed. Size limitations with regard to pump volume (or type, a reciprocating pump would not have a volume limitation) and oven size would be have to be overcome.

2N-1 Mechanism. As the name implies, this mechanism is used for extractions from 11 vessels. The twelfth line is used as a transition line (Figure 5). With this mechanism, three devices are used for actuation, the first is a rotating washer with a dial, which comes with the column selectors. The second is an electric pulse counter and the third is the air actuated solenoid. When the selector is actuated using a solenoid, a rotating dial within the washer in the selector is rotated 30 degrees establishing a conducting connection between two leads extending from the washer. The leads connect to the second solenoid, and thus, actuate the second selector with a fraction of a second delay between the actuations of the two selectors. A counter records the current position of the first selector. When the first selector reaches the sixth column, the transition line, the counter signals the first solenoid to halt. Further actuations will only involve the second solenoid and selector. The next six extractions will involve only the columns of the second selector. The mobile phase, whether modifier or supercritical fluid, will be delivered to the columns via the transition line in the first selector. This mechanism has the sole advantage of maximum extraction flexibility between extraction vessels. This mechanism is complicated by the circuitry involved and was not compatible with the method chaining program of the Model 50 Suprex extractor.

Static/Dynamic Selection Valve. This valve is the key feature of our design in that it eliminates the use of a restrictor. Restrictors are the most common means of controlling the pressure or density of a supercritical process. With no restriction, flow is dead-ended (i.e. restricted) via a switching valve in our invention. Supercritical fluid extractions are then conducted in a static (no flow) mode.

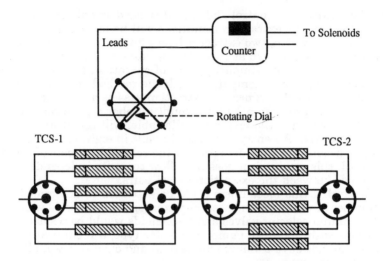

Figure 5. Schematic of Multi-vessel Extractor with 2N-1 Design
and Capability of Eleven Extraction Vessels with Individual
Extraction Programs.

Once the extraction is complete, the static/dynamic selection valve
is repositioned to the dynamic mode to allow flow. Subsequently, pressure and density are rapidly reduced to prevent significant losses of
the supercritical fluid from the syringe pump tank and the extraction
effluent, which is being transferred for collection. With a non-restricted transfer, the flow of supercritical fluid effluent is rapid. This
desire for rapid depressurization led to the development of a delivery
nozzle which would ensure collection of the extracted solutes without
losses. Details of this delivery system can be found in the next section.
Specifically, this sixth feature that has been modified in our design is the static/dynamic selection valve, VALCO SSI 6-port Model
No. E090. The pressure rating of this valve is a function of temperature. This valve was formerly the sample injection valve on the
Suprex Model 50 system. The extractor has been replumbed, so that
the exit position of the six-port column switching valves, now enters
the static/dynamic valve. In one position, the valve dead-ends the
supercritical fluid extract and static extraction is conducted. This is
simply achieved by placing stop pins in the outlet position for the static
mode. The valve is equipped with two volume expansion loops. These
collection loops have been installed on the valve so that losses during
static extraction are not experienced. Each loop has a volume of 19.3
μL. Thus, the total exit collection loop volume is 38.6 μL. In the
dynamic flow position, the valve allows the extraction effluent to pass
unrestricted through the valve. The methods used with this design
are static supercritical fluid extraction followed by a dynamic purge.
Dynamic extraction of the solid phase material is not being conducted
with this system. Figure 6 is an expanded view of this valving setup
for the six vessel multi-vessel extractor.

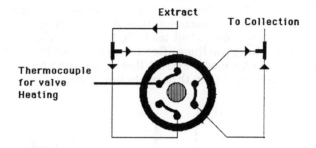

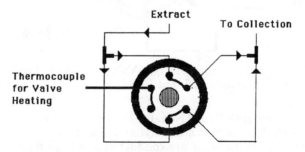

Figure 6. Static/Dynamic Selection Valve Setup for Six Vessel Multi-vessel Extractor. Extraction Effluent Received from One Tandem Column Switching Valve.

The entire valve, located outside the oven is independently heated and insulated from the column oven via 1/2" Cole Palmer flexible heating cord (Cole Palmer Instrument Co., Chicago, IL). Heating is controlled via a thermocouple and an Electronic Control Systems (Model No. 800-262) temperature controller coupled to a Glas-Col variable AC voltage source (Glas-Col Apparatus Co., Terre Haute, IN). In the experiments conducted, the controller setpoint was 40°C.

With the 12-vessel extractor, the 1/8" valve receives the extraction effluent from the vessels in tandem column selectors 1 and 2 (TCS-1 and TCS-2) into two separate ports 1 and 4 as shown in Figure 7. During the static mode, the counter-current valves, i.e. modifier pump valves (MP-3 and MP-4) are closed. Pressure build-up for static extraction then follows. Valves MP-3 and MP-4 are mounted close to the ports so that no accumulation of extract occurs. The valves are connected via a stainless steel tee (T2), to the modifier pump which is also used for flushing the lines after the extractions have been conducted. In the dynamic mode, extract flows from the unblocked ports of 1 and 4 to ports 5 and 6 then through to the delivery nozzles.

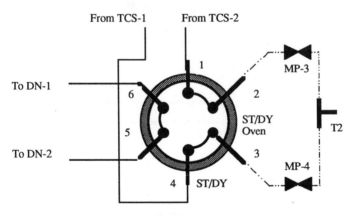

Figure 7. Static/Dynamic Selection Valve Setup for Twelve Vessel Multi-vessel Extractor. Effluent Received from Two Switching Valves.

Delivery of Extract: Delivery Nozzle. Delivery of the extraction effluent is conducted via the six port static/dynamic valve while in the dynamic mode. Generally, extractions are conducted at a high density in the static mode. Once the extraction is complete, the valve is re-positioned into a dynamic evacuation, pressure or density is reduced rapidly to prevent significant losses of the supercritical fluid and the extraction effluent is transferred for collection. The extract leaves through the heated static/dynamic valve to the heated lines then to the delivery nozzle(s). Figure 8 shows a diagram of the delivery nozzle and its components.

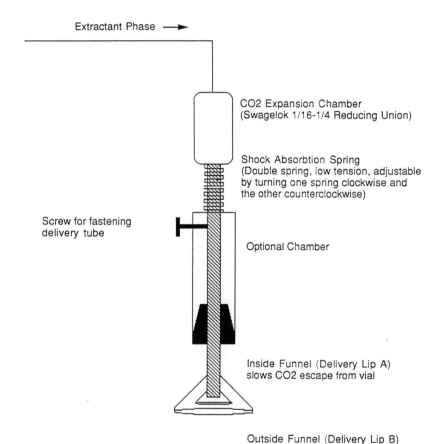

Figure 8. Extraction Effluent Delivery Nozzle Design with Component Parts.

The nozzle consists of an expansion chamber and tube (common stainless steel fittings), a shock absorber (spring) and splash prevention cover (a small inverted polyethylene delivery funnel). In a 12 vessel extraction scheme, two nozzles are used for each pair of vessels.

The nozzle consists of the following components:

A) CO_2 expansion chamber: VALCO (1/4" - 1/16") reducer.
B)   Delivery Tube: 15 cm x 4.6 mm 316 stainless steel tube.
C)   Shock Absorber: made from a 1 1/4" 316 stainless steel low tension spring cut in half to allow for adjustment.
D)   Optional Chamber: Made from two parts:
    1)   7 cm 316 stainless steel hollow cylinder with a top I.D. of 5 mm and a bottom i.d of 1.5 cm.
    2)   1.5 cm 316 stainless steel screw with a 5 mm hole. The screw fists on the bottom end of the hollow cylinder.
E)   Delivery Lip A: This is a polyethylene funnel. 2.5 cm lower I.D. and 5 mm upper I.D., and 1 cm in length.
F)   Delivery Lip B: Polyethylene funnel 3.5 cm lower I.D and 5 mm upper I.D and 1 cm in length..

The funnel arrangement is chosen so that the diameter of Lip A would correspond to the diameter of the collection vial. Lip B is fastened 1 cm above lip A. The typical operation of the funnels is as follows. As expanding carbon dioxide or other supercritical fluid pushes the extraction effluent, predominantly modifier at this point, through the system, the effluent enters the expansion chamber. The shock of this rapidly expanding fluid moves the nozzle, controlled by the spring, downwards. The fluid jet then moves through the delivery tube and exits the nozzle through lip A. After the effluent has been completely dispersed into the collection vial, the delivery nozzle than moves back to its original position, due to the spring retracting. Minimum or no splashing occurs during this period. This collection takes approximately 15 seconds in a normal extraction operation at 300 atm. Following this crucial brief period which typically results in 50-60 % of the recovery, the supercritical fluid begins to evacuate via the nozzle ac-companied by residual modifier in the lines and the remaining recovery is obtained.

System Control. The control unit operates all the functions of the extractor. It is composed of a CRT, keyboard, contact closure outputs and inputs and nine control cards each with special functions. This unit requires almost no modification since its functions are highly specialized and the circuits are delicately assembled. The most important function to the user is its ability to interface with other instruments. This is done via the contact closures. Of primary use in our design is the contact closure which controls the *event end* output pulse. In the Suprex Model 50, this is contact closure B. This *event end* contact closure is the main interface in our design to other instruments and devices such as the fraction collector and the air actuators.

Extraction Conditions

We have used carbon dioxide as the supercritical fluid and methanol followed by water as the modifiers or entrainers. In our typical extraction experiments, which are the extraction of sulfonylurea and phenylmethylurea compounds from soil and plant materials, the solid matrix phase is saturated with the entrainers or modifiers and then pressurized in the extraction vessel with the supercritical fluid. For the sample extractions described here, methanol and water were used as matrix saturants, as opposed to mobile phase modifiers.

The extraction of two typical agricultural products from environmental matrices were chosen as examples for the operation of this system. Diuron, a phenylmethylurea, was freshly spiked onto Tama soil. This soil was characterized and shown to have 3.1% organic material and 14 % clay fraction. In addition, a phenyl metabolite of NUSTAR, a systemic fungicide, on wheat previously unextractable by SFE was extracted. The wheat sample was not classified for its chemical composition. Both samples were treated with radiolabeled compounds (E. I. du Pont de Nemours and Company, Du Pont Agricultural Products, Wilmington, DE) and extraction results are from liquid scintillation counting of the sample extract. Chromatographic evaluation of the Diuron from soil extracts has previously been published (2).

Diuron from Soil. For the extraction of Diuron from soil, the mobile phase that was used was carbon dioxide, the sample size was 4 grams. The 4 gram sample was placed inside an extraction vessel 6.5 cm by 1 cm o.d., any additional volume inside the vessel was filled with reagent grade sand i.e. silica. As is typical practice in our laboratory, reagent grade sand is placed inside the extraction vessel before the sample matrix is introduced. Even though there are frits at the ends of the extraction vessel to prevent sample fines and small particles from leaving the extraction vessel, this extra layer of sand has prevented vessel clogging especially with very fine soil materials e.g. silt. These extractions were conducted at 60°C.

The total extraction time for these samples was 12 minutes per extraction. However, the procedure was conducted twice for aged samples to ensure complete extraction of the desired materials. Therefore, the overall analysis time was 24 minutes per sample. This time estimate can be misleading. The design of our system is such that the extractions are conducted in a combination of serial and parallel fashion, reducing overall analysis time. Illustrated in Table I is the extraction program for Diuron from Tama soil. Prior to the start of this method program, modifier has been introduced to each extraction vessel in a stepwise fashion. This procedure is carried out to insure that there is no pressure build-up where modifier introduction would be prevented. During subsequent extractions during the program, the vessels are filled with the modifier by switching the 4-port selector valve to position 4. At a flow rate of 5 mL/min, each column takes approximately 75 seconds to fill. The modifier pump operates at a pressure of 50 to 60 bar during this procedure. After introduction of the modifier, the vessels are pressurized individually with the chosen supercritical fluid to the desired final pressure (or density) for the extraction.

A time savings is gained because during the static extraction time periods the instrument is programmed to move onto the next sample, introduce the extraction fluid and the modifier, pressurize the vessel and then continue sequentially through all the vessels. In this manner, static periods are overlapped among the samples, when the instrument returns to the original sample vessel, the static extraction time period has been completed. At this point, the vessel can be evacuated in the dynamic mode via the delivery nozzle into a controlled fraction collection device. If a second, or even third or fourth extraction is required for complete recoveries, the sequence is repeated. The additional effluents that may be obtained can be added to the original collection vessel or to separate collection vessels depending on the stage of method development and the desires of the individual.

Phenyl Metabolite of NUSTAR from Wheat. For the extraction of the phenyl metabolite of NUSTAR from wheat, the mobile phase that was used was carbon dioxide, the sample size was 0.1 grams. The 0.1 gram sample was placed inside a 5 mL extraction vessel, any additional volume inside the vessel was filled with reagent grade sand i.e. silica. Varying sample size has not shown any significant variability in recovery as long as the remaining volume inside the vessel has been filled with some inert material. As described above, the reagent grade sand serves a dual purpose. The system temperature was $75°C$. Table II contains the extraction program for this extraction.

Results

Tables III and IV show the results obtained for the two sample types used for our multi-vessel demonstration. As is illustrated, high recoveries and equally high precision resulted.

Diuron from soil. Diuron spiked Tama soil was the extraction example used in the six vessel multi-vessel extractor. As is illustrated in Table III, the average recovery for these samples was 97.3% and the relative standard deviation was 6.6%. When these results are compared to those obtained with the classical extraction techniques, equivalent recoveries were achieved. However, the precision associated with the classical extraction was typically 20%. Acceptable recovery ranges in classical residue analysis are from 70 to 125%. Comparing these precisions with those obtained with a one vessel SFE device, the

Table I

Extraction Program for Diuron from Tama Soil
(12 Columns)

Stage	Ramp	Pressure	Time	Fluid	Select Valve
	Initial	100		CO_2	static
1	Step	200	1.00	CO_2	static
2	Step	250	5.00	CO_2	static
3	Ramp	250	.15	CO_2	dynamic
4	Ramp	150	3.00	Modifier	static
5	Step	250	2.00	CO_2	static
6	Ramp	150	.15	CO_2	dynamic
7	Ramp	150	3.00	Modifier	dynamic

Step 1 is a static extraction at 200 atm.
Step 2 is a static extraction at 250 atm for 5 min.
Step 3 is a sudden depressurization step to 150
Step 4 is the introduction of the modifier for the second static extraction
Step 5 is a second static extraction.
Step 6 is a sudden depressurization.
Step 7 is a cleanout.

Table II

Extraction Program for a Phenyl Metabolite of NUSTAR
Fungicide from Wheat
(12 Columns)

Stage	Ramp	Pressure	Time	Fluid	Select Valve
	Initial	300		CO_2	static
1	Ramp	300	1.00	CO_2	dynamic
2	Step	150	5.00	Modifier	dynamic
3	Step	300	3.00	CO_2	static
4	Ramp	150	.15	CO_2	dynamic
5	Step	150	2.00	Modifier	dynamic
6	Ramp	300	.15	CO_2	dynamic

Step 1 is a dynamic flush of the first static extraction at 300 atm; modifier (methanol/water) was added before the initial extraction began.
Step 2 is the introduction of the modifier for the second static extraction.
Step 3 is a second static extraction.
Step 4 is a sudden depressurization step to 150.
Step 5 is the introduction of the modifier for flushing
Step 6 is an evacuation of the rinse.
Step 7 is a cleanout.

Table III

**Cumulative Results for the Extraction of
Freshly Spiked Diuron From Tama Soil Using
the Multi-vessel Extractor**

Column #	Radiolabeled Recovery
Column 1	90.9%
Column 2	102.3%
Column 3	105.2%
Column 4	91.4%
Column 5	96.5%
Column 6	0 Blank - no samples) no contamination exhibited.
Mean =	97.3%
Std. dev. =	6.4
RSD =	6.6%

Table IV

**Cumulative Results for the Extraction of Phenyl Metabolite of
NUSTAR from Wheat Using the Multi-vessel Extractor**

Set #	Column #	Radiolabeled Recovery
1	Column 1	102.9%
1	Column 2	87.6%
1	Column 3	109.7%
1	Column 4	86.1%
1	Column 5	0 (Blank - no sample) no contamination exhibited.
1	Column 6	87.3%
2	Column 1	101.7%
2	Column 2	75.7%
2	Column 3	80.2%
2	Column 4	Sample Lost
2	Column 5	Sample Lost
2	Column 6	87.9%
	Mean =	91.0%
	Std. dev. =	11.3
	RSD =	12.4%

one vessel precisions generally fall between the multi-vessel and classical results. Average relative standard deviation values ranged from 10 to 15%. It should be noted that the expected precision associated with liquid scintillation counting is +/- 10%.

The soil samples which were extracted in our experiments were saturated with methanol prior to the introduction of the mobile phase. Other samples, especially those which have been field aged, are saturated with water and water/methanol mixtures before the supercritical fluid extraction process begins. Our experience indicates that this saturation step is crucial to obtain high recoveries.

For aged samples, it is often necessary to repeat the extraction steps more than once for complete recovery. The analogous classical extraction comparison would be the introduction of a second liquid phase in a separatory funnel extraction. However, because of our automated system, the introduction of a second or third extraction step is conducted easily. The modifier/entrainer can be independently introduced to the matrix without interrupting the extraction procedure. In the time estimates given for the diuron spiked Tama soil extractions, two extractions per vessel were conducted. This resulted in 3 hours for the extraction of six samples in the six vessel extractor as well as 3 hours for twelve samples in the twelve vessel extractor since the extractions are conducted in pairs.

Phenyl Metabolite of NUSTAR from Wheat. The phenyl metabolite of NUSTAR was the extraction example used for twelve vessel multivessel extractor. As is illustrated in Table IV, the average recovery for these samples was 91.0%, with a relative standard deviation of 12.4%. Two samples were lost in this extraction example. The loss was attributed to overpressurization of the extraction system. The wheat sample which is ground with dry ice in a blender as a standard sample handling procedure for classical as well as the supercritical fluid extraction, frequently contains small fine particles. These particles can readily block the 0.45 μm frits at the end of the extraction vessel. The advantage of our system, is that the loss of one sample in the series does not interfere with the accurate extraction of the rest.

One of the major concerns with this extraction set-up was carryover or contamination from sample to sample. To determine whether or not this occurred, we placed blank sample vessels in series with vessels containing material fortified with a radiolabeled component. No contamination from sample to sample was experienced. This is illustrated for vessel 6 in Table III and for vessel 5 of set 1 in Table IV. In different experiments, these blank vessels have been placed in all positions, no carryover was ever found.

Conclusion

We have designed and constructed two instruments that will allow supercritical fluid extraction from plants and soils with modifiers. Other uses, remain to be investigated. These multi-vessel extractors have demonstrated fast and efficient recovery data from a large number of samples. With this instrumentation and the results that have been generated we have illustrated the applicability of SFE in the routine laboratory. Competition with conventional methodology is no longer a threat.

The use of a restrictor to control pressure and density in supercritical fluid extraction has also been eliminated. Prior to this, restriction and restriction devices have been the Achilles heel of SFE. With this system, this will no longer be a problem.

References

1. McNally, M. E.; Wheeler, J. R.; J. Chromatogr., 1988, 447, 53.
2. Wheeler, J. R.; McNally, M. E.; J. Chromatogr., 1987, 410, 343.
3. Onuska, F. I.; Terry, K. A.; HRC & CC. J. High Res. Chromatogr. & Chromatogr. Comm., 1989, 12, 357
4. Hawthorne, S. B.; Miller, D. J.; Krieger, M. S.; J. Chromatogr. Sci., 1989, 27, 347.
5. Schneiderman, M. A.; Sharma, A. K.; Locke, D. C.; J. Chromatogr. Sci., 1988, 26, 458.
6. King, J. W.; Johnson, J. H.; Friedrich, J. P.; J. Agric Food. Chem, 1989, 37, 951.

RECEIVED January 23, 1992

Chapter 13

Supercritical Fluid Extraction of Polar Analytes Using Modified CO_2 and In Situ Chemical Derivation

Steven B. Hawthorne, David J. Miller, and John J. Langenfeld

Energy and Environmental Research Center, University of North Dakota, Box 8213, University Station, Grand Forks, ND 58202

Polar and ionic analytes can be extracted from solid samples using supercritical CO_2 containing organic modifiers, or alternatively, by performing in-situ chemical derivatization during the SFE step. Static SFE performed with the addition of chemical derivatizing reagents can be used to reduce the polarity of target analytes, which makes the analytes easier to extract and prepares them for direct analysis using capillary GC. Following derivatization, the analytes are extracted using dynamic SFE. Quantitative derivatization (to the methyl ester) and extraction of polar analytes such as 2,4-dichlorophenoxyacetic acid (2,4-D) have been achieved. The use of modified CO_2 and in-situ chemical derivatization for the extraction of polar pesticides, bacterial lipids, ionic surfactants, and wastewater phenolics from real-world samples is discussed.

Analytical-scale supercritical fluid extraction (SFE) has recently been demonstrated to be a rapid and quantitative method for extracting many relatively non-polar organics (e.g., "GC-able" organics) from a variety of sample matrices. However, reports of quantitative extractions of polar, high molecular weight, and ionic analytes have been less frequent, and such extractions have generally required the addition of organic modifiers to CO_2 (1-4), or fluids (e.g., Freon-22) which are less acceptable for routine applications (5). After analytes are extracted, many analytical schemes require that polar and high molecular weight analytes be derivatized so that they can be determined by gas chromatography. For example, EPA methods for acid herbicides require methylation using diazomethane after extraction so that the herbicides can be analyzed using GC with ECD detection.

This paper will discuss the development of SFE techniques for polar and ionic analytes based on two different approaches. The first approach uses the

0097–6156/92/0488–0165$06.00/0

addition of organic modifiers to increase the solubility (and extractability) of polar analytes. The second approach uses in-situ chemical derivatization under SFE conditions to decrease the analyte's polarity, and thus to increase its extractability and ease of analysis by conventional GC techniques.

Experimental

SFE extractions were performed using an ISCO model 260D syringe pump (for pure fluids). Organic modifiers were mixed with CO_2 using two of these pumps according to the manufacturer's directions, except for the linear alkylbenzenesulfonate (LAS) extractions which were performed as previously described (6).

In-situ chemical derivatizations were performed by adding 0.5 to 2 mL of the derivatization reagent, trimethylphenyl ammonium hydroxide (TMPA) in methanol (Eastman Kodak Company, Rochester, NY) directly to the sample cell of an ISCO model SFX extraction unit. The sample and reagent were pressurized with CO_2 (typically 400 to 500 atm), and heated to an appropriate temperature (typically 80°C). The derivatization was performed in the static SFE mode for 5 to 15 minutes, then the derivatized analytes were recovered by dynamic SFE using a typical flow rate of 0.6 to 1.0 mL/min (measured as liquid CO_2 at the pump). No other sample preparation of the derivatized extracts was performed prior to GC analysis. All GC analyses were performed using Hewlett-Packard 5890 GCs with appropriate detectors and HP-5 columns (25 m X 250 μm i.d., 0.17 μm film thickness).

The steps in the SFE/derivatization/extraction were:

1. The sample was weighed into the extraction cell, and the derivatizing reagent was added.

2. The cell was placed into the extraction unit (which was pre-heated to the extraction temperature) and immediately pressurized with 400 to 500 atm CO_2.

3. The sample was derivatized and extracted under static SFE conditions.

4. The outlet valve (restrictor end) of the extractor was opened and the derivatized analytes were extracted using dynamic SFE (400 to 500 atm) and collected in ca. 3 mL of methanol.

SFE and Derivatization/SFE of Linear Alkylbenzenesulfonates (LAS)

Development of extraction conditions for LAS was performed using sludge from a municipal wastewater treatment plant, and agricultural soil from a field which had been used for disposal of sewage sludge several months prior to sample collection. The structure of LAS is shown in Figure 1. As an ionic surfactant, LAS has very low solubility in pure supercritical CO_2, and no detectable amounts of LAS could be extracted with the pure fluid. Supercritical N_2O has previously been reported to yield better extraction efficiencies than CO_2 for some analytes from environmental matrices (*1,2*), but also showed no detectable extraction of LAS. However, when organic modifiers were added to the CO_2, extraction efficiencies increased dramatically, and the extraction with methanol-modified CO_2 (ca. 40 mole %) at 125°C gave good recoveries as shown in Figure 2 (*6*).

An alternate approach for extracting LAS from solid samples is now being developed using in-situ chemical derivatization/SFE with the methylating reagent, trimethylphenylammonium hydroxide (TMPA) in methanol. A 500-mg sample of sludge was placed in a 2.5-mL extraction cell, 1 mL of 5% (wt) TMPA in methanol was added, and the cell was pressurized to 500 atm CO_2 and heated to 80°C. The static derivatization/extraction step was performed for 10 minutes followed by 5 minutes of dynamic extraction at ca. 0.9 mL/min with trapping of the extracted analytes in 3 mL of methanol. The extract was then analyzed using GC without any further treatment. As shown in Figure 3, derivatization of the sewage sludge results in a very complex extract. However, a comparison of an LAS standard (spiked on sand and derivatized in the same manner) with the sludge extract shows good agreement for peaks eluting between ca. 15 and 20 minutes, although the sludge LAS peaks had a slightly higher molecular weight distribution than the commercial LAS peaks. The identity of the methylated LAS species in the sludge extract has been confirmed by GC/MS analysis of the derivatized sludge extract and the derivatized LAS standard. Although only a few of the non-LAS peaks in the sludge extract have been identified, GC/MS analysis shows that many of the major species are methyl esters of biological carboxylic acids.

Since the GC/FID chromatogram of the derivatized sludge extract appears too complex to allow reliable quantitation of the individual LAS homologs (Figure 3), GC with atomic emission detection (GC/AED) in the sulfur-selective mode was also used to analyze the sludge extract (note again that the extract was analyzed without any additional preparation after the derivatization/SFE procedure). As shown in Figure 4, the use of GC/AED in the sulfur mode virtually eliminates the detection of non-LAS peaks in their retention time window, and thus makes quantitation of the LAS derivatives practical. As was shown by GC/MS analysis, the LAS in the sludge sample was shifted to higher molecular weight homologs compared to the standard LAS.

Only preliminary quantitative evaluations for the derivatization/SFE of LAS have been performed, and it appears that the conditions described above yield only ca. 30 to 40% recovery of the native LAS with one derivatization/extraction step. It is apparent from the GC/FID chromatogram that the matrix contains a very high concentration of materials that react with the TMPA and it is likely that the reaction is reagent limited. This idea is further supported since three derivatization/SFE steps on one sample reduces the LAS to undetectable levels.

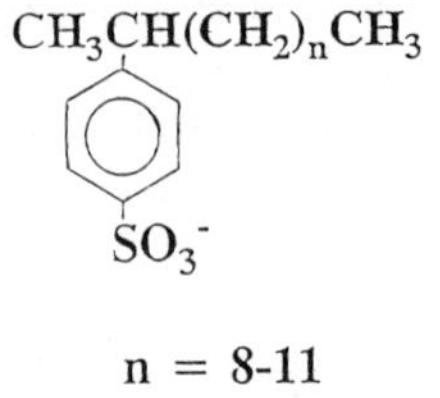

$$CH_3CH(CH_2)_nCH_3$$

$$SO_3^-$$

n = 8-11

Figure 1. Structure of linear alkylbenzenesulfonate (LAS).

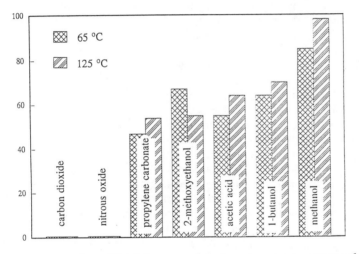

Figure 2. Extraction of LAS from municipal wastewater treatment sludge using pure CO_2, pure N_2O, and organically-modified CO_2. Each sample (50 mg) was extracted for 15 minutes at 380 atm and a flow rate of ca. 0.5 mL/min. Recoveries are based on the average of two extractions. Results are adapted from reference 6.

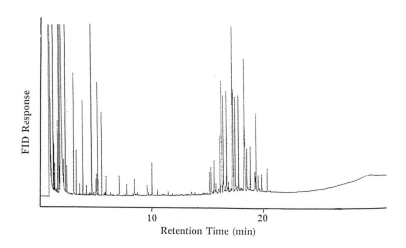

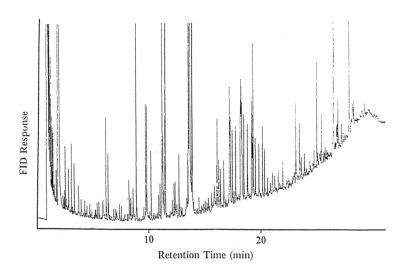

Figure 3. GC/FID separation of the derivatization/SFE extract from municipal wastewater treatment sludge (bottom) and commercial LAS (top).

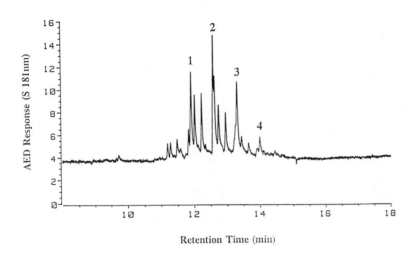

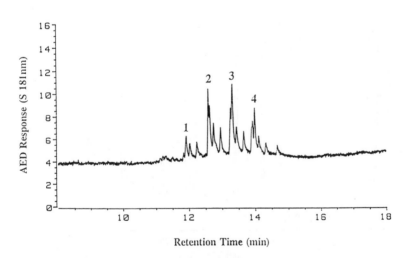

Figure 4. GC/AED   sulfur-selective   chromatogram   of   the
derivatization/SFE extract from municipal wastewater treatment
sludge (bottom) and commercial LAS (top). Numbers on the
peaks correspond to LAS homologs that show the same mass
spectra.

Bacterial Phospholipids

Phospholipid-derived fatty acids are often used to identify bacteria by capillary GC analysis after liquid solvent extraction, concentration steps, and chemical derivatization to their methyl esters. Our initial investigations attempted to extract the intact phospholipids, but no significant recoveries were achieved using pure CO_2. Even if SFE conditions were developed that could extract intact phospholipids, an additional derivatization step would be required before GC analysis of the fatty acid components. For these reasons, chemical derivatization/SFE was investigated in an effort to eliminate the lengthy conventional liquid solvent extractions as well as to combine (and shorten) the extraction and derivatization steps. The derivatization/SFE procedure was performed on samples of whole bacteria using 0.5 mL of 1.5% TMPA in methanol. The static derivatization step was performed for 10 minutes at 80°C and 400 atm CO_2, followed by dynamic SFE for 15 minutes at a flow rate of ca. 0.5 mL/min of the pressurized CO_2. Extracts were collected in ca. 3 mL of methanol and immediately analyzed by capillary GC without any further sample preparation.

Typical results of the derivatization/SFE procedure are shown in Figure 5 for a 20-mg sample of the bacteria Bacillus subtilis. The chromatogram shows the typical distribution of the fatty acid methyl esters expected for this bacteria (back chromatogram). A second derivatization/SFE extraction was performed in an identical manner on the same sample to determine the completeness of the first extraction. Based on the lack of chromatographic peaks for the 2nd extraction (front chromatogram), the first 25-minute derivatization/extraction procedure appeared to be quantitative. Preliminary comparisons of the SFE/derivatization method with the conventional liquid solvent extraction method have also shown good quantitative agreement (7).

Chlorinated Acid Pesticides

SFE/derivatization of several chlorinated acid pesticides (those listed in EPA method 515.1) have been performed using conditions similar to those used for the bacterial phospholipids. The derivatized products from the SFE procedure for several representative organics are shown in Figure 6. As would be expected using the TMPA/methanol reagent, the carboxylic acids form the methyl esters (2,4-D and dicamba) while the phenols form the methyl ethers (pentachlorophenol). Esters of the carboxylic acids (e.g., the di-isopropyl amine ester of 2,4-D) also form the methyl esters. For ethers, two derivatized products resulted since the ether linkage could be cleaved on either side of the oxygen and methylated as shown by acifluorfen.

Quantitative recovery of the pesticides using the SFE/derivatization procedure was found to be matrix dependent, with matrices that contained higher organic content requiring longer derivatization time and higher concentration of the TMPA reagent to obtain good spike recoveries. For example, 2,4-D acid was efficiently derivatized and extracted from a 2-gram

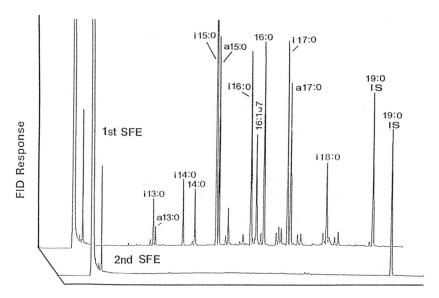

Retention Time

Figure 5. GC/FID analysis of the phospholipid-derived fatty acid methyl esters from derivatization/SFE extraction of <u>Bacillus subtilis</u>. The front chromatogram shows the second extraction of the same sample (I.S. = internal standard).

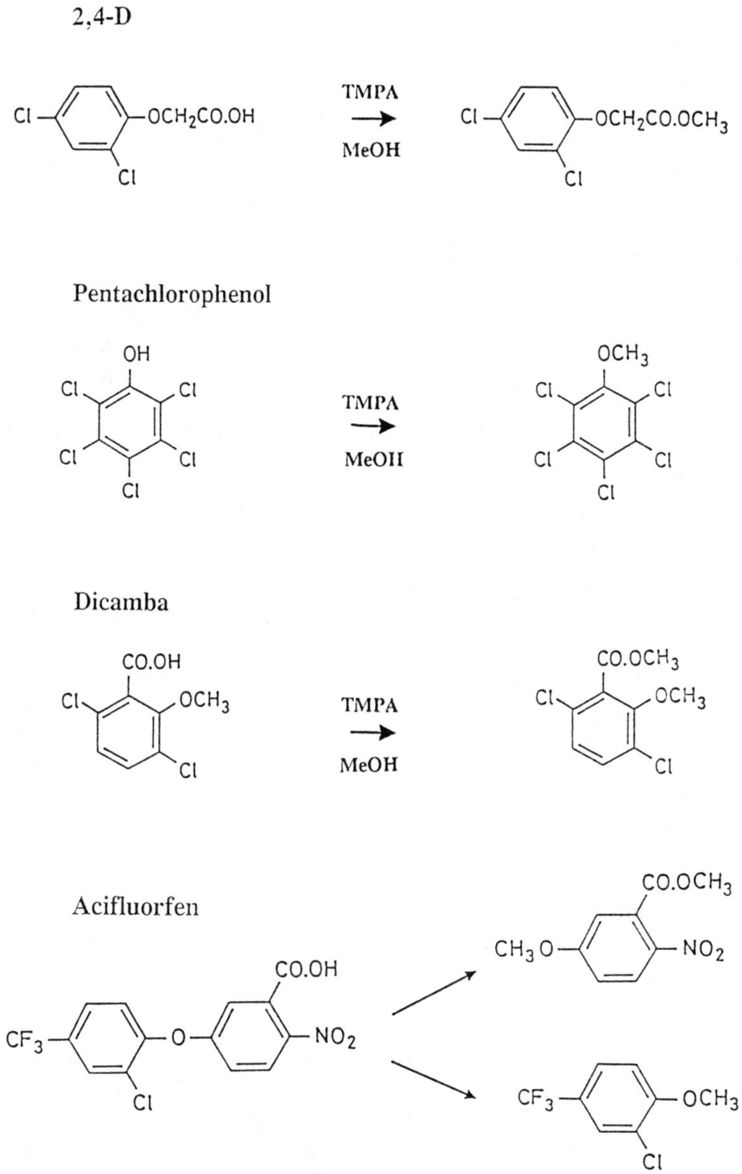

2,4-D

Pentachlorophenol

Dicamba

Acifluorfen

Figure 6. SFE/derivatization products of representative pesticides spiked on sand using the TMPA/methanol reagent.

sample of sand using a 5-minute derivatization procedure (400 atm, 80°C) followed by a 10-minute dynamic extraction (Table I) with collection into 3 mL of methanol. However, when the same extraction conditions were used to derivatize and extract 2,4-D from river sediment, the extraction efficiencies dropped dramatically. The extracts from the river sediment were bright yellow while the extracts from the sand were colorless, and we expect that the higher organic content in the sediment (ca. 4%) was responsible for reacting with and using up the TMPA reagent, as was previously discussed for the sewage sludge. Fortunately, when the concentration of TMPA was increased to 20%, and the derivatization time was increased to 15 minutes, reasonably good recoveries of the 2,4-D were achieved from the sediment sample (Table I).

Phenolic Wastewaters from "Empore" Sorbent Discs

SFE and SFE/derivatization for the extraction of phenols from a coal gasification wastewater have been performed by first collecting the wastewater organics on "Empore" C-18 sorbent discs (8). The sorbent discs were first prepared by washing with methanol and water as per the manufacturer's instructions. A 100-mL sample of wastewater was acidified to a pH <2, 4 mL of methanol was added, and the waters were filtered through the sorbent discs. The discs were then placed in a 2.5 mL cell and extracted at 80°C and 500 atm using pure CO_2, CO_2 with 1.0 mL methanol modifier placed in the cell, or CO_2 with 1 mL of 0.1 M TMPA in methanol placed in the cell. (Extractions using methanol modifier and TMPA derivatization were each performed statically for 5 minutes followed by dynamic extraction for 10 minutes.) No attempt was made to dry the discs before SFE.

Figures 7a to 7d show coal gasification wastewater phenolics extracted from "Empore" discs using pure CO_2 (Figure 7a) and methanol-modified CO_2 (Figure 7c). Note that the two extracts appear virtually identical, which demonstrates that the addition of the methanol modifier was not necessary to recover the phenolics from the disc. The disc that was extracted with pure CO_2 was extracted a second time by shaking for 16 hours in methanol (Figure 7b). The lack of significant peaks in the methanol extract also demonstrates that pure CO_2 efficiently recovers the phenolics from the "Empore" disc. The SFE/derivatization extract of the same phenolics from the disc is shown in Figure 7d. As expected, the phenolics were all derivatized to their methyl ethers (e.g., phenol goes to anisole). None of the parent phenols could be found in the TMPA extract using GC/MS, indicating that the 5-minute SFE/derivatization step was sufficient for quantitative derivatization of the phenols. The weakly acidic hydrogen on indole also was exchanged for a methyl group to form the 1-methylindole derivative. The TMPA extract also contained several higher molecular weight species (not shown) which have been identified by GC/MS as methyl esters of biological carboxylic acids (primarily C-14, C-16, and C-18 acids). As discussed above, bacterial lipids are efficiently derivatized and extracted using the SFE/TMPA procedure, and it seems likely that these species come from bacteria in the wastewater that were collected onto the "Empore" disc during filtration.

Table I. SFE/Derivatization Recoveries of 2,4-D Acid

Matrix	TMPA conc.	Derivatization time	Recovery ± SD[a]
sand	1.5%	5 min	98 ± 8%
sediment	1.5%	5 min	22 ± 2%
sediment	20.0%	15 min	90 ± 4%

[a] Standard deviations were based on triplicate extractions.

The results shown in Figure 7 demonstrate that the filtration of waters through "Empore" discs followed by recovery using either SFE or SFE/derivatization is a viable approach for the extraction and/or derivatization of polar organics from water samples. After the sample is filtered, the disc can immediately be extracted (no drying is required), the discs yield reasonable blanks, and they appear to be stable to the derivatization procedure using TMPA. The primary disadvantage of the TMPA derivatization procedure is the large N,N-dimethylaniline chromatographic peak that results from the TMPA reagent. Fortunately, although the reagent peak obscures a small part of the chromatogram, no negative effects of the TMPA on the chromatographic column or the resulting chromatograms have been observed.

Selection of Derivatization Reagents

Although the majority of our SFE/derivatization studies have been performed using TMPA/methanol, a wide range of reagents is available that should be useful for SFE/derivatization procedures. Reagents could be selected to reduce interferences from the sample matrix. For example, preliminary results using BF_3/methanol for the methylation of the LAS from sewage sludge and 2,4-D from river sediment show much less dependence on interfering matrix organics than the TMPA procedure. BF_3/methanol also has the advantage over TMPA in that no large chromatographic peak (such as shown in Figure 7d) results from the derivatization procedure. Derivatizations can also be performed to increase analytical sensitivity and selectivity by adding easily-detected functional groups to target analytes. For example, derivatizations of the coal gasification wastewater shown in Figure 7 using TMPA with CF_3CH_2OH yields the trifluoroethyl derivative (e.g., 1,1,1-trifluoroethylphenyl ether from phenol), which can then be selectively detected at very low concentrations using GC/ECD.

Summary and Conclusions

Polar and ionic analytes can be extracted from a variety of sample matrices using in-situ chemical derivatization under SFE conditions followed by SFE extraction of the derivatized analytes. Derivatization of the analytes during a

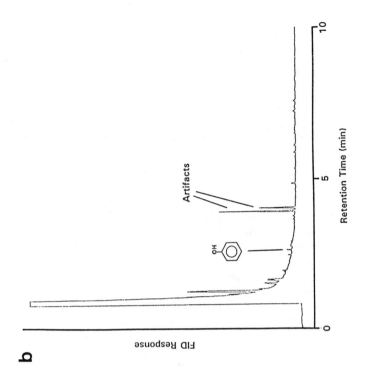

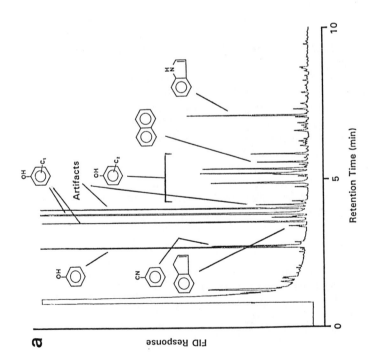

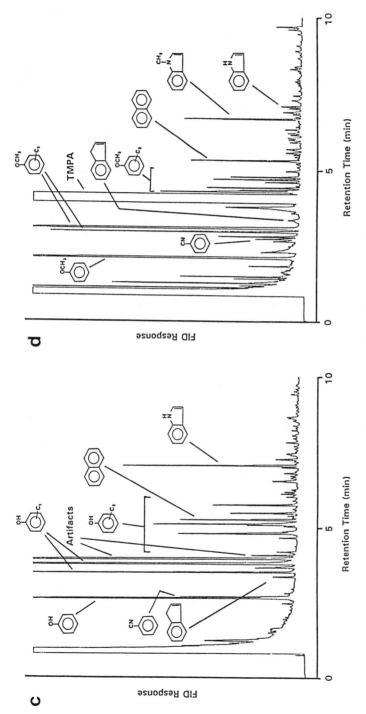

Figure 7.   SFE of coal gasification wastewater organics extracted from "Empore" discs using pure CO_2 (Figure 7a) followed by a second extraction of the same disc with liquid methanol for 16 hours (Figure 7b). The SFE extract obtained using methanol-modified CO_2 is shown in Figure 7c. Figure 7d shows the methyl ethers from the SFE/TMPA derivatization extract.

static SFE step reduces the polarity of the analytes making them easier to extract by SFE and analyze using conventional GC techniques.   The derivatization and extraction steps are completed in < 30 minutes, and many extracts can be analyzed without further treatment. With the use of the TMPA reagent, quantitative recoveries of target analytes can be dependent on the amount of reactive matrix components, so careful evaluation of the reaction/extraction conditions is necessary.   Fortunately, a wide range of potentially useful derivatizing reagents is available that have potential to both reduce the matrix dependence of the derivatization step, as well as to increase analytical sensitivity and selectivity of the SFE/derivatization procedure.

Acknowledgements

The authors would like to thank the U.S. Environmental Protection Agency, EMSL, Cincinnati, for partial financial support. Instrument loans from ISCO are also gratefully acknowledged.

References

1. Hawthorne, S. B. Anal. Chem. **1990**, 62, 633A.
2. Proceedings of the International Symposium on Supercritical Fluid Chromatography and Extraction; Park City, Utah, January 1991.
3. Wheeler, J. R.; McNally, M. E. J. Chromatogr. Sci. **1989**, 27, 534.
4. Ramsey, E. D.; Perkins, J. R.; Games, D. E.; Startin, J. R. J. Chromatogr. **1989**, 464, 353
5. Li, S. F. Y.; Ong, C. P.; Lee, M. L.; Lee, H. K. J. Chromatogr., **1990**, 515, 515.
6. Hawthorne, S. B.; Miller, D. J.; Walker, D. D.; Whittington, D. E.; Moore, B. L. J. Chromatogr. **1991**, 541, 185.
7. White, D. C.; Nivens, D. E.; Ringelberg, D. B.; Hedrick, D. B. Proceedings of the International Symposium on Supercritical Fluid Chromatography and Extraction; Park City, Utah, January 1991, p 43.
8. Markell, C.; Hagen, D. F.; Bunnelle, V. A. LC-GC **1991**, 9, 331.

RECEIVED December 2, 1991

Chapter 14

Supercritical Fluid Extraction in Environmental Analysis

V. Lopez-Avila[1,3] and W. F. Beckert[2]

[1]Mid-Pacific Environmental Laboratory, 625–B Clyde Avenue, Mountain View, CA 94043
[2]Environmental Monitoring Systems Laboratory, U.S. Environmental Protection Agency, 944 East Harmon Avenue, Las Vegas, NV 89109

Samples of sand spiked with 36 nitroaromatic compounds, 19 haloethers, and 42 organochlorine pesticides, and a standard reference soil (certified for 13 polynuclear aromatic hydrocarbons, dibenzofuran, and pentachlorophenol) were extracted with supercritical carbon dioxide in a two- or four-vessel supercritical fluid extractor to establish the efficiency of the extraction and the degree of agreement of the parallel extraction recoveries. Furthermore, the many variables that influence the extraction process (e.g., flowrate, pressure, temperature, moisture content, cell volume, sample size, extraction time, modifier type, modifier volume, static versus dynamic extraction, volume of solvent in the collection vessel, and the use of glass beads to fill the void volume) were investigated.

In this paper, the supercritical fluid extraction (SFE) of organic compounds from sand spiked with 36 nitroaromatic compounds, 19 haloethers, and 42 organochlorine pesticides, and from a standard reference material certified for 13 polynuclear aromatic hydrocarbons (PAH), dibenzofuran, and pentachlorophenol was examined using a two- and a four-vessel extractor. Although the results achieved by SFE for the sand and the standard reference soil samples were very encouraging, previous data obtained in our laboratory on the standard reference soil and a few other standard reference marine sediments were less favorable. It was therefore decided that an investigation of seven variables for their influence on the analyte recoveries from the standard soil sample would be useful. Two tests were conducted in which these variables were investigated. In Test 1, the seven variables selected were pressure, temperature, moisture content, cell volume, sample size, extraction time, and modifier volume. In Test 2, the seven variables were pressure, temperature, volume of toluene added to the matrix, volume of solvent in the collection vessel,

[3]Current address: Midwest Research Institute, 625–B Clyde Avenue, Mountain View, CA 94043

moisture content, presence/absence of glass beads, and static extraction time. For each variable we chose two values. Eight experiments were performed per test, and the results of these experiments are summarized in this paper as the relative changes in recoveries when going from low to high values across the 15 target compounds. In addition to these method optimization experiments, we investigated the effect of varying flowrate, pressure, and temperature using the standard reference soil and a Hewlett-Packard extractor.

Experimental

Apparatus.

- Supercritical fluid extraction system -- Suprex Model SE-50 consisting of a 250-mL syringe pump with the necessary valves and connecting lines to the extraction vessel, a control module containing a microprocessor for controlling the SFE system and able to store up to 25 extraction programs, and an oven module consisting of the extraction oven, the extraction vessel, a 4-port and a 12-port valve configured with electronic actuators for automated operation. The system was set up either with two or four 2-mL extraction vessels in a horizontal position for simultaneous extractions. The extraction vessels (0.9-cm ID x 3-cm length) were obtained from Alltech Associates (Deerfield, Illinois). Supercritical pressures were maintained inside the extraction vessels by using 60-cm length of uncoated fused-silica tubing (50-μm ID x 375-μm OD) from J&W Scientific as restrictors. Collection of the extracted material was performed by inserting the outlet restrictors into 15-mm ID x 60-mm glass vials (Supelco Inc., Bellefonte, Pennsylvania) containing hexane spiked with a known amount of an internal standard (terphenyl-d_{14}). Figures 1 and 2 show schematic diagrams of the two- and the four-vessel extraction systems. For the four-vessel extraction system, we used eight restrictors mounted in the 12-port valve to allow collection of two fractions per sample. Fractions 1A, 2A, 3A, and 4A were collected when the 12-port valve was in the load position. Fractions 1B, 2B, 3B, and 4B were collected when the 12-port valve was in the inject position. The SFE conditions are given in Table I.

- Supercritical fluid extraction system -- Hewlett Packard Model 7680A totally automated system with unlimited-capacity reciprocating pump, specially designed extraction chamber with safety interlocks, a variable restrictor nozzle and analyte collection trap. The operation of the extractor is controlled by a personal computer which is a Microsoft Windows-based system. An animated status screen provides real-time monitoring of the extraction process. Table II gives the SFE conditions for the HP extractor.

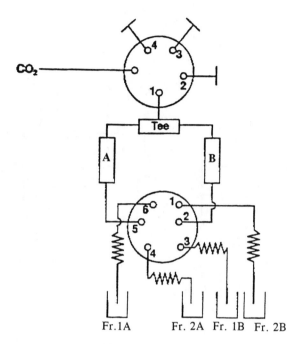

Figure 1. Schematic representation of two-vessel suprex extraction system.

Table I. SFE Conditions for the Suprex Extraction System

Parameter	Value
Pressure (atm)	Varied[a]
Temperature (°C)	Varied[a]
Flowrate (mL/min)	Not known unless volume of compressed CO_2 used in extraction was recorded
Number of extraction vessels	2 to 4
Position of extraction vessel	
Extraction time (min)	30 to 60 per fraction
Extraction vessel volume (mL)	2
Extraction vessel dimensions	9 mm ID x 30 mm length
Restrictor temperature (°C)	Room temperature
Restrictor dimensions	50 μm ID x 60 cm length
Collection solvent	Hexane (5 mL)
Temperature of collection vial	Room temperature[b]

[a]The pressure and temperature were varied as indicated in the tables of results.
[b]Temperature decreased during collection due to cooling of carbon dioxide upon expansion.

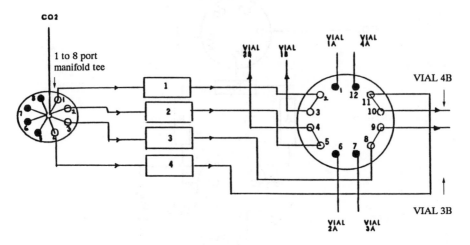

Inject Position

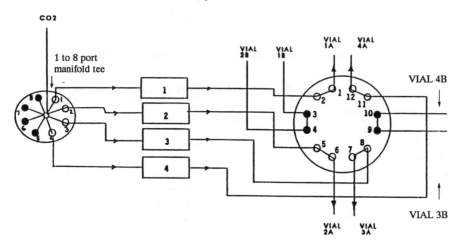

Load Position

Figure 2. Schematic representation of four-vessel suprex extraction system.

Table II. SFE Conditions for HP Extractor Evaluation[a]

Parameter	HP-1	HP-2	HP-3	HP-4	HP-5	HP-6	HP-7	HP-9	HP-10	HP-11	HP-12
Fluid density (g/mL)	0.85	0.85	0.85	0.85	0.6	0.75	0.85	0.90	0.90	0.85	0.80
Pressure (bar)	335	335	335	335	149	218	329	281	350	329	314
Flowrate (mL/min)	1.0	2.0	3.0	4.0	2.0	2.0	2.0	2.0	2.0	2.0	2.0
Temperature (°C)	60	60	60	60	60	60	60	40	50	60	70
Extraction time (min)	30	30	30	30	30	30	30	30	30	30	30
Thimble volumes swept	4.7	9.3	14.0	18.7	13.7	10.6	9.3	8.8	8.8	9.3	9.9
Nozzle temperature (°C)[b]	60	60	60	60	60	60	60	60	60	60	60
Trap temperature (°C)[b]	20	20	20	20	20	20	20	20	20	20	20
Rinse solvent	Hexane	Hexane	Hexane	Hexane	Hexane	Hexane	Hexane	Hexane	Hexane	Hexane	Hexane
Volume (mL)[c]	1.0	1.0	1.0	1.0	1.0	1.0	1.0	1.0	1.0	1.0	1.0
Rate (mL/min)[c]	1.0	1.0	1.0	1.0	1.0	1.0	1.0	1.0	1.0	1.0	1.0
Nozzle temperature (°C)[c]	30	30	30	30	30	30	30	30	30	30	30
Trap temperature (°C)[c]	25	25	25	25	25	25	25	25	25	25	25
Extraction vessel volume (mL)	7	7	7	7	7	7	7	7	7	7	7
Extraction vessel dimensions	10 mm ID x 90 mm length	10 mm ID x 90 mm length	10 mm ID x 90 mm length	10 mm ID x 90 mm length	10 mm ID x 90 mm length	10 mm ID x 90 mm length	10 mm ID x 90 mm length	10 mm ID x 90 mm length	10 mm ID x 90 mm length	10 mm ID x 90 mm length	10 mm ID x 90 mm length
Position of extraction vessel	Vertical	Vertical	Vertical	Vertical	Vertical	Vertical	Vertical	Vertical	Vertical	Vertical	Vertical
Equilibration time (min)	1.0	1.0	1.0	1.0	1.0	1.0	1.0	1.0	1.0	1.0	1.0
Trap material	ODS[d]	ODS	ODS	ODS	ODS	ODS	ODS	ODS	ODS	ODS	ODS
Number of rinses per trap	2	2	2	2	2	2	2	2	2	2	2

[a] All experiments were performed with SRS103-100; the sample size was 1.0 g. An additional experiment identified as HP-8 was planned to be performed at 381 bars and 60°C; however, the prototype system which we used did not allow us to reach 381 bars.
[b] During the extraction step.
[c] During the rinse step.
[d] ODS—octacecyl-bonded silica.

- Gas Chromatograph -- A Varian 6000 equipped with two constant-current/pulsed-frequency electron capture detectors, a 30-m x 0.53-mm ID DB-5 fused-silica open-tubular column (1.5-μm film thickness), and a 30-m x 0.53-mm ID DB-1701 fused-silica open-tubular column (1.0-μm film thickness), both connected to a press-fit Y-shaped fused-silica inlet splitter (Restek Corporation, Bellefonte, Pennsylvania), was used to analyze for the nitroaromatic compounds. The columns were temperature-programmed from 120°C (1.0-min hold) to 200°C (1-min hold) at 3°C/min, then to 250°C (4-min hold) at 8°C/min; injector temperature 250°C; detector temperature 320°C; helium carrier gas 6 mL/min; nitrogen makeup gas 20 mL/min.

- Gas Chromatograph -- A Varian 6500 equipped with two constant-current/pulsed-frequency electron capture detectors, a 30-m x 0.53-mm ID DB-5 fused-silica open-tubular column (0.83-μm film thickness), and a 30-m x 0.53-mm ID DB-210 fused-silica open-tubular column (1.0-μm film thickness), both connected to an 8-in injection tee (Supelco, Bellefonte, Pennsylvania), was used to analyze for the haloethers. The columns were temperature-programmed from 180°C (0.5-min hold) to 260°C (1-min hold) at 2°C/min; injector temperature 250°C; detector temperature 320°C; helium carrier gas 6 mL/min; nitrogen makeup gas 20 mL/min.

- Gas Chromatograph -- A Varian 6000 equipped with two constant-current/pulsed-frequency electron capture detectors, a 30-m x 0.53-mm ID DB-5 fused-silica open-tubular column (0.83-μm film thickness), and a 30-m x 0.53-mm ID DB-1701 fused-silica open-tubular column (1.0-μm film thickness), both connected to a press-fit Y-shaped fused-silica splitter (Restek Corporation, Bellefonte, Pennsylvania), was used to analyze for the organochlorine pesticides. The columns were temperature-programmed from 140°C (2.0-min hold) to 270°C (15-min hold) at 2.8°C/min; injector temperature 250°C; detector temperature 320°C; helium carrier gas 6 mL/min; nitrogen makeup gas 20 mL/min.

- Gas Chromatograph/Mass Spectrometer -- A Finnigan 4510B (Finnigan Mat, San Jose, California) interfaced with a data system for data acquisition and processing and equipped with a 30-m x 0.32-mm ID DB-5 fused-silica open-tubular column (1-μm film thickness) was used to analyze the extract of the standard reference material SRS103-100. The column was temperature-programmed from 40°C (4-min hold) to 300°C (6-min hold) at 8°C/min; injector temperature 270°C; interface temperature 270°C.

Materials.

- Standards -- analytical reference standards of the PAHs, nitroaromatics, haloethers, and organochlorine pesticides were obtained from the U.S. Environmental Protection Agency, Pesticides and Industrial Chemicals Repository (Research Triangle Park, North Carolina); Aldrich Chemical (Milwaukee, Wisconsin); UltraScientific Inc. (Hope, Rhode Island); Chem Service (West Chester, Pennsylvania), and University of Illinois. Purities were stated to be greater than 98 percent. Stock solutions of each test compound were prepared in pesticide-grade hexane at 1 mg/mL. Working calibration standards were prepared by serial dilution of a composite stock solution prepared from the individual stock solutions.

- Carbon dioxide, SFC-grade, liquid (Scott Specialty Gases, Plumsteadville, Pennsylvania).

- Sample matrices -- sand and PAH-contaminated soil SRS103-100 (Fisher Scientific, Pittsburgh, Pennsylvania). The SRS103-100 PAH-contaminated soil was certified by Fisher Scientific for 13 PAHs, dibenzofuran, and pentachlorophenol. The certified values are presented in the Results and Discussion section.

Sample Extraction for the Two-Vessel Setup. The extraction of the 36 nitroaromatic compounds from sand spiked at 600 ng/g (per analyte) was begun at 150 atm/50°C/10 min (dynamic), continued at 200 atm/60°C/10 min (dynamic), and completed at 250 atm/70°C/10 min (dynamic) using carbon dioxide only. Sample size was 2.5 g. For all the experiments reported in this paper, the sample was sandwiched between two plugs of silanized glass wool to fill out the void volume. Two spiked samples identified as Experiments 1 and 2 in Figures 3a and 3b were extracted in parallel using the same conditions. Two additional sand samples spiked at the same concentration were extracted in parallel at 300 atm/70°C/30 min (dynamic) and are identified in Figures 3a and 3b as Experiments 3 and 4. To verify the completeness of the SFE technique, we collected an additional fraction for Experiments 1 and 2 at 300 atm/70°C/30 min (dynamic). No compounds were detected in the second fraction.

The extraction of the 19 haloethers from sand spiked at 600 ng/g (per analyte) was performed at 250 atm/70°C/60 min (dynamic) using carbon dioxide only (Experiments 5 through 8 in Figure 4). Sample size was 2.5 g. In Experiments 5 and 6, two sand samples (spiked in an aluminum cup with 150 μL of 10 ng/μL spiking solution) were extracted in parallel. In Experiments 7 and 8, two other sand samples (spiked directly in the extraction vessel to avoid losses due to compound volatilization and sample transfer) were extracted in parallel. In Experiments 9 and 10, two soil samples (spiked in an aluminum cup as in Experiments 5 and 6) were extracted in parallel at 150 atm/50°C/60 min (dynamic).

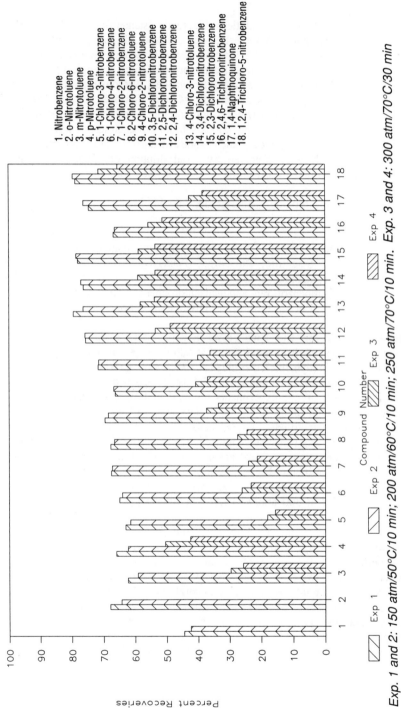

Exp. 1 and 2: 150 atm/50°C/10 min; 200 atm/60°C/10 min; 250 atm/70°C/10 min. Exp. 3 and 4: 300 atm/70°C/30 min. The experimental conditions are given in the experimental section.

Figure 3a. Percent recoveries of 18 nitroaromatics (compounds 1 through 18) extracted from spiked sand (2.5 g). The experimental conditions are given in the experimental section.

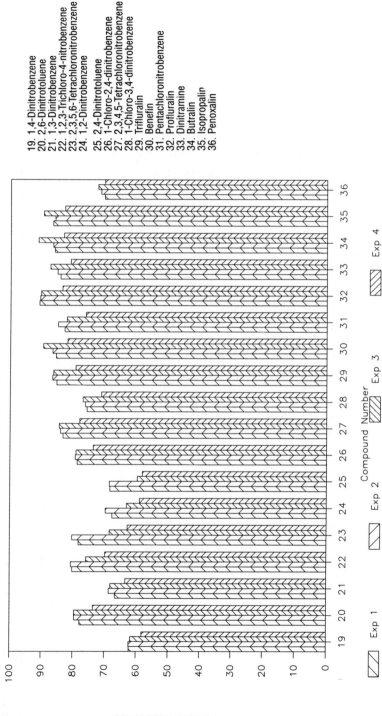

19. 1,4-Dinitrobenzene
20. 2,6-Dinitrotoluene
21. 1,3-Dinitrobenzene
22. 1,2,3-Trichloro-4-nitrobenzene
23. 2,3,5,6-Tetrachloronitrobenzene
24. 1,2-Dinitrobenzene
25. 2,4-Dinitrotoluene
26. 1-Chloro-2,4-dinitrobenzene
27. 2,3,4,5-Tetrachloronitrobenzene
28. 1-Chloro-3,4-dinitrobenzene
29. Trifluralin
30. Benefin
31. Pentachloronitrobenzene
32. Profluralin
33. Dinitramine
34. Butralin
35. Isopropalin
36. Penoxalin

Figure 3b. Percent recoveries of additional 18 nitroaromatics (compounds 19 through 36) extracted from spiked sand (2.5 g). The experimental conditions are given in the experimental section.

Exp. 1 and 2: 150 atm/50°C/10 min; 200 atm/60°C/10 min; 250 atm/70°C/10 min. Exp. 3 and 4: 300 atm/70°C/30 min.

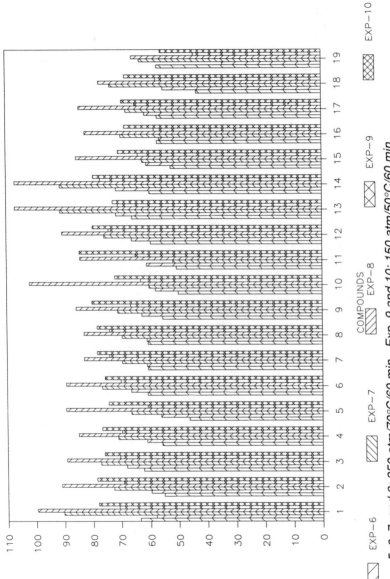

Figure 4. Percent recoveries of 19 haloethers (compounds 1 through 19 in Table V) extracted from spiked sand (2.5 g). The experimental conditions are given in the experimental section.

Exp. 5, 6, 7, and 8: 250 atm/70°C/60 min. Exp. 9 and 10: 150 atm/50°C/60 min

Sample Extraction for Four-Vessel Setup. All extractions were conducted with supercritical carbon dioxide. The extraction of the 19 haloethers from the four spiked sand samples (2.5 g each, spiked at 600 ng/g per compound) was performed at 250 atm/60°C/60 min (dynamic).

The extraction of 42 organochlorine pesticides from the four spiked sand samples (3 g each, spiked at the levels indicated in Table VI) was begun at 150 atm/50°C/10 min (static), continued at 200 atm/60/10 min (dynamic), and completed at 250 atm/70°C/30 min (dynamic) using 2.5-g samples.

The SRS103-100 standard reference soil (2.5 g) was extracted at 300 atm/70°C/30 min (dynamic) for Fraction 1 and 350 atm/70°C/30 min (dynamic) for Fraction 2.

Method Optimization. The experimental conditions for method optimization focused on seven variables, each chosen at two levels (high and low). Two separate tests were performed to see how these variables influence the method performance. The seven variables and their values chosen for each test are presented in Tables III and IV. The group differences for Test 1 (V_P through V_C) were calculated using Equations 1 through 7; those for Test 2 were calculated in a similar manner.

$$V_P = 1/4(w + x + y + z) \quad - 1/4(s + t + u + v) \quad = p - P \tag{1}$$
$$V_D = 1/4(u + v + y + z) \quad - 1/4(s + t + w + x) \quad = d - D \tag{2}$$
$$V_M = 1/4(t + v + x + z) \quad - 1/4(s + u + w + y) \quad = m - M \tag{3}$$
$$V_V = 1/4(u + v + w + x) \quad - 1/4(s + t + y + z) \quad = v - V \tag{4}$$
$$V_S = 1/4(t + v + w + y) \quad - 1/4(s + u + x + z) \quad = s - S \tag{5}$$
$$V_T = 1/4(t + u + x + y) \quad - 1/4(s + v + w + z) \quad = t - T \tag{6}$$
$$V_C = 1/4(t + u + w + z) \quad - 1/4(s + v + x + y) \quad = c - C \tag{7}$$

The relative changes as 100 x change/average recovery at low level were calculated and the data were sorted by variable and compound. The variables were ranked in the increasing order of absolute relative change for each compound and the sum of ranks was calculated.

The effect of varying flowrate, pressure, and temperature was investigated using the SRS103-100 standard reference soil and the Hewlett Packard extractor. The experiments were performed under the conditions given in Table II.

Results and Discussion

Reproducibility of Two-Vessel-System Extractions. Figures 3a and 3b show the recoveries of the 36 nitroaromatic compounds that were spiked on sand at 600 ng/g per compound and extracted with supercritical carbon dioxide using the Suprex extractor. The data indicate that 17 of the 36 compounds (Compounds 1 through 17 in Figure 3a) exhibited higher recoveries when extracted at 150 to 250 atm and 50 to 70°C. For the remaining 19 compounds, the recoveries were above 60 percent, with most of the values around 80 to 85 percent. It is quite possible that the more volatile nitroaromatic compounds get extracted in the first 5 or 10 min of the extraction; however, at higher pressures, the flowrate of the carbon dioxide is higher and these compounds are lost from the collection vessel by volatilization. The

Table III. Experimental Conditions for Method Optimization (Test 1)

Code Letter	Variable	Exp. 1 s	Exp. 2 t	Exp. 3 u	Exp. 4 v	Exp. 5 w	Exp. 6 x	Exp. 7 y	Exp. 8 z
P	Pressure (atm)	300	300	300	300	150	150	150	150
D	Temperature (°C)	70	70	50	50	70	70	50	50
M	Moisture content (%)	10	0	10	0	10	0	10	0
V	Extraction vessel vol. (mL)	4.7	4.7	2	2	2	2	4.7	4.7
S	Sample size (g)	2.5	1.0	2.5	1.0	1.0	2.5	1.0	2.5
T	Time (min)	60	30	30	60	60	30	30	60
C	Modifier volume[a] (µL)	250	50	50	250	50	250	250	50

SOURCE: Reprinted with permission from reference 1. Copyright 1990 Preston Publications.

[a] Hexane was used as modifier. This experiment was initially set up to investigate not only the polynuclear aromatic hydrocarbons, but also a set of organochlorine pesticides, which was spiked on the standard reference material at a low and a high concentration. The low and high concentrations were obtained by spiking 50 µL and 250 µL, respectively, of a hexane solution containing the compounds. Since the organochlorine pesticides could not be detected due to high background, the parameter "analyte spike" was changed to "modifier volume". The eight experiments were performed with the Suprex extractor.

Table IV. Experimental Conditions for Method Optimization (Test 2)[a]

Code Letter	Variable	Exp. 1 *s*	Exp. 2 *t*	Exp. 3 *u*	Exp. 4 *v*	Exp. 5 *w*	Exp. 6 *x*	Exp. 7 *y*	Exp. 8 *z*
P	Pressure (atm)	300	300	300	300	150	150	150	150
D	Temperature (°C)	70	70	50	50	70	70	50	50
F	Volume of toluene (μL)	500	0	500	0	500	0	500	0
G	Collection volume (mL)	5.0	5.0	1.0	1.0	1.0	1.0	5.0	5.0
M	Moisture (%)	10	0	10	0	0	10	0	10
B	Glass beads	w	w/o	w/o	w	w/o	w	w	w/o
E	Static extraction time(min)	30	15	15	30	15	30	30	15

a The eight experiments were performed with the Suprex extractor.

agreement between the duplicate extractions performed in parallel was excellent (within 5 percent).

Figure 4 shows the recoveries of the 19 haloethers. Overall, the recoveries were quite good (averaging around 70 percent across the 19 compounds) but the agreement between the duplicate extractions performed in parallel was not as good as in the case of the nitroaromatics, however, it was within 15 percent for most compounds. To determine if the pressure had any effect upon recovery, we compared experiments 5,6 with experiments 9,10 because they were pairs (performed in parallel at different pressures) and they were performed with sand samples spiked under identical conditions. All recoveries were slightly higher when extractions were performed at lower pressures. This seems to be in contradiction with what we obtained for experiments 7,8; however, in this case the sand was spiked directly in the extraction vessel and therefore, the data from experiments 7,8 cannot be compared with the data from experiments 5,6.

Reproducibility of Four-Vessel-System Extractions. Tables V and VI summarize the data for 19 haloethers and 42 organochlorine pesticides, respectively, that had been spiked onto sand (at the levels indicated in the tables) and extracted with supercritical carbon dioxide using the Suprex extractor. The experimental conditions (pressure, temperature, extraction time) had been established previously using our one-vessel and two-vessel extraction arrangements (1,2). This time, we were primarily interested in establishing the degree of agreement of the four parallel extraction recoveries using the four-vessel extraction system.

The results presented in Table V indicate that of the 19 haloethers extracted from spiked sand, 10 compounds exhibited recoveries greater than 84 percent, five compounds were in the range of 75 to 84 percent, and four compounds were in the range of 46 to 74 percent. The agreements of the four parallel extraction recoveries, expressed as the percent relative standard deviations (RSDs) of the four extraction recoveries, ranged from 1.2 to 36 percent, with 15 compounds having RSDs under 10 percent. In reality, the agreements of these parallel extraction recoveries are better than the RSD values in the table indicate because these tabulated values include not only variations due to the extraction, but also variations associated with matrix spiking, transfer of the sample extracts to the autosampler, extract injection, and peak quantitation.

The results presented in Table VI for the 42 organochlorine pesticides extracted from spiked sand indicate that SFE with carbon dioxide worked reasonably well for 35 of the 42 compounds (recovery > 50 percent); from those 35 compounds, 30 compound had recoveries > 70 percent. Among the seven compounds with recoveries < 50 percent, two are very volatile (DBCP and hexachlorocyclopentadiene). Three of the remaining five compounds (chlorthalonil, captan, and endosulfan sulfate) gave very poor recoveries with carbon dioxide alone, probably because of their polarity; however, in other experiments we have recovered them quantitatively from spiked sand with carbon dioxide modified with 10 percent methanol (1). Chlorobenzilate recoveries were poor, even when we used carbon dioxide with 10 percent methanol, and captafol was not tested with the carbon dioxide/methanol combination.

Table V.   **Average Recoveries and Percent RSDs for 19 Haloethers Extracted
from Spiked Sand with Supercritical Carbon Dioxide[a]**

Compound no.	Compound	Average recovery	% RSD
1	4-Bromophenyl-phenyl ether	93.8	1.2
2	Phenyl-4'-nitrophenyl ether	84.5	3.3
3	2-Chlorophenyl-4'-nitrophenyl ether	66.4	5.8
4	3-Chlorophenyl-4'-nitrophenyl ether	85.5	3.7
5	4-Chlorophenyl-4'-nitrophenyl ether	81.7	6.5
6	2,6-Dichlorophenyl-4'-nitrophenyl ether	85.3	4.6
7	3,5-Dichlorophenyl-4'-nitrophenyl ether[b]	89.4	3.9
8	2,5-Dichlorophenyl-4'-nitrophenyl ether[b]	89.4	3.9
9	2,4-Dichlorophenyl-4'-nitrophenyl ether	80.6	4.1
10	2,3-Dichlorophenyl-4'-nitrophenyl ether	78.8	7.5
11	3,4-Dichlorophenyl-4'-nitrophenyl ether	87.8	8.6
12	2,4,6-Trichlorophenyl-4'-nitrophenyl ether	91.5	1.7
13	2,3,6-Trichlorophenyl-4'-nitrophenyl ether	84.7	12
14	2,3,5-Trichlorophenyl-4'-nitrophenyl ether	88.8	2.0
15	2,4,5-Trichlorophenyl-4'-nitrophenyl ether	78.8	3.3
16	2,4-Dibromophenyl-4'-nitrophenyl ether	78.7	7.7
17	3,4,5-Trichlorophenyl-4'-nitrophenyl ether	59.2	29
18	2,3,4-Trichlorophenyl-4'-nitrophenyl ether	74.0	10
19	2,4-Dichlorophenyl-3'-methyl-4'-nitrophenyl ether	46.2	36

[a] The number of samples extracted in parallel was four. The spiking level was 600 ng/g. The experiments were performed with the Suprex extractor.

[b] These compounds coelute.

**Table VI.   Average Recoveries and Percent RSDs for 42 Organochlorine
Pesticides Extracted from Spiked Sand with Supercritical Carbon
Dioxide[a]**

Compound no.	Compound	Spike level (ng/g)	Average recovery	% RSD
1	DBCP	62.5	7.0	31
2	Hexachlorocyclopentadiene	62.5	22.6	10
3	Etridiazole	62.5	76.8	4.5
4	Chloroneb	1250	85.2	4.1
5	Propachlor	1250	56.8	11
6	Hexachlorobenzene	62.5	73.5	5.3
7	Trifluralin	125	86.7	4.7
8	Diallate	1875	89.2	4.5
9	alpha-BHC	62.5	79.4	5.4
10	PCNB	62.5	75.1	13
11	gamma-BHC	62.5	80.8	6.9
12	Heptachlor	62.5	82.2	5.2
13	Chlorthalonil	125	49.0	10
14	Alachlor	62.5	65.1	10
15	Aldrin	62.5	82.5	5.1
16	beta-BHC	62.5	77.4	5.4
17	delta-BHC	62.5	87.3	4.8
18	DCPA	62.5	77.8	6.8
19	Isodrin	62.5	107	4.3
20	trans-Permethrin	625	69.1	23
21	Heptachlor epoxide	62.5	90.2	12
22	Endosulfan I	62.5	84.1	4.7
23	gamma-Chlordane	62.5	86.1	5.2
24	alpha-Chlordane	62.5	85.5	4.6
25	trans-Nonachlor	62.5	85.4	4.3
26	p,p'-DDE	62.5	89.1	21
27	Captan	125	45.8	7.6
28	Dieldrin	62.5	88.6	3.1
29	Perthane	1875	78.1	5.4
30	Chlorobenzilate	125	0	--
31	Endrin	62.5	83.3	5.0
32	Chloropropylate	62.5	72.5	5.5
33	p,p'-DDD	125	83.3	10
34	Endosulfan II	62.5	74.2	14
35	p,p'-DDT	62.5	74.0	15
36	Endrin aldehyde	125	75.9	3.7
37	Endosulfan sulfate	62.5	0	--
38	Kepone	62.5	64.9	6.1
39	Mirex	125	86.1	4.8
40	Methoxychlor	62.5	60.7	20
41	Captafol	62.5	36.2	19
42	Endrin ketone	62.5	70.1	6.1

[a]The number of samples extracted in parallel was four. The experiments were performed with the Suprex
extractor.

Overall, the agreement of the extraction recoveries (as percent RSDs) of the 42 organochlorine pesticides, obtained for the spiked sand with the four-vessel extraction system, was better than what we had achieved earlier (1) by sequential extraction with the one-vessel extraction system. For example, 26 compounds have RSDs under 10 percent with the four-vessel extraction system while only 12 compounds had RSDs under 10 percent with the one-vessel extraction system. **Standard Reference Material.** The standard reference soil SRS103-100 certified for 13 PAHs, dibenzofuran, and pentachlorophenol at concentrations ranging from 19.1 to 1618 mg/kg was extracted with carbon dioxide. The results are presented in Table VII. These results are from two experiments in which we collected two fractions per sample. The data show clearly that under the conditions used a 30-min extraction will not completely remove the target compounds from the matrix. Even after 60 min, the recoveries of benzo(a)anthracene, chrysene, benzo(b+k)fluoranthene, and benzo(a)pyrene were still below 40 percent (in Fractions 1 and 2 combined). The percent RSDs for this soil were higher than those obtained for the spiked sand samples; RSDs for the soil ranged from 9.9 to 29 percent for Fraction 1 in Trial 1 and from 4.9 to 36 percent for Fraction 1 in Trial 2; the ranges were higher for Fractions 2 in both trials because much smaller amounts were recovered in Fraction 2.

Method Optimization. The experimental conditions focused on two sets of seven variables listed in Tables III and IV, each chosen at two levels. A full factorial design would have required 128 experiments. Instead, we performed only eight experiments. This design allowed us to estimate the main effects of the two sets of seven variables; however, we could not test for statistical significance of any of these effects because there were no degrees of freedom for the error terms. Tables VIII and IX summarize the recovery data for the eight experiments performed for each test. From these tables, one can calculate the effect on the recovery of each variable at its low and its high value using Equations 1 through 7. To get an overall picture of which variables affect the recovery of the highest number of compounds, the group differences were ranked and the ranks across the 15 compounds were summed up.

From these sums, we concluded that in Test 1 recovery was most affected by time (sum of ranks 88) and least affected by the volume of modifier (sum of ranks 31). Pressure and moisture ranked second (sum of ranks 70) and third (sum of ranks 69) in importance after time. The sum of ranks for sample size, temperature, and cell volume are 65, 50, and 47, respectively. Figure 5 shows the relative change in recovery for each compound and for each of the seven variables.

When the experiments were performed at the same pressure, temperature, and moisture content but with toluene as modifier and with a static extraction time of 15 or 30 min prior to the dynamic extraction step, then recovery was most affected by moisture content (sum of ranks 88) followed by pressure (sum of ranks 70) and the toluene volume (sum of ranks 68). The fourth variable to influence was the static extraction time (sum of ranks 57). Temperature, volume of collection solvent, and the presence/absence of glass beads were the least important. Figure 6 shows the relative changes in recovery for each compound and for each of the seven variables investigated in Test 2.

Table VII. Average Recoveries and Percent RSDs for 15 Target Analytes Extracted from a Certified Soil (SRS103-100) with Supercritical Carbon Dioxide (Fraction 1)[a]

Compound	Certified value (mg/kg)	Fraction 1				Fraction 2			
		Trial 1		Trial 2		Trial 1		Trial 2	
		Average recovery	% RSD	Average recovery	% RSD	Average recovery	% RSD	Average recovery	% RSD
Naphthalene	32.4 ± 8.2	54.0	12	56.9	13	0	--	13.5	23
2-Methylnaphthalene	62.1 ± 11.5	79.0	12	66.4	4.9	5.9	28	17.1	43
Acenaphthylene	19.1 ± 4.4	52.4	10	44.2	28	0	--	13.1	20
Acenaphthene	632 ± 105	96.3	10	75.9	15	9.7	17	19.4	24
Dibenzofuran	307 ± 49	105	9.9	80.6	14	9.8	23	19.9	27
Fluorene	492 ± 78	81.2	12	63.2	13	10.4	21	18.9	15
Phenanthrene	1618 ± 348	92.1	23	72.1	17	21.0	18	34.0	24
Anthracene	422 ± 49	79.4	17	62.0	23	16.7	27	24.7	15
Fluoranthene	1280 ± 220	73.1	29	52.0	29	16.0	15	26.0	26
Pyrene	1033 ± 289	48.3	23	44.7	36	16.2	8.8	27.1	22
Benzo(a)anthracene	252 ± 38	32.4	22	27.8	34	11.6	7.8	19.2	28
Chrysene	297 ± 26	31.1	22	27.6	31	10.2	6.0	19.0	29
Benzo(b+k)fluoranthene	152 ± 22	0	--	0	--	7.3	12	12.3	46
Benzo(a)pyrene	97.2 ± 17.1	0	--	0	--	4.5	24	9.3	80
Pentachlorophenol	965 ± 374	52.1	24	52.3	34	14.8	17	31.3	30

[a]The number of samples extracted in parallel in each trial run was four. The experiments were performed with the Suprex extractor.

Table VIII. Extraction of the SRS103-100 by SFE with Carbon Dioxide under Various Conditions (Test 1)[a]

Compound No.	Compound	Percent Recovery							
		Exp. 1	Exp. 2	Exp. 3	Exp. 4	Exp. 5	Exp. 6	Exp. 7	Exp. 8
1	Naphthalene	3.1	1.5	1.5	1.5	2.5	4.6	1.5	4.3
2	2-Methylnaphthalene	0.8	13.1	2.4	7.2	8.1	7.1	2.9	27.1
3	Acenaphthylene	13.1	20.9	10.5	31.4	30.4	34.6	6.3	54.5
4	Acenaphthene	19.9	35.9	16.5	45.7	45.9	55.4	7.5	77.5
5	Dibenzofuran	31.4	44.6	16.6	51.1	50.8	64.2	6.5	80.8
6	Fluorene	37.8	39.4	16.7	47.8	45.9	63.4	7.3	71.5
7	Phenanthrene	74.5	66.4	42.0	91.5	59.2	107.0	19.3	123.0
8	Anthracene	76.8	37.0	27.7	14.9	44.5	63.5	20.1	80.1
9	Fluoranthene	60.9	28.8	39.5	60.9	46.6	46.3	20.3	71.6
10	Pyrene	55.7	30.8	44.2	65.8	43.2	42.8	21.1	61.9
11	Benzo(a)anthracene	35.5	15.7	37.9	43.7	29.3	19.4	18.9	29.8
12	Chrysene	32.0	15.0	36.0	41.1	25.9	17.6	18.0	25.0
13	Benzo(b+k)fluoranthene	21.4	9.5	27.6	30.9	17.0	10.0	15.5	6.2
14	Benzo(a)pyrene	11.3	4.6	18.5	17.0	9.5	5.6	8.2	7.6
15	Pentachlorophenol	40.3	16.4	37.8	50.8	44.1	40.2	18.4	37.1

SOURCE: Reprinted with permission from reference 1. Copyright 1990 Preston Publications.
[a]The experimental conditions are given in Table III.

Table IX. Extraction of the SRS103-100 by SFE with Carbon Dioxide under Various Conditions (Test 2)[a]

Compound No.	Compound	Percent Recovery							
		Exp. 1	Exp. 2	Exp. 3	Exp. 4	Exp. 5	Exp. 6	Exp. 7	Exp. 8
1	Naphthalene	67.9	30.9	18.5	3.1	77.2	49.4	52.5	12.3
2	2-Methylnaphthalene	90.2	24.2	67.6	1.6	101	74.1	75.7	33.8
3	Acenaphthylene	57.6	26.2	47.1	5.2	62.8	57.6	47.1	47.1
4	Acenaphthene	98.6	4.7	80.9	8.9	108	62.7	80.2	88.0
5	Dibenzofuran	105	53.7	87.9	7.5	110	67.4	85.3	104.0
6	Fluorene	87.0	48.4	78.0	10.8	95.1	64.0	70.3	93.1
7	Phenanthrene	124	78.5	129.0	49.2	153	130	92.1	174
8	Anthracene	85.1	59.5	90.8	42.2	101	96.4	60.7	123
9	Fluoranthene	81.3	51.4	97.7	74.4	109	113	50.3	137
10	Pyrene	57.5	37.9	74.7	58.7	74.2	85.0	35.6	96.8
11	Benzo(a)anthracene	36.5	24.2	59.1	38.1	46.4	54.8	19.4	59.9
12	Chrysene	37.7	23.6	63.6	36.7	46.1	49.8	21.9	56.9
13	Benzo(b+k)fluoranthene	30.3	9.2	52.0	17.8	29.6	7.9	7.2	13.2
14	Benzo(a)pyrene	16.5	2.1	31.9	7.2	17.5	2.0	8.2	15.4
15	Pentachlorophenol	89.4	46.5	55.9	69.2	147	117	42.7	144

[a]The experimental conditions are given in Table IV.

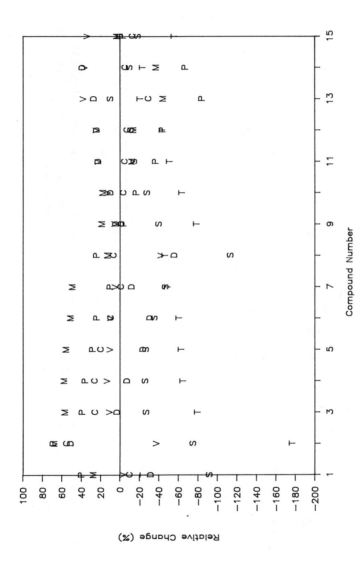

Figure 5. Relative change in percent recovery (when going from low to high) for the 15 target compounds identified in Table VIII (test 1). The variables are identified as follows: pressure (P), temperature (D), moisture (D), cell volume (V), sample size (S), time (T), and volume of modifier (C). The 15 compounds are those listed in Table IX. (Reproduced with permission from reference 1. Copyright 1990 Preston Publications.)

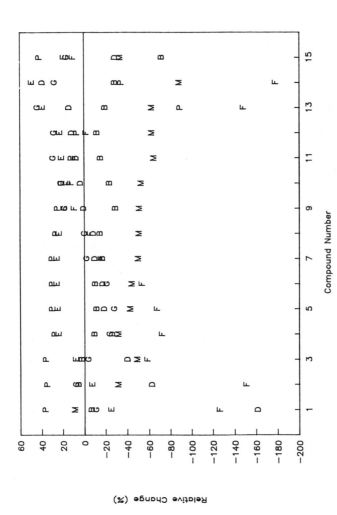

Figure 6. Relative change in percent recovery (when going from low to high) for the 15 target compounds identified in Table IX (test 2). The variables are identified as follows: pressure (P), temperature (D), volume of toluene as modifier (F), volume of collection solvent (G), moisture (M), glass beads (B) and static extraction time (E). The 15 compounds are those listed in Table IX.

The effect of varying flowrate, pressure, and temperature was investigated using the HP extractor. This is the only commercially available extractor that allows flowrates of up to 4 mL/min (as compressed fluid) through the extractor. The recovery data for the 15 target compounds extracted from SRS103-100 as a function of flowrate (experiments HP-1 through HP-4), pressure (experiments HP-5 through HP-7), and temperature (HP-9 through HP-12) are presented in Figures 7 through 9, respectively. These data indicate that the highest recoveries (except for naphthalene and 2-methylnaphthalene) were obtained at flowrates of 4 mL/min; this was quite evident for higher-molecular-weight PAHs. Since we are seeing increased recoveries by increasing the flowrate (or the amount of carbon dioxide used for extraction), it appears that the SFE rates in this case are at least to some extent controlled by compound solubility in carbon dioxide. Pressure affected compound recovery in several ways: for naphthalene and 2-methylnaphthalene, we observed a slight decrease in recovery by increasing the pressure from 149 to 329 bars. Acenaphthylene, acenaphthene, dibenzofuran, and fluorene recoveries were not affected at all by the change in pressure, and the rest of compounds, except benzo(a)pyrene, showed higher recoveries at 218 bars than at 149 and 329 bars. It is possible that we may have reached a maximum for compound solubility at that intermediate pressure. Finally, recoveries were almost identical at 40°, 60°, and 70°C, but they were consistently lower by about 30 percent at 50°C. We repeated the experiment and were able to duplicate the original data at 50°C. Thus, we concluded that pressures of 350 bars and temperatures of 50°C give lower recoveries regardless of the fact that the density of the supercritical fluid was at its maximum under these conditions.

Conclusion

We have demonstrated in this paper that two and four samples can be extracted in parallel with supercritical carbon dioxide without significant impact on data quality. Modifications made to an off-line extractor involved addition of a multiport manifold for the distribution of supercritical fluid to four extraction vessels and of a 12-port, two-way switching valve that allowed collection of two fractions per sample in unattended operation. The only limitation that we have experienced with the four-vessel extraction system was in the duration of the extraction. When working with 2-mL extraction vessels and 50-μm restrictors, and using the pressure/temperature conditions mentioned above, the 250-mL syringe pump allows us a maximum extraction time of 60 min. During this time, two 30-min fractions can be collected with the present arrangement.

Notice

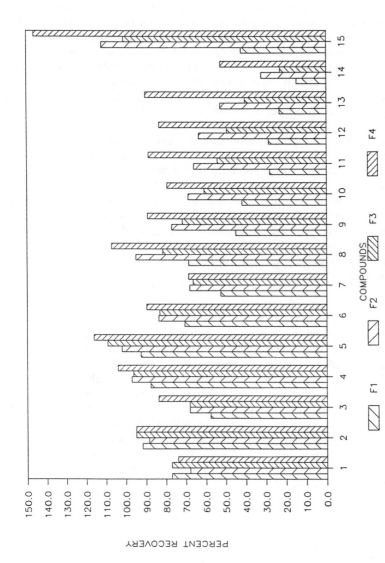

Figure 7. Recovery as a function of flowrate for the 15 target compounds identified in Table IX; F1 is 1 mL/min; F2 is 2 mL/min; F3 is 3 mL/min; and F4 is 4 mL/min. Experimental conditions are given in Table II.

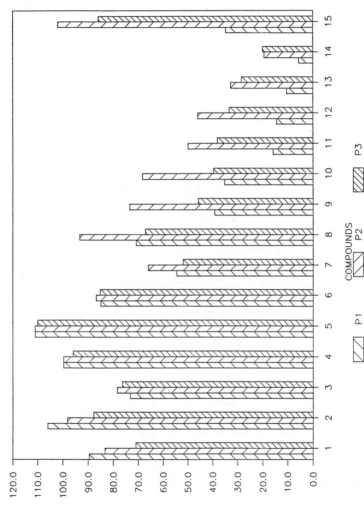

Figure 8. Recovery as a function of pressure for the 15 target compounds identified in Table IX; P1 is 149 bar; P2 is 218 bar; and P3 is 329 bar. Experimental conditions are given in Table II.

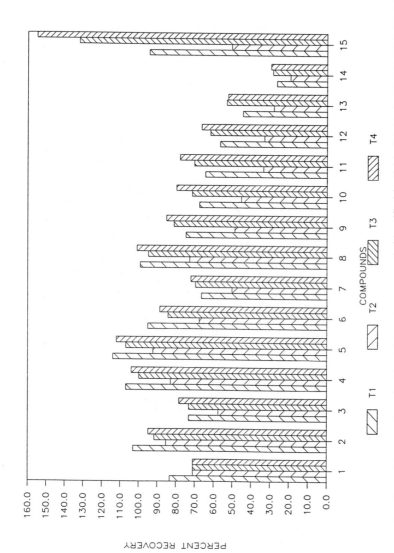

Figure 9. Recovery as a function of temperature for the 15 target compounds identified in Table IX; T1 is 40 °C; T2 is 50 °C; T3 is 60 °C; and T4 is 70 °C. Experimental conditions are given in Table II.

Acknowledgment

We acknowledge the assistance of Nikhil S. Dodhiwala who performed most of the SFE work, of Janet Benedicto and Lisa Balch, who performed all gas chromatographic and gas chromatographic/mass spectrometric analyses reported in this manuscript, and of Karin Bauer of Midwest Research Institute, who provided input on the statistical analysis of the preliminary results from our method optimization study. We would also like to acknowledge the assistance of Ashok Shah and Carl Stadler of Suprex Corporation who helped with the design of the dual-extraction setup on the Suprex SE-50 system, and of Hewlett-Packard Company who made available to us a Hewlett Packard extractor.

Literature Cited

1.   *Lopez-Avila, V.; Dodhiwala, N.S.; and Beckert, W.F.* J. Chrom. Sci. **1990**, 28, 468.

2.   *Lopez-Avila, V.; Dodhiwala, N.S.; and Beckert, W.F.* Off-line Supercritical Fluid Extraction Technique for Difficult Environmental Matrices Contaminated with Compounds of Environmental Significance. Paper presented at the Sixth Annual Waste Testing and Quality Assurance Symposium, Washington, D. C., July 1990.

RECEIVED December 2, 1991

Chapter 15

Supercritical Fluid Extraction of Phenols from Water

J. L. Hedrick, L. J. Mulcahey, and L. T. Taylor[1]

Department of Chemistry, Virginia Polytechnic Institute and State
University, Blacksburg, VA 24061-0212

The feasibility of extracting substituted phenols from an aqueous
solution with supercritical CO_2 is reported. A special extraction
vessel was used in order to overcome the mechanical difficulty in
retaining the liquid matrix in the extraction vessel. Solid phase
trapping was utilized with a diol silica bonded phase. Methanol was
used to rinse the trap. Below 300 atm extraction recovery paralleled
CO_2 pressure at fixed temperature. Phenol was least extractable;
while, 2,4-dichlorophenol yielded the greatest percent recovery.
Above 300 atm extraction yield declined with pressure. It is
theorized that at high CO_2 density there is less mixing with the
aqueous phase because of increased fluid-fluid interaction.

Analytical technology is perhaps no more challenged than in the area of polar
compound analysis. Most of these materials are high boiling, have low solubility in
traditional chromatographic/spectrometric nonpolar solvents, are prone to
decomposition at high temperatures and strongly adduct to both other polar analytes as
well as components in the matrix. If the polar materials in the mixture are numerous
and chemically similar such as substituted phenols, polychlorinated biphenyls, dioxins
and furans, the analysis problem is further compounded. Adding to this dilemma is the
fact that many polar components are multi-functional with both acidic and basic parts. It
therefore is not surprising that modern analytical techniques are only beginning to be
seriously applied to polyfunctional, high molecular weight, slightly soluble, polar
materials.

Current analytical methodology for many polar compounds is time consuming,
insufficiently sensitive, labor intensive, costly and not suitable for use in laboratories
where high sample volume testing is required. Furthermore, many procedures lack
specificity and involve modification (i.e. derivatization) of the material prior to
separation and/or detection. Preparation of the sample, especially if it is found in trace
quantity in a complex matrix such as foods, drugs, chemical waste, polymers and
propellants is perhaps the greatest obstacle. Many sampling (e.g. isolation) methods
have proven to be very time-consuming, cumbersome, include numerous laboratory
operations and may result in chemical alteration of the analyte.

Supercritical fluid extraction (SFE) on the other hand, is fast because

[1]Corresponding author

0097-6156/92/0488-0206$06.00/0
© 1992 American Chemical Society

supercritical fluids have lower surface tension, higher diffusivity and lower viscosity than liquid solvents (1). The solvent strength of a supercritical fluid (SF) can also be varied by adjusting the temperature and/or pressure; while, the solvent strength of a liquid is essentially constant. At ambient conditions the most common supercritical fluid (e.g. CO_2) is a gas and concentration of analyte by evaporation of the CO_2 is simple; whereas, a liquid extract has to be heated to evaporate the liquid in order to achieve a higher concentration for trace organic analytes. From an economic point-of-view liquids are equally expensive to both purchase and dispose (i.e. 55 gallon drum, approximatley $1,000 for disposal). On the other hand, clean CO_2 is readily available and relatively (SFE grade, $150/39 lbs.) inexpensive. The conclusion to be drawn from this rationale is that supercritical fluids offer considerable potential for sample preparation prior to trace organic analysis. In this regard, SFE has been found to be useful in a qualitative sense on both large and small scales. A wide variety of "chemical types" have been extracted from a multitude of matrices (2-4).

   If the principle of "like dissolves like" carries over to "like extracts like", polar analytes present unique challenges for SFE. While several other fluids have been marginally investigated, only CO_2 has been seriously considered to date for analytical SFE. Although once thought to exhibit the solvating power of chloroform and methanol, CO_2 at 500 atm and 100°C is probably no better than hexane. Therefore, CO_2 at high density will probably never exhibit high solubility for a polar compound. In addition, analyte-matrix adduction which must be disrupted to achieve extractability is expected to be high since most "real world" matrices are equally polar. The situation may, however, not be as bleak as depicted here. SFE is normally carried out in the dynamic mode wherein fresh CO_2 is continually passed through the matrix. If the analyte distribution coefficient between supercritical fluid CO_2 (SF-CO_2) and the polar matrix is finite, analytical extractions may be feasible given ample extraction time and CO_2 mass flow. For example, the solubility of ethanol in CO_2 at 1100 psi and 40°C is only approximately 0.5 weight percent, yet ethanol can be readily extracted from an aqueous matrix (5). One additional complicating feature that can be envisioned during the extraction of polar material from a polar matrix is that both material and matrix components may extract under the required high density conditions. In this situation, success will only be realized if a certain degree of selectivity can be achieved in either the extraction, accumulation, or recovery steps. The extraction of phenols with SF-CO_2 from water constitutes a rather challenging and timely problem in this regard. For example, both analyte and matrix are polar and each exhibits some solubility in SF-CO_2. This report discusses our results to date with a variety of substituted phenols.

Experimental

The extraction vessel (Figure 1) used for liquid-fluid extractions was acquired from Keystone Scientific (Bellefonte, PA) and measured 10.0 cm in length with an internal diameter of 0.94 cm (6.94 mL volume). The vessel was subsequently modified for use with liquid samples in the following manner. The zero dead volume channel at either end of the stock vessel was drilled out to a diameter of 1/16". The stainless steel frits that were contained as an integral part of the fitting seal were punched out so that 1/16" stainless steel could pass freely through either end of the vessel. The stainless steel inlet and outlet tubes were 0.01" internal diameter and extended to within 1 cm of the top and bottom of the vessel. The locking ferrule of a Kel-F finger tight fitting (Keystone Sceintific) was swaged manually, once the inlet and outlet tubings were in the proper position. Finger tight fittings, with Kel-F ferrules, were used instead of a standard stainlees steel nuts and ferrules at both ends of the vessel in order to provide a more reliable seal that could be broken and resealed numerous times. Screens were placed at either end of the vessel so that solid material would not be transported through the system and possibly foul valves and restrictions.

The vessel is conceptually the same as one reported earlier by Ong et al. (6) (i.e. SF is introduced from the bottom by passing the tubing in from the top) with a few major differences. The Ong vessel required many different seals, most of which are stainless steel. More importantly the sample volume of the Ong vessel is extremely limited in that a sample size of 0.5 mL would be quite large. This same vessel was also previously employed for the extraction of phosphonates from water in our laboratory (7).

Extraction Profiles of Phenols. Extraction of phenols from water was performed on 3 mL of aqueous sample. Each sample was 160 ppm per component of phenol, 2-chlorophenol, 2,4-dichlorophenol, 2,4,6-trichlorophenol, 3-methyl-4-chlorophenol and 2,4-dimethylphenol. All extractions were performed with 1OO% C02 (SFC grade, Scott Specialty Gases, Plumbsteadville, PA) and a Suprex Model 50 SFE (Pittsburgh, PA). A solid phase trapping apparatus was used for collection of the extracted phenols as previously described (8). Standard solid phase (diol, cyano, amino, phenyl, C_{18}) extraction tubes (Supelco Inc., Bellefonte, PA) were used. Diol-bonded silica (40 um) proved to be the optimum trapping and recovery material for these phenols (i.e. recoveries >90%). In order to obtain the extraction time profile for each phenol, a number of SPE tubes were used sequentially. Each collection tube was used for a specific amount of time (i.e. mass of supercritical fluid). The flow from the extraction chamber could be stopped momentarily for the replacement of collection tubes. Figure 2 shows the plumbing scheme used. The fluid, after exiting the extraction vessel, enters a six port valve. By switching the valve position, the fluid from the extraction vessel either traveled on-line to the collection device or was stopped by a plug in the valve. The rate of flow from the pump (~ 1 mL/min) was monitored and corrected for differences in density between the oven (supercritical fluid), where the extraction took place and the pump head (liquid). Once the SFE was complete, the analyte was removed from each SPE tube with 2 mL of methanol (i.e. four 0.5 mL aliquots). Extraction profiles were obtained for all phenols at 100, 150, 200, 250, 300 and 350 atm at 50°C.

Analysis of the analytes was done by gas chromatography on a HP 5890 equipped with 7673 auto-sampler and DOS series chemstation (Hewlett Packard Co., Avondale, PA). A 25m HP-5 column with 200 um internal diameter and 0.33 um film thickness was used for all separations reported herein. The injection mode was purged splitless (1 uL injection) and the temperature ramp used for all separations was 60°C for 1 minute ramping then to 300°C at 20°C/minute.

Equilibrium Extraction System. An equilibrium extraction system (7) was used to determine distribution coefficients of a phenol/water/C0$_2$ system, as well as to perform static SFE/SFC (Suprex Model 200A, Pittsburgh, PA). The system used (Figures 3a and 3b) consisted of three six-port valves, A, B and C (Rheodyne Inc., Cotati, CA), a recirculating pump, R, (Micropump, Inc., Concord, CA), extraction vessel, E, and associated plumbing in a temperature controlled oven. A threeport switching valve, (VICI, Houston, TX) was used to allow for easy conversion of the instrument back to a conventional SFC. A one-meter length of 100 um i.d. deactivated fused silica was used to interface the extraction apparatus to a 1 mm x 250 mm DELTABOND CN packed column (Keystone Scientific, Inc., Bellefonte, PA) for chromatography. The total volume of the system was calculated to be 8.84 mL, which included all tubing, dead volume associated with the recirculation pump and the 6.94 mL extraction vessel.

Figure 3a shows the supercritical fluid flow path during charging of the system. Fluid enters the system from the syringe pump (SF in) and travels in a direction

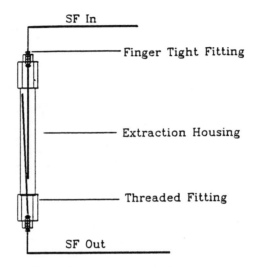

Figure 1. Aqueous extraction vessel.

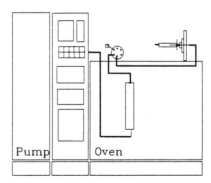

Figure 2. Extraction profile plumbing scheme.

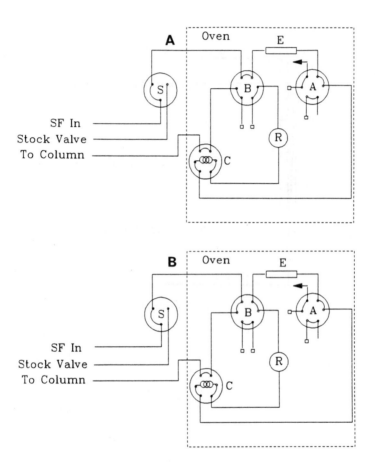

Figure 3.   Static extraction apparatus: (A) pressurization mode and (B)
          equilibration mode

consistent with that which will be induced by the recirculating pump. The system is closed so that no net flow through the system occurs during pressurization. Switching valve "B" to its other position (Figure 3b) creates a closed loop sytem. The recirculation pump moves fluid through the system at a rate of 3-5 mL/min. The exact rate of flow is difficult to determine as the efficiency of the pump with supercritical fluids is not known.

The results stated herein were obtained with a 20 uL sampling loop. Since each 20 uL injection represented only 0.3% of the extracted phase (5.84 mL with a 3 mL sample), multiple injections could be run on the sample without significantly depleting the extracted phase of analyte.

For determination of phenol distribution coefficients the extraction proceeded for 15 minutes in order to reach equilibrium. The time required to reach equilibrium was determined by making five replicate injections of the headspace onto the SFC system. The first injection was after the extraction had proceeded for 15 minutes at 50°C and 100 atm. Following the equilibration time, four further injections at ten minute intervals were made, after which the pressure inside the extraction apparatus was increased and the system was again allowed to equilibrate (i.e. 15 minutes). The five replicate injection process was then repeated. The amount of phenol in each injection was then noted by referring to an external phenol standard calibration curve. As the total volume of the system was known, the amount of phenol in the SF could be calculated. The amount of phenol in the aqueous phase could then be calculated by mass balance.

Results and Discussion

Background. Many different methods have been employed in order to concentrate trace organics from aqueous solution such as liquid-liquid extraction, purge and trap analysis, adsorption on solid adsorbents and passage through capillary polymeric columns. Desirable analyte features for accomplishing this task are low water solubility, low polarity and high volatility. The aqueous solution should also not (a) contain other dissolved solids or particulates, (b) exhibit high/low pH or (c) have other water soluble organics such as methanol or acetone. SFE of analytes from aqueous samples has received little attention as compared to SFE of solid matrices. The mechanical difficulty in retaining the liquid matrix in the extraction vessel is probably to blame. Another problem is the partial solubility of water in SF-CO_2 (i. e. approximately 0.1 wt %) which may present problems during the isolation of the extracted analyte. Water may also act to modify the SF-CO_2 thereby increasing its solvating power. Nevertheless, a few reports have appeared principally concerned with phenols.

Roop et al. (9) showed the extraction of phenol from water using supercritical CO_2. The sample size was 150 mL. The system was charged with CO_2 and then mechanically mixed for one hour. A two-hour separation time was then employed. The water phase was subsequently drawn off and analyzed. The goal of the experiment was to determine the distribution coefficient, K_d, of phenol at 298 K and 323K. Results showed a linear increase in K_d with increasing pressure at fixed temperature. There was a good deal of scatter at the higher temperature, although RSD's of 1.5% were quoted. In another paper, Roop et al. (10) reported the extraction of creosote from water using the same system. The creosote data indicated a reversal in the distribution coefficient at high temperature and fixed pressure. That is, K_d increased with increasing temperature, but beyond 35°C, K_d decreased with a temperature rise for the same pressure.

Departing from the "conventional" extraction cell design, Theibolt et al. (11) used a novel phase segmentor and subsequent phase separator in order to extract 4-chlorophenol and phenol from water. On-line SFC was used for subsequent analysis. Recoveries and reproducibilities for the system were not reported, although with a single pass system 100% recovery would not be expected. We too have demonstrated the feasibility of phenol extraction, albeit from 6M H_2SO_4 (8). Several other organic-water $SF-CO_2$-separations have been screened in feasibility tests. A partial list includes acetone (12), ethanol (13) and formamide (5). Ehntholt et al. (14) also reported the extraction of 23 organics from water at the ppm and ppb level with varying success. Sample sizes were 400 mL. Recoveries of material varied widely with the analyte being observed. Reproducibility of the extraction process varied from 10-50% RSD. Clearly the precision needed for analytical work was not achieved. The lack of reproducibility may have been caused by the relatively crude way in which the analytes were trapped after extraction. The trap was a series of u-tubes in a cooled water bath at 0°C.

Extraction of Phenols. Extraction profiles as a function of time for each phenol (3 mL of solution) at fixed pressure and temperature were obtained at 100, 200 and 250 atm (Figure 4). The same restrictor was employed in each case; therefore, the flow rate of supercritical fluid through the vessel differed as follows: 100 atm, 0.73 mL/min; 200 atm, 1.04 mL/min; etc.). At low pressure, two extraction vessel volumes (8 mL) of CO_2 (100 atm, 50°C) gave less than 40% recovery for each phenol. The lowest recovery was for phenol (<10%) followed by 3-methyl-4-chloro-, 2,4-dimethyl-, 2-chloro- and 2,4-dichloro-. For pressures below 250 atm each substituted phenol extraction behaves as expected with greater recovery occurring for each component with higher $SF-CO_2$ pressure (e.g. phenol: ~5% (100 atm) vs 15% (150 atm) for two vessel volumes (8 mL)). The extraction profiles not only reveal information regarding relative extractability with different substituents attached to the aromatic ring but also data concerning kinetics of extraction. Table I lists various physical and chemical properties of the studied phenols. The strongest acid of the group, no doubt, is 2,4-dichloro-. Since CO_2 (100 atm) saturated water is known to have pH~3, each phenol should be ˙ fully protonated under these extraction conditions and extractability should not correlate with acidity. In other words, if any phenol were deprotonated in part, only the protonated form should extract. If one excludes phenol and considers only data at 100 atm, volatility appears to correlate with extractability since the lowest boiling phenols (2-chloro and 2,4-dichloro-) exhibit the highest recovery. At 200/250 atm, the lowest boiling (phenol) and highest boiling (3-methyl4-chloro) phenols yield recoveries between 60% and 80%; whereas, the remaining three phenols cluster near 100 % recovered. It is interesting to also note that phenol is at least 4X more soluble in water than the other phenols, probably because of hydrogen bonding. The slope of the phenol extraction profile plot at fixed pressure appears to be single valued. Upon increasing pressure, the phenol profile slope also increases. At 200/250 atm the other phenols give rise to plots of varying slope (e.g. relatively high initially, but after approximately 10 mL of compressed CO_2 the profile slope decreases). It has been suggested that the early extracted material may represent removal of bulk analyte, while later more difficult-to-extract material reflects matrix-bound (or diffusion limited) analyte (15). It should be noted however that these comments were made in reference to a semisolid or solid matrix. It might be hypothesized that at 100 atm where $SF-CO_2$ has poor solvating ability only purging of the water is taking place. At high pressures, however, solvation of the substituted phenols becomes dominant which accounts for the rapid early removal. No doubt the following equilibria exist in solution. As indicated previously, equilibrium 1 should lie far to the

Protonated Phenol $\longleftrightarrow$ Non-protonated Phenol   (1)

Water Bonded Phenol $\longleftrightarrow$ Non-water Bonded Phenol   (2)

left. In the case of equilibrium 2, the shift is to the left for phenol, but apparently more of an equal distribution exists for the other substituted phenols.

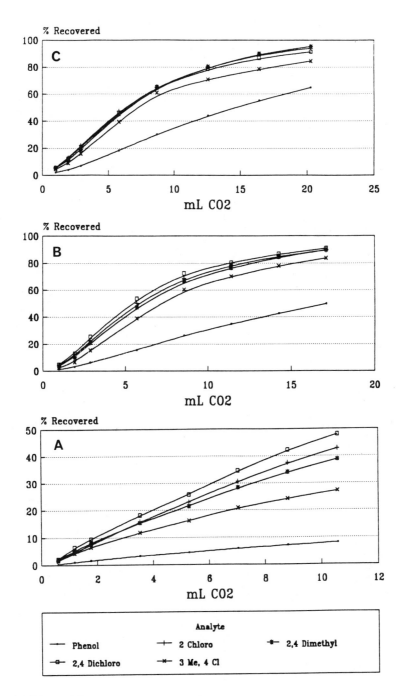

Figure 4. Extraction profile of phenols at 50°C and (A) 100 atm, (B) 200 atm, (C) 250 atm.

Table I

Phenol Solutions Subjected to SFE

Analyte	b.pt. (°C)	M.Wt.	pk_a	H_2O Solubility gm/100 gms pH=7	CO_2 Solubility (16) (wt-%) (25° C/900 psi)
	174.9	128.6	8.48	2.8	∞
	181.8	94.1	9.95	9.3	3*
	210.0	163.0	7.85	< 2	14
	210.0	122.2	----	< 2	----
	235.0	142.3	----	< 2	----

*Approximately 12% at 4000 psi CO_2 and 100°C

Surprisingly, percent recovery decreased upon going to 300 atm (1.17 mL/min SF-CO$_2$) and 350 atm (1.70 mL/min SF-CO$_2$) at 50°C for each phenol. The profile for phenol at various pressures is shown in Figure 5. Notice that for five extraction vessel volumes, percent recovery for phenol is now less than 40% at 350 atm. The slope of each phenol extraction profile plot is nevertheless single-valued at a fixed pressure. Figure 5 also reveals the same general decrease in extractability at pressures greater than 250 atm for 2-chloro and 3-methyl-4-chloro-phenol. The lower recoveries at higher pressure were validated by measuring the phenols in the raffinate. A complete mass balance for each extraction was possible. It is also interesting to note that the extraction profile plots become nearly fixed slopes at the higher pressures (>250 atm). It should also be noted that in all cases trapping was accomplished after CO$_2$ decompression.

An optimized extraction pressure in this study was a surprising result. To aid in understanding this phenomenon which was observed for all five phenols, static extraction of the aqueous solutions with headspace sampling was performed to determine the distribution coefficient of phenol in the SF-CO$_2$/water system. Three mL of 200 ppm phenol was allowed to come to equilibrium at various pressures and temperatures. A portion of the CO$_2$ was then sampled onto an SFC column and the amount of phenol in the CO$_2$ phase was determined by comparing to an external standard. The amount of phenol in the aqueous phase was determined by mass balance from which the distribution coefficient was calculated. It was assumed that H$_2$0 and CO$_2$ did not change appreciably in density or volume due to the presence of the other components in the system. This assumption was thought to be valid because of the extremely limited solubility of H$_2$0 in CO$_2$ and the incompressibility of water. Figure 6 shows the effect of pressure on K$_d$ at four different temperatures. The distribution coefficient was found to steadily increase (more phenol in the CO$_2$ phase) as pressure increased as expected since CO$_2$ density (solvating power) increases. Likewise, at constant pressure the higher temperature yielded the lower K$_d$ suggesting that the decreased solvating power of the CO$_2$ is more of a limiting factor here than the increased phenol vapor pressure at the higher temperature. Our K$_d$ values show the same trend, but are somewhat lower than values reported by Roop et al.(12) (i.e. approximately 0.2 (vol/vol basis) vs 0.7 (wt/wt basis) at 50°C and 25 MPa (250 atm)). K$_d$ as a function of temperature at constant pressure (Figure 7) shows approximately a 50% decrease in going from 35°C to 55°C at four different pressures. From 55°C to 75°C the change in K$_d$ is much less at constant pressure. A similar decrease between 50°C and 75°C was noted by Roop (12) for creosote in a water/CO2 system. For supercritical extraction Kd is a function of both the fluid solvating power and the analyte volatility. In the 35-55°C temperature range, the loss in solvating power apparently outweighs the increase in volatility. Beyond 55°C the volatility and solvation factors are believed to be comparable. K$_d$ as a function of temperature at fixed density was also determined. Surprisingly, a similar plot was obtained. This result was unexpected since at fixed density (solvating power) an increase in temperature was anticipated to enhance volatility and therefore extractability. This notion, however presumes that the solubility of phenol in water is itself independent of temperature. In other words, if the coefficient of phenol solubility in water is much greater than in SF-CO$_2$, K$_d$ would be expected to decrease with an increase in temperature. A definite minimum appears in each of these temperature-density plots around 60°C which indicates the operation of several competing factors. At the present time there are thought to be two possible reasons for the decrease in net extraction with increasing pressure (density). The first deals with the size and number of CO$_2$ droplets that percolate through the aqueous phase. As pressure increases leading to lower CO$_2$ diffusivity, it is possible that the size of the CO$_2$ droplets increase (Figure 8). The total area of CO$_2$ that the analyte sees per unit time would therefore decrease resulting in a slower (less efficient) extraction per

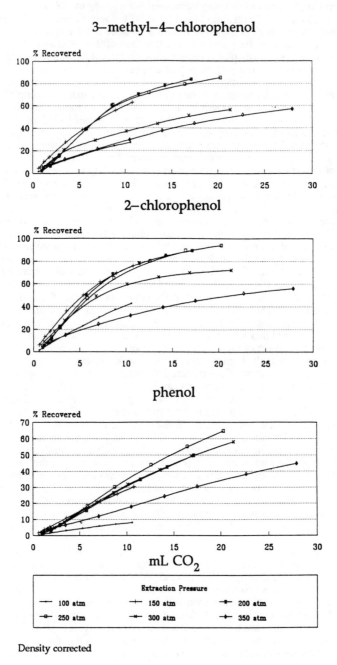

Density corrected

Figure 5.   Extraction behavior of phenol, 2-chlorophenol and 3-methyl-4-chlorophenol at different extraction pressures.

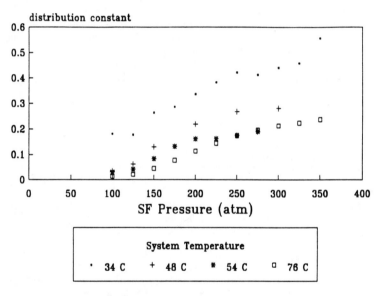

Figure 6.   Phenol distribution coefficient as a function of temperature at constant density.

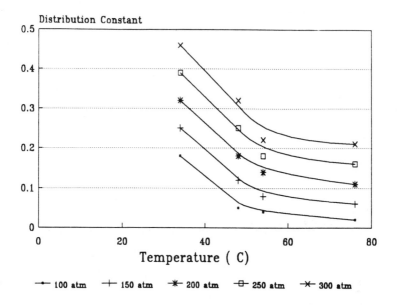

Pressure Constant

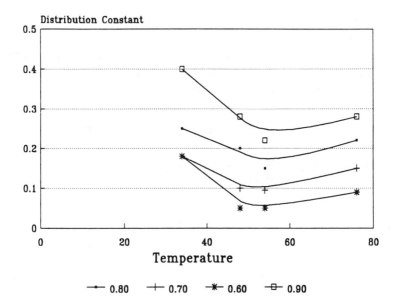

Constant Density

Figure 7.   Phenol distribution coefficient as a function of pressure at four temperatures.

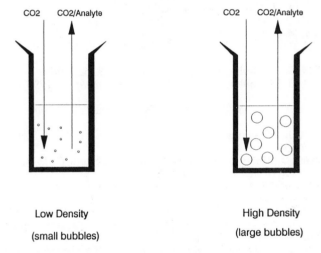

Figure 8.   Representation of liquid-fluid extraction at low and high CO_2 density.

unit time. Another possible explanation for the reduced net extraction is that the CO_2/H_2O boundary has become more defined as the density of CO_2 increases, thereby making mass transfer more difficult. Another plausible account for the decrease is that phenols may react with $SF-CO_2$ at high pressure. The resulting product would presumably be less soluble (or show slower mass transfer rates) in the fluid phase than phenol.

Acknowledgment

The financial assistance of the U. S. Environmental Protection Agency, BP America Inc. and Burroughs Wellcome Co. is greatly appreciated.

Literature Cited

1. Gere, D. R. *Science* **1983**, *222*, 253.
2. Krukonis, V. J. *ACS Symp. Ser.* **1988**, *366*, 26.
3. Hawthorne, S. B. *Anal. Chem.* **1990**, *62*, 633A.
4. King, J. W. *J. Chromatogr. Sci.* **1989**, *27*, 355.
5 McHugh, M.; Krukonis, V. *Supercritical Fluid Extraction*, Butterworths: Boston, MA, 1986.
6 Ong, C. P.; Ong, H. M.; Yau, S. F.; Lee, H. K. *J. Microcol. Sep.* **1990**, *2*, 69.
7. Hedrick, J. L.; Taylor, L. T. *Anal. Chem.* **1989**, *61*, 1986.
8. Hedrick, J. L.; Taylor, L. T. *J. High. Res. Chromatogr.* **1990**, *13*, 312 (1990).
9. Roop, R. K.; Akgerman, A.; Dexter, B. J.; Irvin, T. R. *J. Supercrit. Fluids* **1989**, *2*, 51.
10. Roop, R. K.; Akgerman, A.; Irvin, T. R.; Stevens, E. K.; *J. Supercrit. Fluids* **1988**, *1*, 31.
11 Thiebaut, D.; Chervet, J-P; Vannoort, R. W.; DeJong, G. J.; Brinkman, U. A. Th.; Frei, R.W. *J. Chromatoqr.* **1989**, *477*, 151.
12. Panagiotopoulos A. Z.; Reid, R. C. *Amer. Chem. Soc. Div. Fuel Chem. Prepr.* **1985**, *30*, 46.
13. Kuk, M. S.; Montagna, J. C. In *Chem. Enq. at Super. Fluid Cond.*; Paulatitis, M. E.; Penninger, J. M. L.; Gray, R. D.; Davidson, P. Eds., Ann Arbor Science: Ann Arbor, MI, 1983, pp. 101-111.
14. Ehntholt, D. J.; Thrun, K.; Eppig, C. *Intern. J. Environ. Anal. Chem.* **1983**, *13*, 219.
15. Bartle, K. D.; Clifford, A. A.; Hawthorne, S. B.; Langenfeld, J. J.; Miller, D. J. and Robinson, R. *J. Supercrit. Fluids* **1990**, *3*, 143.

RECEIVED January 16, 1992

Chapter 16

Supercritical Fluid Extraction versus Soxhlet Sample Preparation

A Comparative Study for Extraction and Analysis of Organics in Solid Matrices

L. J. D. Myer, J. H. Damian, P. B. Liescheski, and J. Tehrani[1]

Isco, Inc., 4700 Superior Street, Lincoln, NE 68504

Supercritical fluid extraction (SFE) has emerged as an important analytical-scale sample preparation technique that is applicable to a large variety of analytes and is especially suitable for the analysis of organic compounds in solid matrices. Conventional liquid solvent extraction methods (such as Soxhlet) are time-consuming and labor-intensive. They also require large volumes of toxic organic solvents that are expensive to purchase, expensive to dispose of, and hazardous to work with. SFE of organic compounds from solid matrices can be performed at moderate temperatures using non-toxic, non-flammable supercritical fluids such as CO_2. Extraction times are often less than one hour with analyte recoveries comparable to those of the Soxhlet method. This presentation will discuss the SFE of selected herbicides, pesticides, hydrocarbons, waxes, and fats from a wide variety of solid matrices. Also included is a comparative study of SFE vs. Soxhlet extraction methods.

SFE, although a newly developed technique, has shown great promise as Soxhlet extraction's modern day successor for a wide variety of analytes in many different matrices (1). In fact, interest in SFE has undergone what appears to be exponential growth during the last few years. This interest can be attributed to the speed of SFE and the potential for replacement of the large volumes of toxic organic solvents currently being used with low-cost non-toxic CO_2.

The bulk of the SFE experiments performed to date were executed with systems typically consisting of a syringe or reciprocating pump, a high-pressure containing sample vessel comprised of HPLC column hardware, and a fused silica capillary restrictor. Extraction vessel temperatures of 40-80°C were usually accomplished using a converted oven or with the use of a thermostatted tube heater (2,3). Instrument manufacturers now offer a variety of commercially available SFE systems that vary in design, operation, features, ease of operation, and limitations.

[1]Corresponding author

0097–6156/92/0488–0221$06.00/0

SFE Instrumentation

SFE systems including continuous pumping, dynamic modifier addition, and a portable unit were from Isco, Inc. (Lincoln, NE).

Basic SFE System. The base system used, (Isco SFE System 1200), consisted of a dual-chamber extraction module, 2.5 ml and 10 ml volume extraction cartridges, a 260 ml capacity syringe pump, a pump controller, and all the hardware used to connect the system to a CO_2 supply and collect the extracts. The syringe pump used was capable of 90 ml/min liquid flow at pressures up to 4500 psi and 45 ml/min flow at pressures between 4500 and 7500 psi. The extraction module was capable of withstanding pressures up to 10,000 psi and temperatures up to 150°C.

The extraction module utilized a unique chamber design that eliminated the use of wrench-tightened fittings during sample loading into the cartridge and extraction of the sample. Five-piece cartridges consisting of a metal body, upper endcap, lower endcap, and two frits with polymer sealing rings contained the sample and were hand-tightened and subsequently clipped into the extraction chamber cap. When the cartridge/chamber cap assembly was hand-tightened into the extraction chamber a sealing surface on the chamber cap passed through two circular lip seals in the receiving chamber. During extraction, both the inside and outside areas of the extraction cartridge was filled with fluid, equalizing the pressure and eliminating the need for high-pressure cartridge seals.

There are three needle valves and a check-valve controlling the fluid flow for each extraction chamber of the module. The supply valve regulates the flow into the chambers, the extractant valve allows the flow of fluid through the cartridge, and the vent valve is an alternate flow path for quick depressurization of the chambers through 0.005" I.D. X 20cm lengths of steel tubing once extraction was complete. The check valves were installed in the supply lines, allowed the chambers to be used simultaneously or independently, and prevented the possibility of intra-chamber communication between inside and outside areas of the cartridge.

Temperature regulation of the extraction module was accomplished through the use of a solid state controller regulating two cartridge heaters imbedded in an aluminum heater block that surrounded the extraction chambers. Fluid entering the extraction chambers was pre-heated in coils of tubing imbedded in the heater block and the extractant valves as well as the extractant lines were directly attached to the heater block to maintain temperature.

Since the pump was operated in constant pressure mode, 30cm lengths of 50µm I.D. X 375µm O.D. fused silica capillary tubing was installed on the extract outlet ports and used as flow restrictors as well as a means of directing the extractant flows through teflon™-faced septa into the 150mm X 20mm screw-cap culture tubes containing collection solvent.

Modifier Addition System. In the case where a continuous supply of modifier was desired, the base system was upgraded to perform continuous modifier addition with the addition of a second syringe pump module and a valving package (Isco SFE System 2200). The valving package consisted of two check-valves, a mixing tee, and the hardware necessary to install them in the system. Both pump modules were operated from the same controller. Percent modifier concentrations of up to 50% (v/v) were selectable from the controller keypad.

Continuous Pumping System. In the case where an unlimited supply of pressurized supercritical fluid was desired for extraction, the base system was upgraded to perform continuous pumping by the addition of a second syringe pump module, a valving package, and two cooling jackets (Isco SFE System 2250). The valving package consisted of two check valve housings, tubing, and the hardware necessary to install them in the system. Both pump modules were operated from the same controller. When operated in continuous pumping mode, one pump delivered pressurized fluid while the other pump refilled. During fat extractions, the fused silica capillary restrictors were replaced by 20cm lengths of 0.005" I.D. steel tubing directed into a 1 liter Erlenmyer waste flask through a polyurethane foam plug.

Transportable System. In the case where a transportable system was desired, the SFE system was installed on a portable hand cart, along with a gas-powered generator and a size D-1 cylinder containing the extraction fluid supply (Isco Transportable SFE System).

General Sample Preparation and Analysis Procedures

The following basic procedures generally apply to all extractions performed. In cases where the extract did not undergo further analysis, steps related to analyte collection do not apply.

SFE Procedure. An appropriate temperature was selected with the temperature controller on the extraction module before the sample was placed into the sample cartridge and weighed. If static modifier addition was performed, an organic modifier was added to the sample at this time. An appropriate volume of collection solvent was placed into the collection vessel before inserting the capillary restrictor through the teflon-faced septum of the vessel cap and positioned just below the liquid level of the solvent. A needle was inserted through the septum to act as a gas vent. The pump was placed in the constant pressure mode, filled with CO_2, and pressurized to the desired extraction pressure with all valves of the extractor unit closed. The sample cartridge was clipped into the chamber cap, the cap was hand-tightened into the instrument, and the supply valve was opened.

Either a static extraction was performed by allowing the sample to soak in the supercritical fluid before opening the extractant outlet valve or a totally dynamic extraction procedure was performed by immediately opening the extractant outlet valve. Once the extractant outlet valve was opened, supercritical fluid carrying soluble components from the sample flowed through a frit in the bottom of the sample cartridge, through the capillary restrictor, and into the collection solvent where the extracted material was deposited. The extraction was stopped by closing the extractant outlet and supply valves. The vent valve was then opened to quickly depressurize the chamber and the cartridge was removed.

After SFE was complete, the collection vessel containing the extract was warmed and placed in an ultrasonic bath for 30 seconds to drive residual CO_2 from solution before diluting to volume for chromatographic analysis.

Soxhlet Extraction Parameters. The parameters used for Soxhlet extraction are summarized in Table I.

Table I. Parameters Used for Soxhlet Extraction

Analyte	Hydrocarbons	Fat	Atrazine
Matrix	rock	snack foods	soil
Solvent	1:1 MeOH:CH₂Cl₂	CHCl₃	ethyl acetate
Solvent Volume	300ml	100ml	150ml
Extraction Time	48 hours	7 hours	4 hours
Solvent Reduction	to dryness	to dryness	to 10ml

Extract Analysis Procedures.

GC Analysis Parameters. The parameters used for GC analysis are summarized in Table II.

Table II. GC Conditions Used for Extract analysis

Analyte	PAHs	Hydrocarbons	Waxes	Pesticides	Atrazine
Column	DB-5ᵃ	DB-5ᵃ	AQ3/HT5ᶜ	DB-5ᵃ	DB-17ᵉ
Inj. Type	splitless	splitless	PTV	splitless	splitless
Inj. Temp.	275°C	300°C	40°C-399°Cᵈ	275°C	270°C
Det. Type	MS	FID	FID	FID	NPD
Det. Temp.	285°Cᵇ	300°C	450°C	275°C	300°C
Iso Time 1	4min	3min	2min	2min	2min
Iso Temp. 1	40°C	40°C	80°C	40°C	50°C
Ramp Rate 1	18°C/min	15°C/min	30°C/min	30°C/min	8°C/min
Iso Temp. 2	300°C	140°C	200°C	120°C	270°C
Iso Time 2	19min	-----	-----	-----	10min
Ramp Rate 2	-----	4°C/min	10°C/min	8°C/min	-----
Iso Temp. 3	-----	300°C	430°C	265°C	-----
Iso Time 3	-----	15min	10min	-----	-----

ᵃJ&W, 0.25mm by 30m, 0.25μm film.   ᵈPTV injector with temperature ramp
ᵇMass Spec interface temperature.   ᵉJ&W, 0.53mm by 30m
ᶜSGE 12 AQ3/HT5 0.1 column.

SFE of Polycyclic Aromatic Hydrocarbons from Soil

The SFE of polycyclic aromatic hydrocarbons (PAHs) from environmental type solids has been investigated using various supercritical fluids and mixtures of supercritical fluids plus modifier (1,4-7). Invariably, the higher molecular weight PAHs, many of which are carcinogenic pollutants, have been difficult to quantitatively extract from all but the most simple matrices. Here we will describe the quantitative recoveries of 14 PAHs from a certified soil sample using only CO_2, but at elevated extraction temperatures.

Procedure. Three fractions were collected. Two were extracted at 90°C for 30 minutes each, followed by one at 120°C for 60 minutes. The SFE System 1200 and a 10 ml cartridge was used to extract duplicate 2.5 g samples at 4980 psi.

Results. Tables III&IV show the SFE-GC/MS results obtained for duplicate extractions of 14 PAHs plus pentachlorophenol from a EPA standard reference material soil sample. Table III lists the certified values of the analytes as determined by a standard method as well as the SFE recoveries for the individual fractions of sample 1, total recovery from SFE, and total percent recovery from SFE in relation to the certified values. Table IV shows the repeatability of the experiment by comparing total SFE recoveries from two identical sample extractions.

The more volatile PAHs, from naphthalene through pyrene were extracted at 85-125% recovery during the first of the three fractions collected. The higher molecular weight PAHs required additional extraction time for quantitative recovery, but all the heavier PAHs except benzo(a)pyrene were extracted at >95% recovery by the end of the second fraction. Fraction #3 did not contribute significantly towards recovering any of the compounds except benzo(a)pyrene (total=74%). Percent recoveries were very repeatable between the two sample duplicates. It was interesting to note that pentachlorophenol was recovered in duplicate at >239% of its certified value, suggesting that the method used to certify the sample left more than half the pentachlorophenol in the soil matrix.

SFE of Atrazine Herbicide from Soil

Since many herbicides are crop-specific, herbicide concentrations in farmer's fields are routinely monitored to determine herbicide carry-over. Residual herbicides from the previous year adversely affect alternate crops grown to be grown during the subsequent season. There has been a limited amount of literature describing the successful SFE of herbicides such as sulfonylureas, diruron, linuron, and s-triazines spiked on various solid matrices with CO_2 and methanol modified CO_2 (1,8-11). Summarized here are the differences in recovery of atrazine from an actual farmer's soil sample as a function of extraction temperature and pressure using both CO_2 and methanol modified CO_2. Also shown are comparisons of recoveries from real vs. spiked samples and also static vs. dynamic modifier addition techniques.

Procedure. Seven extractions of the same soil sample were performed at temperatures between 40°C and 80°C, pressures between 4000 psi and 7000 psi, and also with and without modifying the CO_2 with methanol. Both static and dynamic modifier techniques were used. CO_2 alone and static modifier addition procedures were performed with a SFE System 1200 while dynamic modifier addition procedures were executed with a SFE System 2200. The spiked sample was prepared by pipetting a much higher concentration of atrazine in ethyl acetate onto the soil matrix

Table III. Recoveries of PAHs from a Certified Soil Sample by SFE

Analyte	Certified (mg/kg)	Frac 1[a] (mg/kg)	Frac 2[b] (mg/kg)	Frac 3[c] (mg/kg)	Total (mg/kg)	Percent Recovery
Naphthalene	32+/-8.2	31	0	0	31	97
2-Methylnaphthalene	62+/-11.5	67	0	0	67	108
Acenaphthalene	19+/-11.5	17	0	0	17	89
Acenaphthene	632+/-105	748	5	0	753	119
Dibenzofuran	307+/-49	384	0	0	384	125
Fluorene	492+/-78	481	4	0	485	99
Phenanthrene	1618+/-348	1500	68	5	1573	97
Anthracene	422+/-49	466	22	5	493	117
Fluoranthene	1280+/-220	1110	197	19	1326	104
Pyrene	1033+/-289	947	215	25	1187	115
Benzo(a)anthracene	252+/-38	168	73	18	259	103
Chrysene	297+/-26	194	93	27	314	106
Benzo(b+k)flouranthene	152+/-22	97	58	25	180	118
Benzo(a)pyrene	97+/-17.1	37	21	14	72	74
Pentachlorophenol	965+/-374	1870	403	83	2356	244

[a]CO_2, 350atm, 90°C, 30min
[b]CO_2, 350atm, 90°C, 30min
[c]CO_2, 350atm, 90°C, 60min

Table IV. Recoveries of PAHs from a Certified Soil Sample by SFE
Comparison of Extraction #1 / Extraction #2

Analyte	Certified Value (mg/kg)	Sample #1 Total (mg/kg)	Sample #1 Recovery (%)	Sample #2 Total (mg/kg)	Sample #2 Recovery (%)	Average Recovery (%)
Naphthalene	32+/-8.2	31	97	28	90	94
2-Methylnaphthalene	62+/-11.5	67	108	65	105	106
Acenaphthalene	19+/-11.5	17	89	16	84	87
Acenaphthene	632+/-105	753	119	715	113	116
Dibenzofuran	307+/-49	384	125	366	119	122
Fluorene	492+/-78	485	99	462	94	96
Phenanthrene	1618+/-348	1573	97	1471	91	94
Anthracene	422+/-49	493	117	500	118	118
Fluoranthene	1280+/-220	1326	104	1316	103	103
Pyrene	1033+/-289	1187	115	1128	109	112
Benzo(a)anthracene	252+/-38	259	103	252	100	101
Chrysene	297+/-26	314	106	281	95	100
Benzo(b+k)flouranthene	152+/-22	180	118	181	119	119
Benzo(a)pyrene	97+/-17.1	72	74	77	79	76
Pentachlorophenol	965+/-374	2356	244	2273	236	240

prior to extraction.   Extraction parameters used for individual samples are summarized in Table V.

Table V. Parameters used for the SFE of Atrazine with CO_2 and Methanol Modified CO_2

Extraction Fluid	CO_2 only	CO_2+2ml methanol (static method)	CO_2+5% methanol (continuous method)
Matrix	soil	soil	soil
Pressure/Temp	a,b,c	b,c	b
Static Time	------	10min	------
Dynamic Time	15-20min	15-20min	15-20min
Fluid Volume	30ml	30ml	30ml
Sample Weight	9.0g	9.0g	9.0g
Cartridge size	10ml	10ml	10ml
Collection Solvent	d	d	d

[a]4000psi, 40°C.   [b]5000psi, 60°C.   [c]6000psi, 80°C.   [d]Iso-propyl Alcohol

Results. The results for the recovery of atrazine from real and spiked samples by SFE are shown in Table VI. It should be noted that all SFE recoveries listed are reported using the assumption that conventional extraction recovers 100% of the analyte.

The effect of raising temperature is shown by the recoveries of atrazine from identical real samples with CO_2 only at 40°C/4000psi, 60°C/5000psi, and 80°C/6000psi. Raising the pressure along with the temperature maintains a constant CO_2 density, therefore differences in recoveries must be attributed to extraction temperature. Recoveries ranged from a low of 24% at 40°C to a high of 40% at 80°C during this part of the experiment.

Unlike the real samples, a high level spike pipetted onto the same identical sample just prior to SFE was much easier to extract using mild conditions. The same extraction parameters of CO_2 only at 40°C/4000psi, when used for a spiked sample, gave a recovery of 88%.

The common practice of placing modifier directly into the cartridge and performing a static extraction was investigated at 60°C/5000psi and 80°C/6000psi. A real sample statically extracted with modifier at 60°C/5000psi had only a 46% recovery of atrazine. Increasing the pressure and temperature to 80°C/6000psi gave virtually the same results of 47% recovery, therefore no temperature effect was observed during static modifier addition.

When a continuous flow of 5% methanol modified CO_2 was used for extraction at 60°C/5000psi, an 85% recovery of atrazine was observed, an increase of 38% over the static modifier method and a recovery very close to that observed with a spiked sample.

SFE of Organochlorine Pesticides From Soil

The concentration of pesticides in a variety of matrices such as grain, soil, sediment, vegetables, and processed foods are of intense interest to environmental agencies as well as consumers. The SFE of a seemingly vast number of pesticides from widely

Table VI. Recovery of Atrazine from Soil by SFE

Sample Type (real or spiked)	Modifier	Temp (°C)	Pressure (psi)	Recovery[c] (%)
real	none	40	4000	24
spiked	none	40	4000	88
real	none	60	5000	32
real	methanol[a]	60	5000	46
real	methanol[b]	60	5000	85
real	none	80	6000	40
real	methanol[a]	80	6000	47

[a]Static method, 2ml methanol directly into cartridge, 10min static
[b]Dynamic method, $CO_2 + 5\%$ methanol
[c]In comparison to conventional extraction with ethyl acetate, Table III for parameters

varying matrices has been reported (1,12-15), but there are many whose SFE have yet to be studied.   In this report we demonstrate the SFE of 18 common organochlorine pesticides from soil, using un-modified CO_2 and a field-portable SFE system.

Procedure. The field-portable system was taken to a grassy field where pesticides were extracted from a soil sample spiked with 100 μg each of 18 organochlorine pesticides. The extraction from the 2.5 ml cartridge was performed at 55°C, 5000 psi, with 30 ml of CO_2 in 15 minutes. The collection solvent was hexane.

Results. Gas chromatograms and results from the SFE of 18 organochlorine pesticides spiked on soil using a transportable system are shown in Figure 1 and Table VII, respectively. The lower tracing in Figure 1 corresponds to the initial extraction while the upper tracing to a follow-up extraction of the same sample. The ninth visible eluting peak is actually an unresolved doublet of Dieldrin and 4,4'-DDE. The recoveries of these two compounds were calculated from the combined integration of the two and therefore they have the same value.
   As indicated in Table VII, 13 of the 18 OCPs were retrieved in excess of 90% recovery during the first extraction. The remaining four OCPs, endrin, endosulfan II, endosulfan sulfate, and methoxychlor were recovered at levels between 10-57%. Since no significant amount of any OCP was recovered during the follow-up extraction, breakdown of some organochlorine pesticides endrin in the injection port of the GC was suspected.

SFE of Alkanes from Rock

Geochemists and the oil industry routinely screen rock samples for bio-markers that indicate the age and source of hydrocarbon constituents. The conventional extraction takes 48 hours and requires extract clean-up by thin-layer chromatography to separate the alkanes from aromatic hydrocarbons and heteroatom containing species (16). Here the selectivity of SFE in extracting only the alkanes from a mature source rock

Table VII. SFE of Pesticides from Soil

Analyte	% Recovery[a]	Analyte	%Recovery[a]
1. alpha-BHC	98	10. 4,4'-DDD	99
2. beta-BHC	91	11. Endrin	10
3. gamma-BHC	100	12. Endosulfan II	49
4. delta-BHC	94	13. 4,4'-DDD	94
5. Heptachlor	102	14. Endrin aldehyde	101
6. Aldrin	102	15. Endosulfan sulfate	57
7. Heptachlor epoxide	98	16. 4,4'-DDT	96
8. Endosulfan I	98	17. Endrin Ketone	17
9. Dieldrin	99	18. Methoxychlor	47

[a]CO_2, 5000psi, 55°C, 30ml, 15min

sample is shown, not only shortening extraction time dramatically, but also eliminating the subsequent preparation of the extract for GC analysis.

Procedure. The SFE System 1200 was used to extract 5.3 g of powdered rock, contained in a 10 ml cartridge, with 30 ml of CO_2 at 60°C and 4000 psi over a period of 21 minutes. The collection solvent was hexane.

Results. Gas chromatograms of both the SFE and Soxhlet extracts of a mature rock sample are shown in Figure 2. The upper tracing is from the Soxhlet method and the lower from SFE. Linear alkanes up through C35 can be seen in both extracts, as well as smaller amounts of isoprenoid and cyclic alkanes located between the linear hydrocarbon peaks. The Soxhlet extract was missing a range of lower molecular weight hydrocarbons, and it must be assumed that they were lost during the extraction or in the solvent reduction steps. No quantitation of the hydrocarbons was attempted.

SFE of Waxes, Mineral Oil, Dimers, and Trimers from Polystyrene

The polymer manufacturing industry monitors extractables in their products for a variety of purposes. Polymers contain a variety of additives such as anti-oxidants, mineral oils, and waxes as well as incompletely polymerized dimers and trimers that have a pronounced effect in the finished product (17). This study illustrates the efficiency of SFE in the extraction of polystyrene samples containing mekon and polystyrene waxes as well as mineral oil, dimers, and trimers.

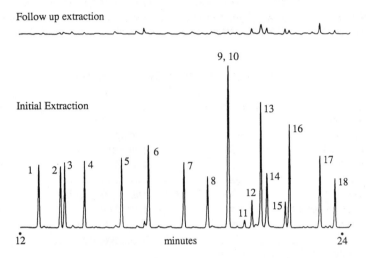

Figure 1. SFE, Pesticides from Soil

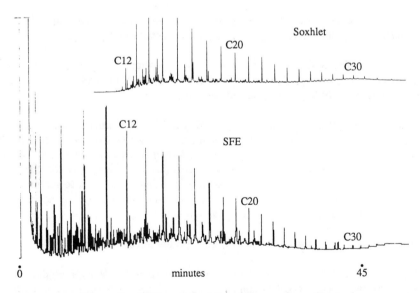

Figure 2. SFE and Soxhlet, Alkanes from Rock

Procedure. Two 4 g polystyrene samples containing different waxes were extracted in 10 ml cartridges with 60 ml of CO_2 at 7000psi and 70°C over a period of 45 minutes. The SFE System 1200 was used for this experiment.

Results. Figures 3-5 and Table VIII display the results for the SFE and Soxhlet extraction of polywax 1000 and mekon waxes from polystyrene pellets.

High-temperature gas chromatograms of both the SFE and Soxhlet extracts of polywax 1000 from polystyrene are shown in Figure 3. The upper tracing is from the SFE extract while the lower is from the Soxhlet extraction. Dimers and trimers (initially eluting peaks) as well as mineral oil (broad band) were extracted along with components of polywax 1000. Soxhlet extraction gave wax components up to C70 and SFE extracted components up to C60, but the chromatograms were otherwise visibly identical.

High-temperature gas chromatograms of the SFE and Soxhlet extracts of mekon wax from polystyrene are shown in Figure 4 and Figure 5, respectively. As was observed in the polywax 1000 extracts, dimers, trimers and mineral oil were extracted along with the mekon wax. Again, Soxhlet extraction gave wax components up to C70 while SFE extracted components up to C60. With the exception of a few unidentifiable artifacts at the end of the SFE extract tracing, the chromatograms were again visibly identical.

Overall extraction efficiencies were calculated by determining the total peak area count for all components, with the exception of the solvent peak. The latter peak area was also noted as this value gives a number proportional to the actual amount of sample reaching the column, (as opposed to the amount injected). By proportionality and normalization, the relative percent of SFE extractables were compared to the Soxhlet method. The results of these calculations are included as Table VIII.

Table VIII. Recovery of Waxes from Polystyrene
Comparison of SFE vs. Soxhlet Extraction

	SFE PW1000	Soxhlet PW1000	SFE Mekon	Soxhlet Mekon
Corrected Peak Area Counts	367,751	532,535	2,771,044	3,384,381
Extraction Efficiency (%)	69.0	100[a]	81.8	100[a]

[a]Based on a single extraction

SFE of Fat from Processed Snack Foods

In processed snack foods, fat content can be defined as the amount of material extractable using a Soxhlet/chloroform extraction technique that takes in excess of seven hours to complete. Such lengthy extractions require large amounts of hazardous solvents that must be completely evaporated during the analysis (18). Using SFE with industrial grade CO_2 at flow rates of 20 ml/min, we present

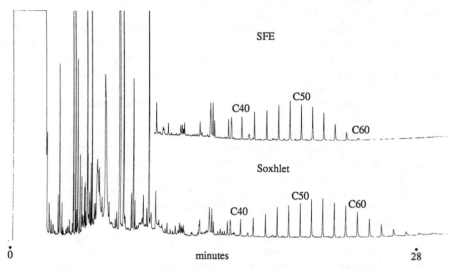

Figure 3. SFE and Soxhlet, Polywax 1000 from Polystyrene

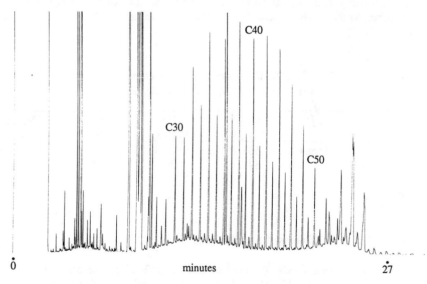

Figure 4. SFE, Mekon Wax from Polystyrene

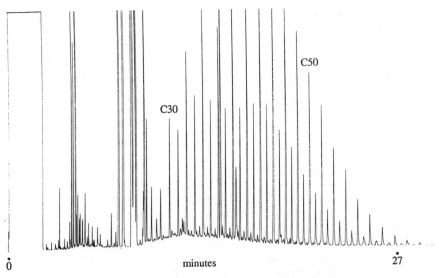

0   minutes   27

Figure 5. Soxhlet, Mekon Wax from Polystyrene

comparable determinations of the fat contained in such samples after only 10 minutes extraction time.

Procedure. A System 2250 was used to perform the SFE of fat from processed snack foods. Unlike other SFE procedures described in this paper, the fat extract was not collected for any subsequent analysis. Since gravimetric determination of weight loss in the sample was used for determination of fat, a large amount of CO_2 was passed through each sample in a very short period of time. Parameters used are summarized in Table IX.

Table IX. Parameters used for the SFE of Fats from Foods with CO_2

Analyte	Fat	Fat	Fat
Matrix	potato chips	cheese curls	corn chips
Temperature	80°C	80°C	70°C
Pressure	7500psi	7500psi	7500psi
Extraction Time	12min	12min	8min
Fluid Volume	200ml[a]	200ml[a]	150ml
Sample Weight	1.6g	3.0g	1.7g
Cartridge Size	2.5ml	5.0ml[b]	2.5ml
Collection Solvent	----------	----------	----------

[a]Methanol (2ml) added to the sample and immediately followed by dynamic extraction.
[b]Custom size cartridge made by machining a 2.5ml cartridge oversize.

Results. Table X lists the results of the fat determinations by both SFE and standard Soxhlet methods. SFE determinations of fat in potato chips and cheese curls were virtually identical to the values obtained by Soxhlet extraction, while the results from SFE of corn chips was a few percent low. The low results can be attributed to the SFE of the corn chips being performed without adding methanol to the sample cartridge and also to the fact that 50ml less CO_2 was used in the extraction.

SFE vs. Soxhlet Cost Comparison

A comparison has been made between SFE and Soxhlet extraction methods in regards to recovery of analyte, speed of extraction, and relative cost per extraction (15). Such comparative studies are dependent, as is our study, not only upon the extraction conditions used but upon the type and cost of the equipment. The investigation described in this paper compares SFE with standard Soxhlet methods in terms of extraction time, analyte recovery, method simplification, environmental hazards, portability, and cost per extraction using a commercially available SFE system.

As an extraction technique, SFE proved to give comparable recoveries to those of Soxhlet extraction. In all cases, SFE dramatically shortened extraction times and minimized most environmental hazards, solvent concentration steps, and waste disposal costs. A summary of this comparison is included as Table XII and is a projected cost comparison of SFE to Soxhlet extraction, based on our experiences with the SFE system used in these studies (Isco SFE System 1200). The projected cost per extraction was determined to be $15.85(SFE) vs. $22.60(Soxhlet).

Table X. Determination of Fat in Snack Foods
Comparison of SFE vs. Soxhlet Extraction

	SFE % Fat	Soxhlet %Fat
Potato Chips	22.6[a]	22.6[c]
Corn Chips	19.6[b]	23.7[c]
Cheese Curls	32.8[a]	33.7[c]

[a]CO_2, 2ml methanol, 7500psi, 80°C, 200ml in 10min
[b]CO_2 only, 7500psi, 70°C, 150ml in 8min
[c]Chloroform, 100ml, 7 hours

Table XI. SFE vs. Soxhlet Extraction

Analyte (Method)	Time	Solvent	Concentr. Step	Disposal Costs	Environ. Hazards
PAHs					
SFE	2h	CO_2	none	none	minimal
Soxhlet[a]	24h	300ml acetone/hexane	yes	yes	yes
Pesticides					
SFE	15min	30ml CO_2	none	none	minimal
Soxhlet[a]	24h	300ml acetone/hexane	yes	yes	yes
Waxes					
SFE	45min	60ml CO_2	none	none	minimal
Soxhlet[b]	16h	150ml IPA	yes	yes	yes
Fats					
SFE	10min	200ml CO_2	none	none	minimal
Soxhlet[b]	7h	100ml $CHCl_3$	yes	yes	yes
Hydrocarbons					
SFE	15min	30ml CO_2	none	none	minimal
Soxhlet[b]	48h	300ml methanol/CH_2Cl_2	yes	yes	yes

[a]Based on EPA methods
[b]Based on conventional methods described in this paper

Table XII. Cost Comparison of SFE vs. Soxhlet Extraction

Component	Cost per Extraction	Component Lifetime (# Extractions)
SFE		
Restrictor	$2.00	40
Collection vial	$1.00	1
Frits	$0.50	100
Extraction cartridge	$1.70	100
Seals	$0.40	200
CO_2	$0.20	1
SFE system (Isco)	$1.55	100,000
Labor	$7.50	
Waste disposal	negligible	
Cleaning	$1.00	
Total	$15.85	
Soxhlet		
Misc.	$1.06	1000
Snyder column	$0.22	1000
Flask and tubing	$0.24	1000
K-D flask	$0.15	1000
Sodium sulfate	$0.20	1
Solvent	$0.28	1
Soxhlet & condenser	$1.66	1000
Labor	$10.80	
Waste disposal	$5.00	
Cleaning	$3.00	
Total	$22.60	

Acknowledgements

We wish to acknowledge the following individuals and their contributions to this paper.

Dr. Viorica Lopez-Avilla (Mid-Pacific Environmental Laboratory). Certified PAHs in soil sample, SFEs, GC-MS analysis of extracts.

Dr. Hwang (Frito-Lay, Inc.). Snack food samples, Soxhlet extraction method, Soxhlet extractions, and analysis of Soxhlet extracts.

Dr. Roger Miller (Huntsman Chemical Company). Polystyrene samples, Soxhlet method, Soxhlet extraction, and High Temperature GC analysis of both SFE and Soxhlet extracts.

Literature Cited

1. Hawthorne, S. B. *Anal. Chem.* **1990**, 62, 633A-642A.
2. Hawthorne, S. B.; Kreiger, M.S.; Miller, D. J. *Anal Chem.* **1988**, 60, 472.
3. Kalinoski, H. T.; Udseth, H. R.; Wright, B. W.; Smith, R. D. *Anal. Chem.* **1986**, 58, 2421-2425.
4. Hawthorne, S. B.; Krieger, M. S.;Miller, D. J. *Anal. Chem.* **1989**, *61*, 736-740.
5. Hawthorne, S. B.; Miller, D. J. *J. Chromatogr.* **1987**, 403, 63-76.
6. Hawthorne, S. B.; Miller, D. J. *Proceedings of the EPA/APCA Symposium on Measurement of Toxic and Related Air Pollutants;* May, 1987.
7. Levy, J. M.; Cavalier, R. A.; Bosch, T. N.; Rynaski, A. F.; Huhak, W. E. *J. Chromatogr. Sci.* **1989**, 27, 341-346.
8. Janda, V.; Steenbeke, G.; Sandra, P. *J. Chromatogr.* **1989**, *479*, 200-205.
9. McNally, M. E.; Wheeler, J. R. *J. Chromatogr.* **1988**, *435*, 63-71.
10. McNally, M. E.; Wheeler, J. R. *J. Chromatogr.* **1988**, *447*, 53-63.
11. Wheeler, J. R.; McNally, M. E. *J. Chromatogr. Sci.* **1989**, *27*, 534-539.
12. Campbell, R. M.; Meunier, D. M.; Cortes, H. J. *J. Microcol. Sep.* **1989**, *1*, 302-308.
13. King, J. W. *J. Chromatogr. Sci.* **1989**, *27*, 355-364.
14. Raymer, J. H.; Pellizzari, E. D. *Anal. Chem.* **1987**, *59*, 1043-1048.
15. Lopez-Avila, V.; Dodhiwala, N. S.; Beckert, W. F. *J. Chromatogr. Sci.* **1990**, *28*, 468-476.
16. Philp, R., University of Oklahoma, personal communication, 1990.
17. Miller, R., Huntsman Chemical Co., personal communication, 1990.
18. Hwang, J., Frito Lay, personal communication, 1991.

RECEIVED January 21, 1992

Chapter 17

Relative Effect of Experimental Variables

Analytical Supercritical Fluid Extraction Efficiencies and Correlations with Chromatographic Data

Kenneth G. Furton and Joseph Rein

Department of Chemistry, Florida International University at University Park, Miami, FL 33199

The relative effects of supercritical carbon dioxide density, temperature, extraction cell dimensions (I.D.:Length), and cell dead volume on the supercritical fluid extraction (SFE) recoveries of polycyclic aromatic hydrocarbons and methoxychlor from octadecyl sorbents are quantitatively compared. Recoveries correlate directly with the fluid density at constant temperature; whereas, the logarithms of the recoveries correlate with the inverse of the extraction temperature at constant density. Decreasing the extraction vessels' internal diameter to length ratio and the incorporation of dead volume in the extraction vessel also resulted in increases in SFE recoveries for the system studied. Gas and supercritical fluid chromatographic data proved to be useful predictors of achievable SFE recoveries, but correlations are dependent on SFE experimental variables, including the cell dimensions and dead volume.

Supercritical fluid extraction (SFE) was first described over two decades ago (1). Since then, it has developed into an important industrial-scale technique as an alternative to distillation and traditional solvent extractions. More recently, the application of SFE on an analytical-scale has received a great deal of attention. Analytical SFE can provide more efficient and selective extractions with as much as an order of magnitude increase in the rate of extraction compared to conventional methods (2-7). Although the numerous applications which have been reported recently have begun to demonstrate the advantages of SFE, the development of analytical SFE is still in the early stages. Future development of analytical SFE, including the use of polarity modifiers and the development of selective extractions, will depend upon a better understanding of the complex mechanisms governing SFE as well as the interrelationships between the numerous experimental variables which can effect SFE recoveries.

0097–6156/92/0488–0237$06.00/0
© 1992 American Chemical Society

Variables affecting SFE of Analytes from Solid Matrixes

The phenomena in SFE are of an extremely complex nature, and, although theoretical models for supercritical fluid extraction have been reported (8-10), they generally require a great deal of physicochemical data and are often of limited value in predicting optimum experimental parameters for SFE. Other approaches to predicting optimal extraction conditions have been based on calculating solubility parameters for analytes in supercritical fluids at a given temperature and pressure (11,12). Although solubility parameters can be useful in choosing initial SFE conditions, extraction efficiencies may be dominated by such factors as the vapor pressure and the diffusion coefficient of the analyte in the supercritical fluid, and, particularly, the affinity of the analyte for sorptive sites on the sample matrix and the diffusion of the analyte out of the sample matrix. Recent work has demonstrated that SFE of small quantities of extractable materials from solid matrixes, particularly during the latter portions of the extraction, may be diffusion controlled, rather than solubility limited (13).

The supercritical fluid extraction of analytes from solid sorbents is controlled by a variety of factors including the affinity of the analytes for the sorbent, the tortuosity of the sorbent bed, the vapor pressure of the analytes, and the solubility and the diffusion coefficient of the analytes in the supercritical fluid. Additionally, SFE efficiencies are affected by a complex relationship between many experimental variables, several of which are listed in Table I. Although it is well established that, to a first approximation, the solvent power of a supercritical fluid is related to its density, little is known about the relative effects of many of the other controllable variables for analytical-scale SFE. A better understanding of the relative effects of controllable SFE variables will more readily allow SFE extractions to be optimized for maximum selectivity as well as maximum overall recoveries.

TABLE I.   Some of the Variables Which May Affect Supercritical Fluid
Extraction Efficiencies

analyte concentration	mode of analyte accumulation
analyte type	modifier concentrations
extraction cell agitation (e.g. sonication)	pressure (density)
extraction cell dead volume	restrictor type
extraction cell dimensions (I.D.:Length)	sample condition (humidity, pH, etc.)
extraction cell size	sample matrix (sorbent, co-extractants)
extraction fluid	sample particle size
extraction time	sample size
fluid flow rate	temperature
fluid modifiers	total volume of extraction fluid

To limit the number of variables in the present study, a model series of compounds of equal concentration were prepared on a highly-purified solid. The model series of compounds used in this work include polycyclic aromatic hydrocarbons (PAHs), which comprise the largest class of known chemical carcinogens whose isolation and identification continues to be an important analytical problem. The common herbicide, methoxychlor, was added for comparative purposes and is discussed later in this work. The widely used solid phase extraction (SPE) sorbent, octadecyl-bonded silica, was chosen as the model support. The system studied is also of interest in investigating the use of supercritical fluids for selective trapping and elution of analytes from SPE cartridges. To accurately quantify the effect of variables such as extraction cell dimensions and dead volume on the observed SFE efficiencies, care must be taken to ensure that all of the other controllable experimental variables are kept constant and accurately measured. It is also important to note that SFE processes are highly analyte/matrix dependent; therefore, the results presented here cannot necessarily be related to other analyte/matrix systems including natural samples.

Apparatus. The packed-column supercritical fluid chromatograph (SFC) used to generate the SFC retention data for this work was constructed from a computer (Zenith) controlled Isco SFC-500 high pressure syringe pump (Lincoln, NE), Valco C14W 0.5 ml injection valve, Isco 1mm I.D. microbore C18 column, Fiatron column heater and controller (Oconomowoc, WI), and an Isco V^4 UV-Vis detector with 6000 psi flow cell. The extraction apparatus consisted of an Isco 260D syringe pump, Valco C6W valve, Fiatron heater and controller, a heated Isco V^4 UV-Vis detector with 6000 psi flow cell, and linear restrictors fabricated from fused silica tubing (Polymicro Technologies, Phoenix, AZ) with dimensions of either 41mm I.D. x 362ml O.D. or 17mm I.D. x 368ml O.D. HPLC guard columns with 2mm stainless steel frits (Upchurch Scientific, Oak Harbor, WA) were used for the extraction cells and High Pressure Fingertight II fittings with PEEK ferrules and capillary sleeves were used for connections. SFC grade carbon dioxide (Scott Specialty Gases, Plumsteadville, PA) was used for all of the extractions.

The octadecyl coated silica support from fifty 1000 mg Prep Sep (Fisher) C18 solid phase extractions cartridges was used to prepare the standard packing used throughout this study. A stock packing containing 200 ppm of the standards was prepared from a chloroform solution with evaporation from a packing flask using the rotary evaporator technique. Extractions were performed on aliquots of the standard packing that were weighed directly into the extraction cells. Extractions were performed on the packings contained in cells with several different internal diameter to length ratios (I.D.:Length). Extractions were performed in triplicate with carbon dioxide at an average flow rate of 0.600 ml/min, controlled by varying the length of a linear capillary restrictor. Replicate extractions were carried out with a constant total volume of carbon dioxide. The extracted analytes were collected by inserting the end of the outlet restrictor into a 10 ml volumetric flask containing several mls of absolute ethanol or methylene chloride. 2.5 mg of fluorene was added to the volumetric flask as an internal standard. External standards were used to define 100% of each solute. The accuracy of this approach was confirmed by performing exhaustive extractions of the standards which resulted in calculated recoveries of 97-103%.

Blockages were minimized by heating the length of the fused silica restrictor and agitating the collection vessel in an ultrasonic bath (Branson,

Shelton, CT). Identification and quantitation of the extracted standard PAHs was performed using a Hewlett-Packard Model 5890 Series II GC and a HP5971 MSD or flame ionization detection. To ensure that the restrictor was not limiting observed recoveries for this system, initial studies compared the use of 10 centimeter x 17 micrometer I.D. restrictors with 60 centimeter x 41 micrometer I.D. restrictors, each of which yielded average flow rates of ca. 0.600 ml/min. As no significant difference in the recoveries was observed, and the longer 41 mm I.D. restrictors were chosen for the study as they had less tendency to block compared to the narrow I.D. restrictors.

Correlation Between Cell Dimensions and SFE Recoveries. The term "cell dimensions" in the present work refers to the dimensions (I.D.:Length) of the sorbent bed within the microextraction cell. Triplicate extractions of the model sorbent were compared using microextractor cells of the following dimensions (I.D.xLength): 1.0x1.0 cm and 0.37x10.0 cm. The total volume of the sorbent bed placed in the cells was 0.79 ml for the cells resulting in sorbent bed dimensions of 1:1 (1.0x1.0cm) and 1:20 (0.37x7.3cm). The extraction cells must be carefully designed to prevent channeling from occurring, where the supercritical fluid passes preferentially through the center of the cell, which can result in lower observed recoveries as the cell is made more broad (I.D. increased). Both cell geometries had 8^0 distribution cones leading to stainless steel frits of the same I.D. as the sorbent bed I.D. which evenly distributed the extraction fluid across the bed.

Extraction conditions were purposely chosen to yield less than quantitative recovery of the PAHs to enable comparisons to be made between the different extraction cell geometries. Unfortunately, conditions required to yield reproducible recoveries of coronene ($>1\%$) often resulted in quantitative (90-100%) recoveries of naphthalene and anthracene, and, therefore, often only the last four PAHs were useful for comparison. The average recoveries for pyrene, perylene, benzo[ghi]perylene, and coronene from octadecyl packings using the two different geometries are shown in Figure 1 (data from Reference 14). Each sample was extracted at 100.0^0C and 4500 psi (density = 0.675 g/ml) at an average flow rate of 0.600 ml/min. The total volume of carbon dioxide used for each extraction was 7.5 ml. For all of the PAHs, a markedly higher recovery was achieved with the short broad cell (1:1) versus the long narrow cell (1:20).

Correlation Between Cell Dead Volume and SFE Recoveries. Replicate extractions were compared using microextractor cells of the following dimensions (I.D.xLength): 0.37x3.0 cm and 0.37x10.0 cm. The total volume of the sorbent bed placed in the cells was 0.32 ml for both cells resulting in the same sorbent bed dimension of 1:8. The first cell was completely filled and, therefore, essentially had zero dead volume (ZDV). The second cell was only one-third full and, therefore, had ca. 70% dead volume. Again, extraction conditions were purposely chosen to yield less than quantitative recovery of the PAHs to enable comparisons to be made between the extraction cells with different amounts of dead volume. Again, conditions required to yield reproducible recoveries of coronene ($>1\%$) generally resulted in quantitative (90-100%) recoveries of naphthalene and anthracene, thereby limiting the data set available for comparison. The average recoveries for pyrene, perylene, benzo[ghi]perylene, and coronene from octadecyl packings using the two different cells are shown in Figure 2. Each sample was

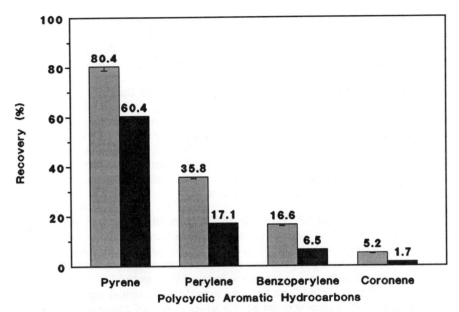

Figure 1.   Graphs of percent recoveries for pyrene, perylene, benzo[ghi]-perylene, and coronene using a 1:1 sorbent bed dimensions (light shading) and a 1:20 sorbent bed dimensions (dark shading).

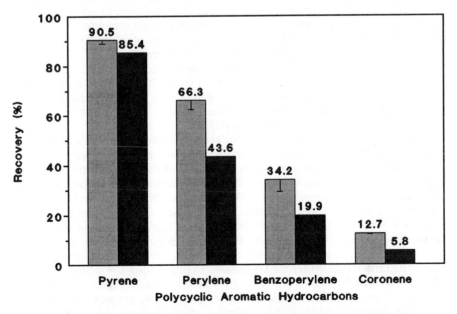

Figure 2.   Graphs of percent recoveries for pyrene, perylene, benzo[ghi]-perylene, and coronene using 1:8 sorbent bed geometries, one containing ca. 70% dead volume (light shading) versus one with zero dead volume (dark shading).

extracted at 100.0°C and 4500 psi (density = 0.675 g/ml) at an average flow rate of 0.600 ml/min. The total volume of carbon dioxide used for each extraction was 3.5 ml. For all of the PAHs, a markedly higher recovery was achieved with the cell containing dead volume versus the cell completely filled with sorbent.

Predictive Information from Chromatographic Retention Data

Many of the processes controlling extraction of analytes from a sorbent, including diffusion out of the sorbent, vapor pressure of the analyte, and affinity of the analyte for the sorbent, are similar to those controlling chromatographic retention. Therefore, the vast compilations of chromatographic retention data may provide information useful in predicting SFE recoveries, provided an analogous mobile phase/sorbent system is chosen for comparison.

Comparison of SFC and SFE Data. Qualitative correlations between SFC retention and supercritical fluid extractability of herbicides have been reported (15,16) and recently, we have discussed quantitative correlations for PAH extraction for the system discussed here (17). Extraction of compounds from solid sorbents is controlled by many of the same factors which control SFC retention. Therefore, SFC retention data can be useful in predicting potential supercritical fluid extraction (SFE) recoveries for compounds or classes of compounds, and may provide insight into the processes involved in supercritical extraction. Although considerable progress has been made in recent years, the relationships between solute properties and their retention in supercritical fluid chromatography are still not well understood. Correlations between various physicochemical parameters of solutes and their chromatographic retention (e.g. capacity factor, k) may provide valuable insights into the dominant retention mechanisms and may provide useful predictive information for establishing optimum experimental parameters (18,19). A variety of physical parameters have been shown to correlate with chromatographic retention. Several physical properties, measured SFC capacity factors, as well as GLC derived retention indices for the PAHs studied are listed in Table II. The capacity factors, k, were calculated from an isoconfertic-isothermal SFC separation of a mixture of the PAHs on an octadecyl bonded packed column using CO_2 as the mobile phase (4500 psi, 100°C).

Since it has been shown that molecular connectivity correlates well with SFC retention data for PAHs (19), one might intuitively expect SFE recoveries to correlate similarly, assuming that similar mechanisms dominate. Plots for the logarithms of the SFC capacity factors and the SFE recoveries for the two different cell geometries are shown in Figure 3. The excellent correlation between molecular connectivity and the SFC retention data is obvious, yielding a linear correlation with r > 0.999. Although there is obvious curvature with the SFE recovery data, the correlation appears to become more linear as the extraction cell I.D.:Length is decreased (made more chromatographic column-like). For example, a linear least squares regression analysis for the 1:1 dimensions cell yields a r = 0.967; whereas, the 1:20 dimensions cell yields a r = 0.984. Therefore, although SFC data derived from a similar system (similar sorbent, analyte, and supercritical fluid) may provide useful qualitative predictive information, any quantitative information may be limited by experimental variables, including the extraction cell dimensions.

Table II.   Physical Properties, SFC Capacity Factors, and GLC Retention Indexes for the PAHs Studied

Solute	Melting Point (°C)	Fused Ring Number	Molecular Connectivity (x)	k(SFC) [C18]	Retention Index OV-101
Anthracene	217	3	4.81	1.97	1846
Pyrene	150	4	5.56	4.07	2139
Perylene	274	5	6.98	14.10	2888
Benzo[ghi]perylene	278	6	7.72	26.03	3185
Coronene	427	7	8.46	50.16	3498

Comparison of GLC and SFE Data. Since the octadecyl system under study represents a non-polar sorbent system, and supercritical carbon dioxide is also non polar, a non-polar GLC system, OV-101, was chosen for comparison. Plots of the observed recoveries versus GLC retention index values for OV-101 (data from Reference 20 shown in Table II) are compared in Figure 4 for three of the cells studied. The data shows excellent linear correlation for the 1:8 dimensions cell with zero dead volume (r = 0.998). The 1:1 dimensions cell with zero dead volume also shows very good linear correlation (r = 0.997); whereas, the cell containing significant dead volume does not correlate well (r = 0.976). This data suggests that, when significant dead volume is introduced in the extraction cell, correlations between SFE recoveries and chromatographic retention data are diminished. Additionally, as the cell I.D.:Length is increased (made less column-like), correlations may again be diminished. Therefore, it is obvious that when attempting to predict potential SFE recoveries from chromatographic data, experimental variables, including dead volume and cell dimensions, may need to be considered.

The Effect of Fluid Density and Temperature on SFE Recoveries

Correlations Between Fluid Density and SFE Recoveries. One of the primary advantages of using supercritical fluids as extraction solvents is the ease with which the density, and effectively the solvating power, of the fluid can be changed by varying the pressure at constant temperature. Data for SFE of the analytes at different fluid densities at a constant temperature, flow rate, and extraction fluid volume are tabulated in Table III. Extractions were performed with a 1:1 dimensions cell at 100°C with an average flow rate of 0.600 ml/min. The sorbent bed volume was 0.79ml and the total volume of carbon dioxide used for each extraction was 3.2 ml. For all of the analytes studied, there was a direct linear correlation between supercritical fluid density (d) and SFE recovery as seen in Figure 5. This data suggests that SFE for the model solute/sorbent studied here

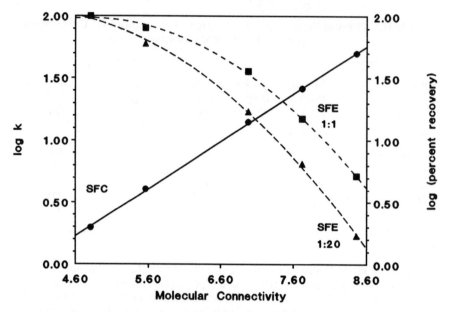

Figure 3. Plot of log k and log (% recovery) vs. molecular connectivity for anthracene, pyrene, perylene, benzo[ghi]perylene, and coronene from C18 SFC retention data (circles), SFE data with a 1:1 extraction cell dimensions (squares), and SFE with 1:20 extraction cell dimensions (triangles).

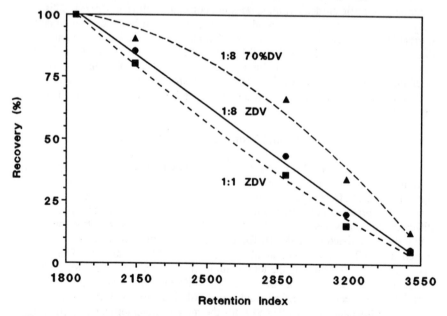

Figure 4. Plot of the average supercritical CO_2 extraction recoveries vs. GLC derived retention indexes for anthracene, pyrene, perylene, benzo[ghi]perylene, and coronene using a extraction cell of two different cell geometries (1:1 and 1:8) completely filled and the 1:8 dimensions cell with ca. 70% dead volume.

may be dominated by solubility limitations other than diffusion controlled processes. Linear least squares regression analysis of this data yields equations 1-5.

$$\text{Recovery}_{\text{(methoxychlor)}} = 186.10(d) - 51.38 \quad [r = 0.995] \tag{1}$$

$$\text{Recovery}_{\text{(pyrene)}} = 140.46(d) - 29.99 \quad [r = 0.991] \tag{2}$$

$$\text{Recovery}_{\text{(perylene)}} = 55.20(d) - 17.46 \quad [r = 0.987] \tag{3}$$

$$\text{Recovery}_{\text{(benzo[ghi]perylene)}} = 33.90(d) - 13.57 \quad [r = 0.989] \tag{4}$$

$$\text{Recovery}_{\text{(coronene)}} = 26.00(d) - 11.10 \quad [r = 0.975] \tag{5}$$

Table III.   Recovery of Analytes from Octadecyl Bonded Packings using Various Densities at a Constant Temperature of 100°C

Density (g/ml)	0.40	0.50	0.60	0.70
Analyte	% Recovery (standard deviation)			
Methoxychlor	25.4(1.2)	38.8(2.5)	59.0(3.1)	80.7(2.0)
Pyrene	26.6(1.1)	41.4(0.5)	50.8(0.1)	70.2(1.1)
Perylene	5.6(0.9)	9.2(1.3)	14.6(0.7)	22.2(0.9)
Benzo[ghi]perylene	0.6(.04)	2.6(0.4)	6.6(0.8)	4.0(0.1)
Coronene	-	1.1(.08)	4.0(0.1)	7.7(0.3)

Correlations Between Fluid Temperature and SFE Recoveries. One very important, yet often underrated, SFE variable is the extraction temperature. SFE recoveries of analytes can be greatly enhanced by increasing the temperature of extraction at a constant fluid density as seen in Table IV. The extractions were performed with extractors with a 1:1 dimensions cell at 100°C and an average flow rate of 0.600 ml/min. The sorbent bed volume was 0.79ml and the total volume of carbon dioxide used for each extraction was 3.2 ml. The logarithms of the observed recoveries correlate with the inverse of the extraction temperature (at constant density) as illustrated in Figure 6. Again, linear least squares regression analysis of this data yields equations 6-9.

$$\log R_{\text{methoxychlor}} = 2.318 - 27.474(1/t) \quad [r = 0.958] \tag{6}$$

$$\log R_{\text{pyrene}} = 2.358 - 38.834(1/t) \quad [r = 0.975] \tag{7}$$

$$\log R_{\text{perylene}} = 2.564 - 88.756(1/t) \quad [r = 0.991] \tag{8}$$

$$\log R_{\text{benzo[ghi]perylene}} = 2.709 - 130.90(1/t) \quad [r = 0.981] \tag{9}$$

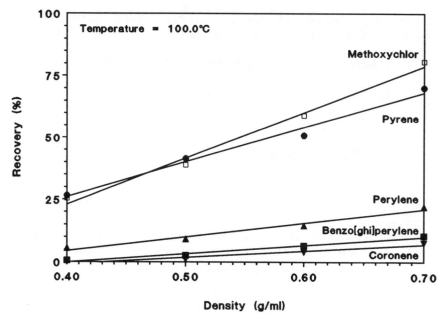

Figure 5. Plot of recoveries for supercritical CO_2 extractions at 100°C vs. the fluid density for methoxychlor, pyrene, perylene, benzo[ghi]perylene, and coronene.

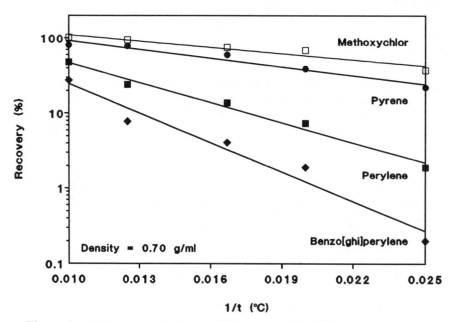

Figure 6. Plot of recoveries (log scale) for supercritical CO_2 extractions at a density of 0.70 g/ml vs. the inverse of the extraction temperature for methoxychlor, pyrene, perylene, benzo[ghi]perylene, and coronene.

Table IV.   Recovery of Analytes from Octadecyl Bonded Packings using
Various Temperatures at a Constant Density of 0.70 g/ml

Temperature (°C)	40.0	50.0	60.0	80.0	100.0
Analyte	% Recovery (standard deviation)				
Methoxychlor	37.6(1.3)	69.1(1.6)	75.2(3.3)	94.3(0.7)	102.1(1.4)
Pyrene	22.1(1.3)	39.2(0.3)	60.1(4.7)	78.7(1.3)	80.9(2.2)
Perylene	1.9(0.4)	7.4(0.4)	13.7(0.7)	24.1(2.5)	47.8(1.8)
Benzo[ghi]perylene	0.2(.06)	1.9(0.2)	4.1(0.6)	7.8(0.2)	27.4(1.9)
Coronene	-	-	-	2.3(0.1)	7.7(0.8)

The Relative Effects of Experimental Variables on Achievable Recoveries

The relative effects of SFE variables on achievable recoveries is important for any systematic design of an extraction protocol for maximum selectivity and, particularly, maximum overall recovery in the minimum analysis time. A comparison of the data presented above reveals that the relative increase in recovery for the analytes studied increases as the extractability of the compounds decreases. For example, the relative increase in recovery for the PAHs increased in an approximately linear fashion with PAH fused ring number. A direct comparison of the four variables discussed here is shown in Figure 7 for pyrene, perylene, benzo[ghi]perylene, and coronene. The density and temperature comparisons were calculated using the appropriate equations (1-9) for each analyte. As expected, the greatest overall effect was generally seen upon increasing the supercritical fluid density. Somewhat surprising, however, was the fact that the other variables studied, particularly temperature, have effects of a similar order of magnitude. The overall effect of extraction cell dimensions and cell dead volume on achievable recoveries was in all cases significant. The results indicate that, for maximum recovery, the minimum I.D.:Length ratio cell should be used. On the other hand, if one is interested in selective extractions, a larger I.D.:Length ratio cell may be desirable. For example, the amount of coronene extracted compared to pyrene for the 1:1 cell under the conditions studied is 6.5%; whereas, the amount is less than half that, 2.8%, for the 1:20 cell. Therefore, for selective extractions, long cell geometries appear to be more desirable, for the system discussed here.

Direct Comparison Between the Variables Investigated. Using equations 1-9, one can calculate the fluid densities and temperatures which would yield similar recovery increases as those seen upon changing the cell dimensions and dead volume. Figures 8 and 9 illustrate the relative increase in recovery upon changes in dead volume, cell dimensions, temperature, and density for pyrene and methoxychlor respectively. It is obvious that the relative effect of each variable is highly analyte dependent. For pyrene, similar recovery increases (increasing recovery from ca. 62% to 82%) were seen upon decreasing the I.D.:L ratio, increasing the temperature 20°C (from 70 to 90°C), or increasing the density 0.15

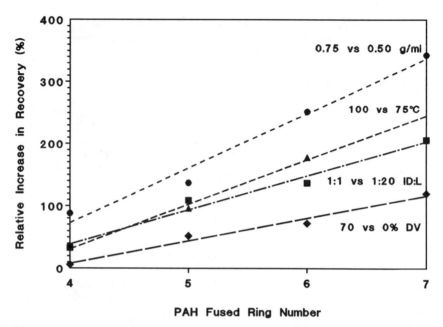

Figure 7. Plot of the relative increase in recovery achieved as a function of the PAH fused ring number as follows: 70 vs. 0% dead volume; 1:1 vs. 1:20 cell dimensions; 100 vs. 75°C; and 0.75 vs. 0.50 g/ml.

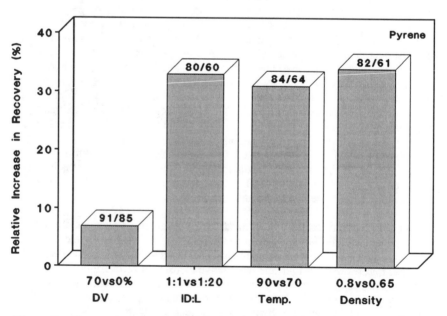

Figure 8. Bar graph comparison of the relative increase in recovery achieved for Pyrene for the following changes in experimental variables: 70 vs. 0% dead volume; 1:1 vs. 1:20 cell dimensions; 90 vs. 70°C; and 0.80 vs. 0.65 g/ml. The actual percent recoveries are indicated in the top of each bar.

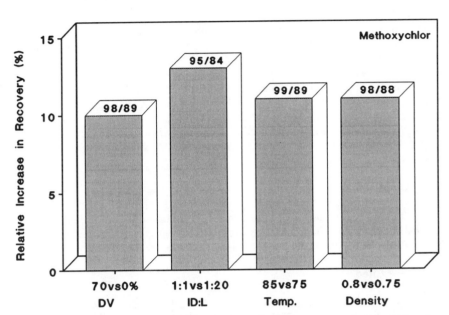

Figure 9. Bar graph comparison of the relative increase in recovery achieved for Methoxychlor for the following changes in experimental variables: 70 vs. 0% dead volume; 1:1 vs. 1:20 cell dimensions; 85 vs. 75°C; and 0.80 vs. 0.75 g/ml. The actual percent recoveries are indicated in the top of each bar.

g/ml (from 0.65 to 0.80 g/ml). In contrast, for methoxychlor, a similar effect as that seen upon changing the cell dimensions or introducing dead volume (increasing recoveries from ca. 88% to 98%) is achieved by increasing the temperature only 10°C (from 75 to 85°C) or increasing the density by just 0.05 g/ml (0.75 to 0.80 g/ml).

Conclusions

Achievable SFE recoveries for PAHs from octadecyl sorbents can be estimated from their corresponding SFC retention data, GLC retention indexes and, to a lesser extent, from basic physical properties such as molecular connectivity. The degree of correlation increases for column-like extraction vessels (long and narrow) and those containing minimal dead volume within the extraction cell. In addition to density and temperature, other SFE variables, including the microextractor cell dimensions and amount of dead volume within the vessel, may have a significant effect on the recoveries achievable by analytical supercritical fluid extraction, depending on the analyte/matrix system under study. For the present system, recoveries correlated directly with the fluid density and the logarithms of the recoveries correlated with the inverse of the extraction temperature. Recoveries were increased by decreasing the I.D.:Length ratio of the extraction vessel and incorporating dead volume into the extraction cell. For the system studied here, maximum SFE recoveries were achieved with a short/broad cell dimensions. For maximum selectivity, long/narrow cell geometries may be more desirable. Comparison of the relative effects of the four variables studied here reveals a large dependence on the type of analyte. The relative effect of temperature on overall recoveries was similar in magnitude to that observed for density changes. In all cases, the effects of cell dimensions and cell dead volume on observed recoveries were significant. Therefore, the effects of variables such as temperature, cell dimensions, and cell dead volume should be considered and combined with the well-established density effects (related to solubility) when designing an analytical-scale SFE technique for optimal selectivity, as well as maximum overall recoveries.

Acknowledgments

This work was supported in part by a FIU College of Arts & Sciences Faculty Development Minigrant. This work was presented in part at the 201st ACS National Meeting, Atlanta, GA.

Literature Cited

1.   Zosel, K. German Patent No. 1.493.190, Berlin, 1969.
2.   Schantz, M.M.; Chesler, S.N. *J. Chromatogr.* **1986**, 363, 397.
3.   Wright, B.W.; Wright, C.W.; Gale, R.W.; Smith, R.D. *Anal. Chem.* **1987**, 59, 38.
4.   McNally, M.E.; Wheeler, J.R. *J. Chromatogr.* **1988**, 435 63.
5.   Hawthorn, S.B.; Krieger, M.S.; Miller, D.J. *Anal. Chem.* **1989**, 61, 736.
6.   Furton, K.G.; Rein, J. *Anal. Chim. Acta* **1990**, 236, 99.
7.   Hawthorne, S.B. *Anal. Chem.* **1990**, 62, 633A.
8.   Ziger, D.H.; Eckert, C.A. *Ind. Eng. Chem. Process Des. Dev.* **1983**, 22, 582.
9.   Mansoori, G.A.; Ely, J.F. *J. Chem. Phys.* **1985**, 82, 406.
10.   Lira, C.T. In *Supercritical Fluid Extraction and Chromatography*; Charpentier, B.A.; Sevenants, M.R., Eds.; ACS Symposium Series 366; American Chemical Society: Washington, D.C., 1988; pp 1-25.
11.   Czubryt, J.J.; Meyers, M.N.; Giddings, J.C. *J. Phys. Chem.* **1970**, 74, 4260.
12.   King, J.W. *J. Chromatogr. Sci.* **1989**, 27, 355.
13.   Bartle, K.D.; Clifford, A.A.; Hawthorn, S.B.; Langenfeld, J.J; Miller, D.J.; Robinson, R. *J. Supercritical Fluids* **1990**, 3, 143.
14.   Furton, K.G.; Rein, J. *Chromatographia,* **1991**, 31, 297.
15.   McNally, M.E.; Wheeler; J.R. *J. Chromatogr.* **1988**, 447, 53.
16.   Wheeler, J.R.; McNally; M.E. *J. Chromatogr. Sci.* **1989**, 27, 534.
17.   Furton, K.G.; Rein, J. *Anal. Chim. Acta* **1991**, 248, 263.
18.   Jinno, K.; Saito, M.; Hondo, T.; Senda, M. *Chromatographia* **1986**, 21, 219.
19.   Rein, J.; Cork, C.M.; Furton, K.G.; *J. Chromatogr.,* **1991**, 545, 149.
20.   Grimmer, G.; Bohnke, H. *J. Assoc. Off. Anal. Chem.* **1975**, 58, 725.

RECEIVED December 2, 1991

Chapter 18

Supercritical Fluid Extraction
Developing a Turnkey Method

C. R. Knipe[1], D. R. Gere[2], and Mary Ellen P. McNally[3]

[1]Hewlett–Packard Company, P.O. Box 900, Route 41 and Starr Road,
Avondale, PA 19311–0900
[2]Hewlett–Packard Company, 39550 Orchard Hill Place Drive,
Novi, MI 48050
[3]Agricultural Products, Experimental Station, E402/3328B, E. I. du Pont de
Nemours and Company, Wilmington, DE 19880–0402

Supercritical fluid extraction (SFE) is a new and promising
technique for sample preparation. Because SFE is so new, and
there are so many control parameters available to the scientist, it
can be difficult to know how to proceed when developing an SFE
method. This paper discusses two samples and describes a
systematic approach for evaluating SFE extraction parameters for
methods development. The effects of modifiers are investigated.

Although supercritical extraction (SFE) has been known for some time, it is still a
relatively new technique to the analytical chemist. Before developing an SFE
method, the chemist must understand the composition of the matrix and the
analyte properties. The key instrumental parameters affecting the extraction of
analytes from the matrix include: fluid density, temperature, and fluid
composition. Both the make-up of the matrix and the analytes must be
considered when selecting the extraction conditions. Consideration of the
extraction parameters must be given with respect to their affect on the analytes of
interest and on the compounds present in the matrix that may either coextract
with the analytes or inhibit their extraction by physical or chemical means.
    Choice of the extraction fluid density will be dictated by the volatility and
polarity of the compounds to be extracted. In general low density CO_2 will
extract volatile non-polar compounds while higher density CO_2 will generally
extract less volatile more polar material.
    The useful extraction temperature range will be influenced by the vapor
pressure and thermolability of the analytes. The extraction temperature can also
alter the physical state of the matrix itself, for example, liquifing fats oils and low
molecular weight polymers. This in turn will affect recoveries and extraction

0097–6156/92/0488–0251$06.00/0
© 1992 American Chemical Society

times necessary to remove the analytes of interest. The temperature will also determine the amount of modifer that can be solubilized by the extraction fluid. The choice of extraction fluid along with selection of modifier and amount of modifier will also affect extraction recoveries and extraction selectivity. In this work the extraction fluid chosen was CO_2 as well as CO_2 with the addition of a variety of organic modifiers.

In addition to these extraction parameters the matrix itself will have a large impact upon the choice of the extraction conditions. The matrix, besides containing coextractable materials will impact the solute-solid interaction kinetics and thermodynamics. The coextractants themselves can also act as modifiers and, as their concentrations change over the course of an extraction, the make-up of the extraction fluid will dynamically change.

Distribution of solutes throughout the matrix can also affect the extraction process. The analytes can be adsorbed on the matrix surface or encapsulated in the matrix. Encapsulated analytes may require harsher extraction conditions or additional preparation steps such as grinding or chopping, prior to extraction. Once this information is collected the chemist can address the problem of developing the extraction method.

Developing a Method

Extraction Parameters. The challenge in SFE is to determine the best extraction parameters for the analytes of interest in a particular matrix. These parameters are listed in Table I. The solvent power of the extraction fluid as controlled by the density, temperature, and composition is the primary parameter which must be controlled to obtain optimal recoveries. The definition of solvent power used here is that of Snyder and Kirkland (1). As previously mentioned low densities will extract non-polar volatile compounds and high densities will extract relatively more polar, less volatile compounds.

A quick way to evaluate the density effect is to do several extractions at different densities (low, medium, and high) to see when the analytes begin to extract (2). Another parameter to control is the extraction temperature. In addition to affecting the density, raising the temperature can add thermal energy to the system. This can either increase the partial pressure of the solutes or reach the melting point of the matrix, thereby freeing up the solutes, and allowing them to be transported from the matrix by the extraction fluid. The increase in temperature will also increase the diffusivity of fluids above their critical temperature.

Another parameter that can have a large effect on the extraction process is the addition of modifiers. The effects of modifiers are still not well understood, but they will change the solvent power of the fluid and/or change the solute/solid interaction (generally both). After the fluid characteristics have

Table I. Key Extraction Parameters

Physical parameter	Experimental parameter
Solvent power	Fluid density Temperature Fluid composition
Solvent contact	Flow rate Extraction time Extraction mode (static/dynamic)

been determined the fluid/sample interaction time and the fluid volume must be specified. This interaction is determined by flow rate, extraction time, and whether or not the extraction is done in a static or dynamic mode.

Analyte Properties. Once the analytes of interest are extracted, they must be collected/trapped for analysis and quantitation. Once collected/trapped they may have to be reconstituted, filtered, or derivatized prior to final analysis. Again the chemical/physical properties of the analytes will determine the collection and reconstitution rinse parameters . During the extraction step, the volatility of the analytes will determine the collection temperature or type of adsorbent material used for collection. If the analytes are volatile, then a cold trap or cooled collection solvent along with a low flow rate should be used. This is because the analytes are volatile and the expansion of the CO_2 can create aerosols or mechanically move the analytes past the collection device. Less volatile analytes can tolerate higher extraction flow rates and higher collection temperatures can be used. If an adsorbent trap is used for collection, the chemist can specifiy an appropriate adsorbent and rinse solvent for optimal recoveries for the analytes of interest[3]. Flow rate and volume are parameters that also need to be specified. The flow rate and rinse volume are determined by the solublity of the analytes in the rinse solvent and the amount of material to be removed from the trap.

Two examples were chosen to show the effect of these SFE parameters and the process of developing a method. The first example is the extraction of paprika and an examination of some of these extraction parameters in a qualitative way. The second example, developing a method for the extraction of several herbicides from soil, shows a more quantitative approach. The herbicide example demonstrates the significant effect that modifiers can have upon extraction recoveries. It also shows that the solute/matrix interactions are as important as the solute/solvent interactions when developing an SFE method.

Qualitative Method

Paprika Analysis. Paprika was selected because it has a wide range of readily extractable compounds. There are essential oils, aroma components, and carotenoids contributing to color and taste. Paprika is readily available and provides various extracts easily detectable by sight and smell. Thus using paprika as a sample yields a large amount of qualitative information that can be quickly and easily gleaned concerning the effect of varying SFE extraction conditions.

Fractionation of Paprika. Since there are so many extractable components in paprika, it helps to simplify the analysis by being somewhat selective during the extraction. The solvent power of the fluid will have the most impact on the selective extraction of certain classes of compounds from the sample. Selective extractions were obtained by extracting the sample multiple times at different densities, commonly called density stepping or fractionation (as discussed by W.S. Miles at the Pittsburgh Conference in New York City 1990, paper 543).

Table II. Paprika Extraction Conditions

Experimental:

Sample Size: 10-300 mg of store bought paprika.
Chamber temperature 40 and 80 °C.
Extraction fluid: CO_2.
Flow rate: 4 ml/min mass flow of liquid CO_2 at the pump.
Thimble volumes: 10
Thimble size: 1.5 ml.
Extraction time: 1.0,   2.43,   3.25,   3.85 min
Density: 0.25,   0.6,   0.8,   0.95 g/ml.
Extractor: HP7680A SFE.

In this case both density and temperature were varied to investigate qualitatively the conditions at which the various compounds would be extracted from the paprika. Table II indicates the extraction conditions used for the paprika sample.

Step I. The first set of experiments involved extraction of paprika using pure CO_2 at three different densities; 0.25, 0.6, 0.95 g/ml and at a temperature of 40 ^{o}C. As expected, the first collection vials came out clear, suggesting that if anything was extracted, it was limited to the more volatile non-polar aromatic components or oils. The second extraction step, at a density of 0.6 g/ml CO_2, produced a pale orange extract. This suggests that some of the less volatile more polar compounds were extracted. The third extraction step, at a density of 0.95 g/ml CO_2, produced a much darker orange color. At this higher density of CO_2 the carotenes and carotenoids were extracted. This confirmed the expectation that extraction of less volatile, more polar compounds occurs at the higher densities.

Step II. Next, the same SFE extractions were rerun using the same flow rates and thimble volumes as for step I, but at a chamber temperature of 80 ^{o}C. The same densities were used except that 0.8 g/ml was substituted for 0.95 g/ml at the new temperature. The density of 0.8 g/ml at 80 ^{o}C was used because of the upper pressure limit of the extractor. The upper pressure limit of the extractor was 400 atm., therefore at 80 ^{o}C, this corresponds to a density of 0.80 g/ml for CO_2. In this case, the extracts collected at 0.6 and 0.8 g/ml densities were darker in color than the corresponding extracts for those densities at 40 ^{o}C. This indicates that an elevated temperature can aid in the extraction of colored components (which are typically more polar, less volatile components) from the paprika with SFE. Without some quantitative results it is difficult to say what the differences are in relation to specific compounds. Nor does this finding supply specific information about how much additional material was extracted at 80 ^{o}C versus 40°C, all other conditions being equal. The results for step II are summarized in Table III.

Table III. Extracted Paprika Fractions

Extraction Density	Extract
Temperature = 40 ^{o}C.	
0.25 g/ml	Colorless extract.
0.60 g/ml	Pale orange extract.
0.95 g/ml	Orange extract.
Temperature = 80 ^{o}C.	
0.25 g/ml	Colorless extract.
0.60 g/ml	Orange extract.
0.80 g/ml	Dark orange extract.

Thimble Volumes. To properly compare extractions while changing the density, the number of thimble volumes swept was kept constant for all extraction steps. In this case the extraction time was changed as a new density setpoint was chosen for each extraction step while the flow rate at the pump head was kept constant. The pump head is where the flow of liquid CO_2 is controlled. Control of the liquid CO_2 controls the mass flow of the system. As the density and the extraction times changed, so did the mass of CO_2 used per step; however, the net volume of solvent seen by the sample did not change. Importantly, as the solvent changed from 0.25 g/ml CO_2 to a solvent of 0.95 g/ml CO_2, more CO_2 was required to displace the same volume element. Since the mass flow was held constant in this set of experiments the extraction time had to be changed to normalize for equivalent volumes of solvent as seen by the sample.

Quantitative Method

Herbicide Extraction. Traditionally one of the more time consuming and tedious sample preparation problems has been the extraction of herbicides and pesticides from soil. Due to Federal regulations these agrochemical products must be monitored, and the number of samples that need to be analyzed can be significant[4]. Along with the large number of samples that must be analyzed there is a need for improved sample preparation methods that minimize the preparation time and solvent usage, such as SFE. Two families of herbicides that require monitoring are the s-triazines and the phenylureas. The triazines selected for extraction were atrazine and cyanazine, and diuron was selected from the phenylureas[5][6]. Table IV shows some of the physical properties of these herbicides. All three have similar melting points and molecular weights. They are relatively small molecules and are not highly polar and so should be

Table IV. Herbicide Properties

	Atrazine	Cyanazine	Diuron
Melting Point	171-174 °C	167-169 °C	158-159 °C
Molecular Weight	215.68g/mol	240.68g/mol	233.10g/mol
Vapor Pressure	0.113 mPa	0.21 µPa	0.25 mPa

amenable to extraction using supercritical CO_2. The vapor pressures show that they are not highly volatile, so a technique such as headspace analysis can be ruled out.

Spiked Celite. To determine extraction conditions, a standard solution of the three herbicides was extracted from celite. The celite provided a relatively inert solid matrix for the solutions to adsorb onto. Since the celite is inert, this eliminates many of the solute/solid interactions, thereby providing information about the extraction conditions that concern primarily just the solvent/solute interactions.

Table V. Herbicide Extraction Conditions

Extraction Conditions

	Step 1	Step 2	Step 3
Density:	0.25	0.60	0.9 g/ml
Flow rate:	1.0	2.4	3.6 ml/min
Temperature:	50 °C		
Fluid:	CO_2		
Extraction time:	9.4 minutes		
Thimble volumes:	5		

$200\mu l$ of a standard solution was pipetted onto approximately 1 gram of celite. The solvent was allowed to evaporate and the celite was placed into the extraction chamber. The spiked celite was then extracted at three densities; 0.25, 0.6, 0.9 g/ml CO_2. The extraction time for the steps was selected to allow five thimble volumes to be swept for each step. The full set of extraction conditions is shown in Table V.

Thimble Volumes. For these experiments the thimble volumes and the extraction times were held constant. To accomplish this, the mass flow of the system had to be varied by changing the flow rate at the pump head for each density step. Controlling the mass flow rate allowed the linear/volumetric flow to be consistent throughout the experiments. This is different from the paprika experiments in which the mass flow was held constant and the extraction times were changed, to keep thimble volumes constant for each extraction step. Flow control is one of the major advantages of variable restrictor based SFE units.

The variable restrictor allows the user to control both the mass flow of the system and the density independently.

Table VI. Analytical Conditions

Chromatographic Conditions:

Gas Chromatograph: HP 5890 Series II
Column: Supelco SPB-5 column (530 μm x 15 m)
Detector: Nitrogen phosphorus detector (NPD)
Injector: Cool on-column
Temperature program: Ramp 35 ^{o}C - 250 ^{o}C
Injection volume: 2μl

Spiked Celite Recoveries. Recoveries were determined by analyzing the extracts by gas chromatography using a nitrogen phosphous detector(NPD). (See Table VI.). A calibration curve was generated using the standard solution that was used to spike the celite. To quantitate the extracts, the collection vials were diluted to volume and the GC/NPD area counts were compared to the calibration curves. Recovery results from this density stepping showed that the atrazine was soluble at all densities of CO_2, while cyanazine and diuron did not extract well until the density of the CO_2 increased to 0.9 g/ml. The results are shown in Table VII. This indicates the need to use a high density to extract these solutes from soil.

Table VII. Recovery from Celite

Density	Atrazine	Cyanazine	Diuron
0.25 g/ml	32.5%	1.0%	1.1%
0.60 g/ml	47.2%	5.3%	7.7%
0.90 g/ml	23.3%	58.2%	60.8%
Total	103.0%	64.5%	69.6%

Spiked Soil Samples. To be sure that the soil matrix did not contribute extractable materials in addition to the analytes of interest, a soil blank was extracted at a density of 0.9 g/ml of CO_2 as a control. Gas chromatography using flame ionization detection (FID), showed that very few co-extractants were present in the matrix. Using an element specific detector to analyze for these herbicides (NPD), can provide a false sense of security when considering co-extractants and their impact upon the extraction method.

The density stepping method was then run on a spiked soil sample. Five grams of soil were spiked with the same solution of herbicide standards used to spike the celite sample and the solvent was allowed to evaporate. The spiked sample was then placed into an extraction thimble and extracted. The results were different than those obtained from the spiked celite. Even at a high density of CO_2 the herbicides were not extracted. Previous experience showed that using water as a modifier aided in the extraction of diuron from soils *(6)(7)*. Therefore 1 ml. of water was added to the spiked soil and the sample rerun at a density of 0.9 g/ml of CO_2. The results showed a significant increase in recovery of the herbicides with the addition of water. This demonstrates that the addition of a modifier added to the extraction cell can have a significant effect upon the extraction recoveries *(8)(9)*. The results are summarized in Table VIII.

Table VIII. Recovery from Soil

Density	Atrazine	Cyanazine	Diuron
0.25 g/ml	0%	0%	0%
0.60 g/ml	1.2%	1.1%	0.5%
0.90 g/ml	1.0%	1.6%	0.9%
Total recovery	2.2%	2.7%	1.4%
0.90 g/ml (1.0ml H2O added)	32.5%	33.5%	28.4%

The results of extracting these herbicides from celite versus soil, display a marked difference (as seen by comparing Table VII to Table VIII). The only difference between the extraction from celite and the extraction from soil was the matrix itself, all other conditions were identical. This raises the question; is the water modifying the solvent or modifying the matrix?

Modifier Effects

As a result of the increased recovery when water was added as a modifier, it was decided to look at the effect of various modifiers upon the extraction recoveries in a systematic manner.

Table IX. Solute Solubilites

Solvent	Atrazine	Cyanazine	Diuron
Water	70 ppm	171 mg/l	42 ppm
Methanol	18,000 ppm	NA	NA
Ethanol	NA	45 g/l	NA
Chloroform	52,000 ppm	210 g/l	low
Hexane	NA	15 g/l	low

NA: Solubility data was not available.

Table IX shows the solublity of the herbicides in various solvents. If the extraction process is based on a solvation mechanism then a non-polar solvent (such as chloroform) should extract these solutes better than a more polar solvent (such as methanol or water).

Modifiers were selected for addition to the spiked soil samples representing a range of solvent polarities. Based on the results of the density stepping experiments, the extraction times were increased to 19 minutes from 9.4 minutes, which increased the thimble volumes from 5 to 10. To be consistent throughout the experiments, 1 ml of modifier was added directly to each extraction cell. Methylene chloride was chosen as a non-polar modifier. Results show that the methylene chloride did little to aid in the extraction process for any herbicide.

Next, more polar modifiers were selected, isopropyl alcohol and methanol. Both alcohols helped increase the recoveries from the soil, with the methanol being slightly better than the isopropyl alcohol. These results indicate that the more polar the modifier, the better the extraction recoveries. It required a very polar modifier (water), to obtain recoveries from the spiked soil comparable to those obtained from the unmodified spiked celite samples. Tables X, XI and XII show the herbicide recoveries and the relative solublities of these herbicides in various solvents.

The results are not what one would expect based upon a solvation model of extraction. Since the herbicides are more soluble in non-polar solvents (Table IX), one would expect that a non-polar extraction fluid, such as CO_2 or CO_2

modified with a non-polar solvent, would produce the greatest recoveries. The results from the spiked soil samples show that the opposite occurs, polar modifiers increase recoveries much more than non-polar modifiers. This indicates that the solute/matrix interactions were the interactions being modified versus the solute/solvent interactions. These results are in opposition to the practice of enhancing recoveries by changing the solvent power of the extraction fluid, using density, temperature or fluid composition.

Table X. Recovery of Atrazine

Modifier (1.0 ml added)	Soil	Celite	Relative Solubility
None	0%	96.8%	-
Methylene chloride	4.1%	117.4%	52,000 ppm chloroform
Isopropyl alcohol	27.2%	15.9%	-
Methanol	49.8%	45.8%	18,000 ppm in methanol
Water	108.9%	94.6%	70 ppm in water

Table XI. Recovery of Cyanazine

Modifier (1.0 ml added)	Soil	Celite	Relative Solubility
None	0%	96.8%	-
Methylene chloride	11.3%	116.6%	210 g/l in chloroform
Isopropyl alcohol	33.0%	15.7%	-
Methanol	52.3%	47.5%	45 g/l in methanol
Water	106.4%	81.8%	171 mg/l in water

Table XII. Recovery of Diuron

Modifier (1.0 ml added)	Soil	Celite	Relative Solubility
None	0%	96.8%	-
Methylene chloride	4.0%	108.7%	low, in hydrocarbon solvents
Isopropyl alcohol	16.8%	12.6%	-
Methanol	35.8%	43.1%	-
Water	91.7%	93.3%	42 ppm in water

The same experiments with modifier were run on spiked celite. The results were similar to those seen with the soil samples except in the case when methylene chloride was added to the celite. With the methylene chloride added as the modifier, the herbicides were totally extracted as in the unmodified experiments. One possible explanation is that the methylene chloride itself was being removed from the matrix more readily than the other modifiers, along with the herbicides.

The alcohols may have inhibited the full recovery of the herbicides under these conditions by providing a liquid phase which could solublize the herbicides. The herbicides could then partition themselves between the liquid alcohol phase and the CO_2/alcohol phase. This would lead to the herbicides being retained on the celite until the all of the alcohol modifier was removed by the CO_2 extraction fluid.

Modifier Addition Using Premixed Cylinders. The use of pre-mixed modifiers and CO_2 was also investigated. A 5% by weight mixture of isopropyl alcohol and CO_2, and a 7% by weight mixture of methylene chloride and CO_2, were used under the same extraction conditions. These mixtures represent approximately a 4 mole % mix of each modifier with CO_2. Under these extraction conditions this corresponds to 3.6 - 4.4 ml of liquid modifier being used per extraction instead of the 1.0 ml volume added directly to the extraction thimble. The recoveries obtained by using modified tanks were generally lower than those obtained by adding the modifier directly to the extraction thimble. (See table XIII).

The premixed modifiers enter the extraction cell already in the CO_2 fluid phase. The modifier stays in the fluid phase during the extraction, therefore has less interaction with the matrix itself compared to the liquid modifier being placed directly on the sample(10). This reduced interaction with the matrix is probably responsible for the lower recoveries. All of these experiments were done in a dynamic mode, the results might have been quite different if the experiments with the modified tanks were run in a static mode. Under static conditions the modified CO_2 would have a longer time to interact with the matrix and potentially produce higher recoveries.

Table XIII. Different Methods of Modifier Addition

Modifier	Atrazine	Cyanazine	Diuron
Celite			
Isopropyl alcohol[a]	15.9%	15.7%	12.6%
5% IPA in cylinder[b]	6.5%	7.9%	7.5%
Methylene chloride[a]	117.4%	116.6%	108.7%
7% MeCl$_2$ in cylinder[b]	69.2%	70.0%	65.6%
Soil			
Isopropyl alcohol[a]	27.2%	33.0%	16.8%
5% IPA in cylinder[b]	0%	0%	0%
Methylene chloride[a]	4.1%	11.3%	4.0%
7% MeCl$_2$ in cylinder[b]	8.7%	18.8%	7.8%

a) 1.0 ml of modifier added directly to the sample
b) (wt/wt)% for modifier in cylinder.

Reliability. The robustness of SFE as a routine technique is shown in Table XIV. These data were generated by loading a sample tray and the samples were run overnight with the help a robotic manipulator. The celite was spiked with the water in batch mode but extracted sequentially. It can be seen from the data that having the samples sitting at room temperatures had no deleterious effects upon the extraction recoveries. The relative standard deviation values include not only the extraction variations but all variations from sample preparation (spiking) and analysis.

Table XIV. Repeatability Data for Recoveries from Celite

Matrix	Atrazine	Cyanazine	Diuron
Dry	96.8%	97.2%	95.3%
Dry	96.0%	97.1%	91.3%
1.0 ml H_2O added	88.0%	82.0%	83.7%
1.0 ml H_2O added	94.6%	81.8%	93.3%
1.0 ml H_2O added	86.7%	86.8%	90.6%
Relative std. dev.	4.6%	7.7%	4.3%

Summary

SFE can be a fast and efficient method for sample preparation. The paprika sample demonstrates how techniques such as density stepping can rapidly give qualitative information concerning starting points for further SFE method development. However to properly develop an SFE method requires a systematic approach for varying the extraction parameters.. The herbicide method was developed by first extracting a standard solution from an inert matrix (celite). These results showed that supercritical CO_2 could extract the herbicides of interest. However when these same SFE conditions were used to extract the herbicides from a spiked soil sample, the recoveries were quite different. Several modifiers were added to the spiked soils. The modifiers were selected to represent a wide range of polarity from non-polar (methylene chloride) to polar (water). These results showed that the modifiers producing the best recoveries, were those that the herbicides were least soluble in, suggesting that these modifiers are modifying the solute/matrix interaction not the solute/solvent interaction. Results also indicate that it is more efficient to apply the modifier directly to the matrix versus mixing the modifier with the CO_2, prior to introducing it into the extraction cell.

Literature Cited

1. Snyder, L. R.; Kirkland, J.J., *Introduction to Modern Liquid Chromatography*; John Wiley & Sons, Inc.: New York, NY, 1979; pp. 257-258.

2. Giddings, J.C.; Myers, M.N.; King, J.W. *J. of Chrom. Sci.* **1969**, *7*, pp. 276-283.
3. Mulcahey, L.J.; Hedrick, J.L.; Taylor, L.T. *Anal. Chem.* **1991**, *63*, pp.22225-2232.
4. Buser, H-R. *Environ. Sci. Technol.* **1990**, *24*, pp. 1049-1058.
5. Janda, V.; Steenbeke, G.; Sandra, P. *J. of Chrom.* **1989**, *479*, pp. 200-205.
6. Hance, R.J. *Weed Res.* **1965**, *5*, pp. 108-114.
7. McNally, M.E.; Wheeler, J.R. *J. of Chrom.* **1988**, *447*, pp. 53-63.
8. Dooley, K.M.; Ghonasgi, D.; Knopf, F.C. *Environ. Prog.* **1990**, *9*, pp. 197-203.
9. Wheeler, J.R.; McNally, M.E. *J. of Chrom. Sci.* **1989**, *27*, pp. 534-539.
10. Berger, T.A.; *HRC* **1991**, *14*, pp. 312-316.

RECEIVED December 2, 1991

Chapter 19

Supercritical Fluid Extraction in the Analytical Laboratory

A Technique Whose Time Has Come

W. S. Miles and L. G. Randall

Hewlett–Packard Company, P.O. Box 900, Route 41 and Starr Road, Avondale, PA 19311–0900

The robustness of a sample preparation technique is characterized by the reliability of the instrumentation used and the variability (precision) of the information obtained in the subsequent sample analysis. Thus, variations in controlled parameters and sequences are to be avoided. In sample preparation methods employing supercritical fluids as the extracting solvents, it has been our experience that minimal variations in efficient analyte recoveries are possible using a fully automated extraction system. The extraction solvent operating parameters under automated control are temperature, pressure (thus density), composition and flow rate through the sample. The precision of the technique will be discussed by presenting replicability, repeatability, and reproducibility data for the extraction of various analytes from such matrices as sands and soils, river sediment, and plant and animal tissue. Censored data will be presented as an indicator of instrumental reliability.

Supercritical fluid extraction (SFE) has been studied as a preparative technique for over 50 years *(1,2)*. In particular, interest in SFE by chemical engineers for use in large, industrial-scale processes grew rapidly in the late 1960's and early 1970's. Early pioneering work exploring analytical-scale applications was proceeding during that time as well -- for example, references *(3-5)*. By the early 1980's studies of SFE as a tool in analytical laboratories had begun to grow very quickly -- as evidenced by the appearance of many review articles *(6-8)* which explored prior work as part of making the transition to the analytical laboratory.

At this time sample preparation procedures using extraction solvents at supercritical or near-critical conditions are becoming more and more

0097–6156/92/0488–0266$06.50/0
© 1992 American Chemical Society

commonplace in analytical laboratories, and there have been many recent references in the literature on feasibility studies of SFE as an alternative sample preparation technique -- including comparisons of recoveries provided by SFE to those afforded by traditional techniques. For example, Schantz and Chesler *(9)* reported recoveries and standard deviations for the extractions of Arochlor 1254 from sediment and PCB's from an urban particulate sample. Two extractions for each sample type were carried out on separate samples and SFE recoveries were compared to those from Soxhlet extractions. Kiran and Li *(10)* reported standard deviations of less than 0.6% for repeatability studies on dissolution and precipitation of wood from a variety of supercritical fluids. Capriel et al *(11)* demonstrated quantitative feasibility of SFE in their assay of bound pesticides from soil and plant tissue. For example, they reported the recoveries of atrazine from an Ottawa, Canada corn to be 95% by SFE and 78% by high-temperature distillation (HTD). The reported relative standard deviation was $\pm 5\%$. Hopfgartner *(12)* et al determined replicability of five extractions of biomarkers in ground rock samples and demonstrated that the standard deviation for SFE (2.34 % for the "% Tri" maturity parameter, which measures the ratio of two compound types) was clearly better than that obtained by solvent extraction (13.85%). They also qualitatively compared the three approaches of SFE, solvent extraction and thermodesorption. Barry et al *(13)* have described their work in fractionating polymers. Although no precision data was reported, Barry et al successfully applied SFE to fractionate commercially available OV-17 into a low molecular weight fraction and an enhanced thermostable high molecular weight fraction. In a similar study in our laboratory we were able to purify a high molecular weight polymer by removing approximately 97% of the low molecular weight oligomers using pure CO_2 after dissolving the polymer in methylene chloride; 5% to 15% of the high molecular weight polymer was lost. However, with the traditional liquid-liquid extraction method essentially all the oligomer is removed, but 95% of the high molecular weight polymer is lost as well. Moreover, the traditional liquid-liquid method requires 3-4 days compared to about 1.5 hours by SFE.

The good recoveries reported above by some earlier workers in SFE clearly demonstrate the feasibility and comparative accuracy of SFE as a sample preparation technique for a variety of analytes and matrices. However, the question of the "robustness" of SFE as an analytical tool easily used on a routine basis by chemists throughout the analytical community remains to be answered. What is the expected variability of results for any given application and how much user interaction is required to maintain functional equipment sample after sample? Such information -- precision and mean-time-between-failures (or mean-time-between-maintenance) -- has not been routinely reported in published literature since the emphasis has heretofore been on initial feasibility experiments.

The main purpose of this paper is to explore the robustness of SFE as an analytical technique. To do this, we have used guidelines published by the AOAC *(14)*, Association of Official Analytical Chemists, as a way to define and measure contributors to method robustness. In particular, method robustness can be characterized by the reliability of the analytical instrumentation employed and the precision (variability) of the results. In the "Results and Discussion" section, anecdotal information will be presented as an indication of instrumentation reliability and many studies will be summarized to provide precision data for the factors of replicability, repeatability, and reproducibility.

An underlying assumption in these discussions is that SFE is a viable alternative for sample preparation procedures for a significant number of samples -- even though equipment more sophisticated than traditional laboratory glassware is necessary. For example, SFE systems can be operated at temperatures up to 150 C and pressures to 600 bar using a variety of fluids. The unique characteristics of supercritical fluids which make them so attractive as solvents have been discussed fully on many occasions elsewhere *(15-17);* a similar discussion is outside the scope of this paper. However, in the next section we will briefly 1. explore the use of supercritical fluids from the perspective of potentially enhanced robustness and 2. outline considerations which are typically considered prior to analytical methods development and which should be employed for SFE as for any other technique.

The Process for Evaluating SFE as an Alternative Sample Preparation Technique

Why SFE is an Attractive Alternative to Established Procedures. The main task of sample preparation is to separate an analyte from a matrix by investing the minimum amount of effort and time necessary to provide quantitation with acceptably high degrees of accuracy and precision. Traditional approaches to sample preparation are often very labor intensive and require manual manipulation of many different pieces of glassware -- as pictorially summarized in Figure 1. The number of major steps in a traditional preparation method can range from fewer than 10 to more than 100. For example, with solid samples the prep method may range from simple "shake-and-shoot" extractions to very complex procedures consisting of extraction, concentration, fractionation, concentration, solvent exchange, and calibration steps as outlined in Figure 1. As noted, typical fractionation processes may consist of preparative column chromatographic clean-ups (e.g. references *18, 19*) or several liquid-liquid back extractions *(20)*. Our laboratory has been studying the problem of automating sample preparation in the analytical laboratory since 1980. Efforts have ranged from defining useful discrete instruments *(21)* to implementing robotic systems *(22, 23)*. During that time we have interviewed hundreds of chemists with sample prep methods ranging from simple to complex. For those simple methods, instrumentation with a technique as complex as SFE is unwarranted; however, robotic automation which emulates the simple process is often highly desirable. For more complex procedures, a common problem (*21*) has been a lack of robustness of established manual methods. Laboratory managers have often described their concern about results which vary from person to person in the same lab. Methods developers have noted that, for moderately complex methods, the time it takes to develop a routine, established manual method can range from several months to 1-2 years. However, such methods exhibit a notorious lack of robustness: it is quite common that procedures are fine-tuned or customized from one person to the next in order to achieve acceptable results. Indeed the variability of any analytical method can be simplistically reduced to two sources:

* Manual technique of the person
* Degree of control of the instrumentation.

If instrumentation with appropriate degrees of control can be engineered, it is a logical step to develop analytical methods with more and more automated control of the preparation and analysis processes to minimize variability. However, one of the disadvantages of traditional methods is the difficulty in directly automating manipulative steps such as outlined in Figure 1. This difficulty serves as an opportunity for SFE as an alternative sample preparation technique, since many of these steps can be easily and reliably automated due to the physicochemical characteristics of supercritical fluids. Therefore, the benefits obtained by applying the SFE technique to sample preparation are 1. a reduction in manual processes and 2. a high degree of amenability for precise, well-known process controls of timed sequences and operating parameters. Extraction fluid density (pressure), temperature, flow rate and extraction time (static or dynamic) are the operating parameters that directly influence variability of recoveries. These parameters can and should be independently and precisely controlled to provide minimum variability and thus a routine, robust method. For example, one significant component found in most SFE instrumentation is the restrictor. Its function is to expand the pressurized solution to separate the "solvent gas" from dissolved extracted components. If a fixed restrictor is used, the mass flow rate of the fluid changes as a function of pressure (density): mass flow can increase by a factor of 25 as pressure is increased from 80 to 400 bar (*24*). Not only are the pressure and flow coupled, the coupling is via a static conduit whose dimensions are imprecisely controlled during an extraction (partial plugging by particulates and precipitated components, temperature) and from component to component during maintenance replacements. This results in a lack of control in operating parameters (density) and timed sequences (via flow rate and time). A variable restrictor whose dimensions are set and adjusted by an electronic feedback control loop is an alternative solution.

In fact control of all the operating parameters is well within the realm of current best engineering practices, making SFE a powerful technique to use in automating sample preparation. While end results of supercritical fluid extraction and manual methods can be similar, the actual interweaving of discrete processes is different. A full discussion comparing the structures of traditional manual methods and those using SFE -- particularly in the context of SFE operating parameters -- is forthcoming (C.R. Knipe, W.S. Miles, F. Rowland, and L.G. Randall, "Designing a Sample Preparation Method which Employs Supercritical Fluid Extraction (SFE)", Hewlett-Packard Publication # (43)5091-2102E (1991), in preparation).

Initial Considerations Prior to Methods Development. When evaluating SFE as a sample preparation technique we must be able to compare SFE to other sample preparation techniques using criteria such as initial methods development considerations and standard quantitative measures used in analytical laboratories, such as the AOAC guidelines. Considerations in designing and developing analytical methods (which encompass both sample preparation and analysis procedures) can be categorized as practical (related to business) or scientific (related to technical) in nature (Figure 2). Some of the practical aspects are time, ease of use, cost per analysis and reliability. Examples of scientific considerations that must be addressed when evaluating method performance are practicality, sensitivity, accuracy and precision as well as dependability (reliability). As an example of a scientific consideration, in the development of an SFE method for the simultaneous determination of chlorpyrifos and pyridinol from grass, it was found that methanol was a better modifier than toluene. However, toluene was a better reconstitution solvent for pyridinol since it was quantitated as the trimethyl derivative. On this basis a mixture of CO_2/toluene was chosen as the extracting solvent. Also note that there is some overlap of the categories of considerations: reference to Figure 2 shows that reliability is both a practical and scientific issue. If a technique or method is to be robust, there must be a high degree of certainty that (all) the instrumentation will perform on a routine basis with minimum user intervention to maintain functioning performance. Maximum "up-time" is also a cost/sample consideration. From a technical perspective, the user must have confidence in the ability of the instrumentation to deliver the expected results of the method time after time, sample after sample. For example, in SFE the instrumentation must transfer analytes from an extraction cell to a collection device quantitatively.

As the potential user of SFE runs through the checklist of considerations for evaluating techniques for specific applications (Figure 2), the area of robustness as characterized by reliability and variability (precision) has been less well-documented than the other items. In looking for a way to evaluate the

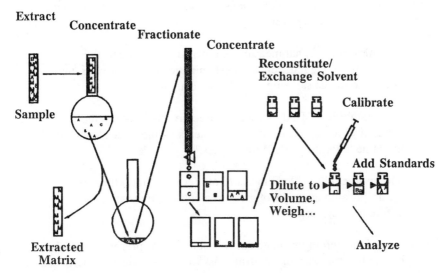

Figure 1.   A typical sequence of sample preparation manipulations for a procedure of Soxhlet extraction followed by preparative column chromatography for a solid sample -- e.g., glycolipids out of wheat flour. Note that equivalently complex processes are encountered with "simple" liquid-solid or liquid-liquid extractions followed by a series of back extractions -- e.g., pesticides from fish tissue. (Reproduced with permission by Hewlett-Packard, HP Publication # (43)5091-2102E, (1991), in preparation.)

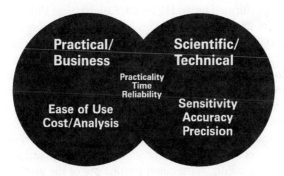

Figure 2.   Initial considerations prior to methods development.

robustness of SFE in the context of existing laboratory practices, we selected the AOAC guidelines as a vehicle to present various sets of experimental results. The following are the AOAC definitions of the sources of variability in measures of precision taken from Reference *14*:

> * **Replicability** - "That measure of precision that reflects the variability among independent determinations on the same sample at essentially the same time by the same analyst."

> * **Repeatability** - "That measure of precision that reflects the variability among results of measurements on the same sample at the same laboratory at different times."

> * **Reproducibility** - "That measure of precision that reflects the variability among results of measurements of same sample at different laboratories."

A summary of the definitions in an easily usable form is presented in Table I.

Reliability, that part of robustness associated with ongoing performance, can be summarized as the "ability of method or technique to successfully perform a required function under stated conditions for a stated period of time." The work presented in this paper in the "Results and Discussion" section focuses predominately on measures of precision. Some anecdotal information -- essentially censored data -- is presented for reliability.

Experimental

Extractions were performed on an HP7680A Supercritical Fluid Extractor. This system is a graphics-driven, PC controlled instrument featuring a combination of extraction, fractionation, concentration, and solvent exchange/reconstitution processes automated within a single method. This instrumentation employs an electronically controlled variable restrictor so that the fluid pressure (density) and flow rate are decoupled, allowing the flow rate to be independently settable over the entire pressure range. The user is at all times knowledgeable of the amount of solvent having passed through the sample for a given density, extraction time, and flow rate via the reported parameter of "thimble volume"; this monitoring of a controlled sequence is necessary for quantitative methods development and, especially, subsequent routine use of the method after it is developed. Also, the independently settable pressure provides the capability of fractionation (i.e., selective removal of groups of components from samples via the corresponding change in extracting fluid solvent strength) while allowing a choice of the flow rates most appropriate to the application.

Table I.   Measures of Precision per AOAC [a]

Sources of Variability	Replicability	Repeatability	Reproducibility
Specimen (Subsample)	Same or Different	Same or Different	Most Likely Different
Sample	Same	Same	Same
Analyst	Same	(At least one	Different
Apparatus	Same	of these must	Different
Day	Same	be different)	Same or Different
Laboratory	Same	Same	Different

[a] *Reprinted with permission from reference 14*

Extraction solvents used were Scott Specialty Gases, Supercritical Fluid Grade CO_2 and premixed molar percentages of CO_2 with selected organic solvents as modifiers. The same organic solvents were used as reconstitution solvents. In all cases these were Spectro-Quality or HPLC quality in purity (Alltech Associates Inc., American Burdick and Jackson).

Pre-extraction sample manipulation involved chopping and in some instances adding an adsorbent and/or modifier to the sample. Whole hops, green onions, lettuce, alfalfa, oranges, and strawberries (as purchased from local product supplier) were chopped in a Sorvall "Omni-mixer" until homogenous. Portions of green onions, lettuce, oranges, and strawberries were mixed with small amounts of Celite 545 (Supelco, Inc.) or sandwiched between layers of Celite or filter paper within the sample-containing extraction vessel (referred to as "thimble"). This was needed to prevent loss of any dissolved analyte during weighing. Dry samples such as ground coffee, grass, alfalfa, sand, and soils were weighed directly into the 7 milliliter extraction thimbles (also fitted with filter paper discs at both ends). Depending upon the application, modifiers may have been used by dispensing measured aliquots of the modifier directly onto these samples prior to tightening the thimble caps. One reason for this approach is that for some applications the extraction efficiency for low volatiles such as benzo(a)pyrene can be greatly increased by wetting the sample with modifier. Small volumes of liquids, for example, the performance evaluation standard (PES), were placed on Celite or filter paper and extracted. The phenylmethyl polymer was first dissolved in modifier before transferring to the extraction thimble with the solution being deposited on filter paper previously installed in the thimble. The only pre-extraction hardware manipulations necessary were to hand-tighten the extraction thimble caps, to place thimble in extractor, and press "start" to begin the method sequence. SFE operating conditions for the above applications are presented in Table II.

Results and Discussion

Reliability. According to AOAC, reliability is the ability of a method or technique to perform a required function under a set of conditions for some pre-determined time. In the work reported here, even though we did not carry out a formal study of reliability, we were able to gain censored estimations of instrument dependability. In Table III we show a diverse list of applications ranging from simple to complex samples requiring rather different operating conditions. The term "censored" is used to indicate that the number of samples run for each application was based on experimental design and experiments in those areas were stopped without encountering any failures in instrumentation. Here, an instrumentation failure is taken to mean some problem has occurred which necessitates maintenance and repair to resume functioning. [A failure

Table II. SFE Operating Conditions

Analyte/Matrix	Solvent	Density (g/ml)	Pressure (psi)	Temp (°C)	FlowRate (ml/min)	Extraction Time(min)	Thimble Volumes
1. PES/filter paper	CO_2	0.75	1,937	40	3.0	14	7.4
2. Bitter acids/hops	CO_2	0.40	1,255	40	3.0	10	6.6
3. PAHs/EPA reference material sediment	CO_2/CH_2C_{12} added thimble[a]	0.85	4,767	60	2.5	20	7.8
4. LMWO/phenylmethyl siloxane polymer	CO_2/dissolved in CH_2C_{12}[b]	0.90	4,082	40	2.0	45	13.2
5. Pesticides/coffee	CO_2/Modifiers added to thimble[c]	0.25 0.90	1,117 4,082	40 40	2.0 2.0	10 18	10.6 5.3
6. Pesticides/green onions	CO_2	0.25	1,117	40	2.0	6	6.3
7. Pesticides/lettuce	CO_2	0.90	5,082	50	2.0	10	2.9
8. Pesticides/strawberries	CO_2	0.75	2,548	50	2.0	6	2.1
9. Pesticides/oranges	CO_2	0.75	2,548	50	2.0	10	3.5
10. Pesticides/alfalfa	CO_2/5M%MeOH[d]	0.40 0.90	1,255 5,082	40 50	2.0 2.0	6 10	4.0 2.9
11. Chlorpyrifos/grass	CO_2/toluene added to thimble	0.95	5,560	40	3.0	6	2.5
12. Pesticides/ freeze-dried fish tissue	CO_2, CO_2/8M%[e] MeOH	0.30 0.85	1,610 4,559	70 70	2.0 2.0	15 10	13.2 3.3

[a] "added to thimble" indicates that modifier was dispensed directly on the sample prior to extracting.
[b] Polymer was dissolved in methylene chloride prior to placing in extraction thimble.
[c] Two density steps, Step I. D = 0.25 and Step II. D = 0.90.
Modifiers added to different samples were 2 ml of water, methanol, acetic acid, acetone, and methylene chloride.
[d] Two density steps, Step I. D = 0.40 and Step II. D = 0.90. 5M% MeOH modifier was pre-mixed for both steps.
[e] Two density steps, Step I. D = 0.30 and Step II. D = 0.80. Modifier used in Step II. was pre-mixed.

often encountered in many SFE systems which use fixed restrictors is "plugged restrictors"; the condition can be resolved by replacing the restrictor, measuring the flow, and adjusting the method to accommodate the new flow before proceeding with the sample or those that follow.]

In the last entry in Table III more than 50 samples were extracted in an experiment surveying the performance of laboratory robotic equipment. These included 2.0 gram samples of soils extracted with pure CO_2, 2.5 gram samples of a reverse phase material extracted with CO_2 and mixtures of CO_2 with various modifiers, 2.0 gram samples of ground coffee plus aliquots of modifiers dispersed on the samples within the sample thimble extracted with pure CO_2, and 20 microliter aliquots of performance evaluation standard (PES, octadecane in iso-octane) on simple matrices extracted with pure CO_2. The variety of analytes and complex matrices along with the number of different runs clearly show that the instrumentation is reliable.

Variability as Measured by Precision. As summarized in Table I, there are three measures of precision used to report variability among results of measurements: replicability, repeatability, and reproducibility. In any experimental work the sources of variability affecting precision are sample, analyst, apparatus, day of analysis and laboratory where the analysis was performed. In establishing a particular measure of precision, these sources are held constant in a prescribed manner. In the results to follow, various analytes were extracted from different matrices, keeping the indicated sources of variability constant according to the measure under scrutiny. For example, in examining the first measure, replicability, all the sources of variability are held constant and all extractions are performed in the same manner. In this discussion our primary interest was to evaluate precision; thus, accuracy will not be discussed in depth for each set of data. However, percent recoveries are reported for all applications. Recoveries in most cases met our methods development goals of $95 \pm 5\%$ at the 95th confidence level.

Replicability. Table IV shows replicability data for five extractions of a hydrocarbon mixture from a simple matrix (filter paper). This was an early application run on preliminary instrumentation (note footnote at bottom of Table IV) in order to establish on-going tracking of the performance of developing instrumentation. Even under these conditions the precision (as indicated by the relative standard deviation) is good, ranging from 5% for C_{20} to 11.5% for C_{12}. Of the reported variability, about 2% is found in the chromatography. Also given in Table IV are recovery values indicating essentially complete recovery of analytes. The significance of precision results for this hydrocarbon application is that one of our earlier goals was to compare SFE to existing automated or semi-automated techniques. When compared to techniques such as headspace and purge and trap these results compare favorably (*25*).

Table III.   Censored[a] Reliability Data

Application	Number of Samples
■ Bitter acids from hops high wax content	5
EPA reference material sediment for high PAHs	12
■ NOAA river sediment Containing 40-50% water	16
■ Phenylmethyl siloxane cyclic oligomers	20
■ HP 7680A Performance Evaluation Sample (PES)	>2 mos, 3-12 per day
■ Soils/reverse phase material/ ground coffee/simple standards	>50

[a] *No User Intervention Sample to Sample*

Table IV.   Replicability: Hydrocarbons from Filter Paper 10 ppm [a]

Run No.	C8	C9	C10	C11	C12	Anthracene	C20
1	152710	166960	181350	188480	187980	288850	184060
2	165750	163720	174280	179070	177380	280740	184690
3	157640	155690	165460	169300	166400	280740	184690
4	142190	139990	149240	199950	200760	315180	204450
5	158130	175040	191990	153440	148060	314260	198830
Av Area Ct	155284	160280	172464	178048	176116	301848	194730
Std Dev	86786	13295	16216	17830	20194	15939	9661
Rel Std Dev	.056	.083	.094	.100	.115	.053	.050
% Recovery	97.1	98.8	99.5	99.7	95.6	99.8	100.0

[a] *Preliminary Instrumentation -- sequencing and control loops not finalized*

In the next example exploring replicability, a slightly more difficult sample was used: more analytes and a large range in concentration levels. The results are given in Table V. Four replicate samples containing three pesticides and eight PAH's (concentrations ranging between 7.5 and 4300 ppb) were extracted from filter paper or Celite. Figure 3 is an FID chromatogram of one of the extracts. The relative standard deviations for these analytes, listed in the fourth column in Table V, indicate low variability, i.e., good precision -- with one exception, anthracene. There was poor precision in quantitating the anthracene. This was due to the anthracene shoulder being an extracted matrix impurity whose concentration varied from sample to sample as the amount of support sorbent varied so that the variability in integrating the anthracene peak ranged from 3% to 15%. For resolved peaks, chromatographic variability is expected to be better than 2%. As the reader looks at Figure 3 he will note two instances of unresolved chromatographic peaks: the first is anthracene plus the variable-level co-extractant and the second, overlapping dieldrin and DDE peaks. In the case of the overlapping dieldrin and DDE peaks, there is no source of a variable-level co-elutant so the precision for these compounds is good. Recoveries for these analytes are better than 90%. This is important since for the higher molecular weight analytes in this sample, much lower results have been reported (26).

As a last determination of precision by replicability the quantitation of pesticides from some real samples is shown in Table VI. The precision for the extraction of diazinon from green onions, lettuce and oranges ranged between 7% and 9%. The range for the extraction of chlorpyrifos from green onions, lettuce, strawberries, oranges and alfalfa was 6% to 11%. Slightly better precision, 4% to 6% was obtained for the extraction of malathion from the same samples. Precision ranges for aldrin extracted from green onions, strawberries and alfalfa were 6% to 10% and 7% to 13% respectively. For these applications an average relative standard deviation of about 7% is very good considering that there was no sample cleanup subsequent to the supercritical extraction to remove chromatographic interferences (27). This is important since in most other techniques, especially for onions, lettuce and other produce with high levels of chlorophyll, considerable time and effort are spent to clean up the extract prior to analysis. It should be noted that possible interference by water was addressed by either adding small amounts of Celite or sandwiching between filter paper plugs. The use of filter paper as plugs or disks proved to be an effective way of confining matrix to the extraction vessel and preventing downstream clogging of the system. No clogging was experienced for any of the runs and the only other notable setpoint value was the use of a somewhat lower flow rate. It was observered that lower flow rates also aided in preventing plugging caused by transport of very small particles.

The results from the above applications indicate that the SFE method has very good replicability when compared to values typically cited by those using existing automated techniques and established manual methods (e.g., 25, 27) and when applied to a range of sample types of varying complexity and analyte levels.

Table V. Replicability Pesticides and PAHs from Celite

	Concentration (ppb)	Av Area Count	Std Dev	Rel Std Dev	% Recovery
Naphthalene	280	390	39	.101	92.9
Acenaphthene	293	318	34	.107	92.4
Fluorene	525	350	36	.103	99.7
Phenanthrene	2580	198	12	.060	94.6
Anthracene	615	440	90	.200	95.5
Pyrene	4287	256	13	.051	94.6
Chrysene	2393	146	9	.061	94.7
Benzo (a) pyrene	2087	234	11	.047	90.1
Dieldrin	20.0	114	9	.077	101.8
DDE	15.0	88	6	.065	98.3
DDT	7.5	203	9	.046	97.5

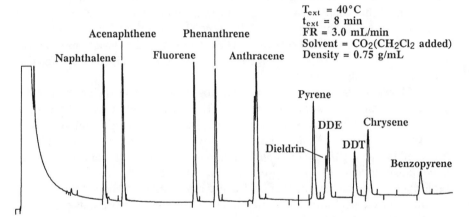

T_{ext} = 40°C
t_{ext} = 8 min
FR = 3.0 mL/min
Solvent = $CO_2(CH_2Cl_2$ added)
Density = 0.75 g/mL

Figure 3.   Recovery of pesticides and PAH's from Celite for a one-step extraction. Note that to achieve a one-step extraction of components ranging up to benzo(a)pyrene, the sample was wet with CH_2Cl_2 prior to extraction.

Table VI. Replicability Pesticides from Fruits and Vegetables

	n=3				
	Diazinon 64ppb	Chlorpyrifos 286ppb	Malathion 32ppb	Aldrin 41ppb	Diuron 21ppb
	RSD %R	RSD %R	RSD %R	RSD %R	RSD %R
Green onions	.07 101	.06 104	.04 87	.09 90	.07 82
Lettuce	.07 94	.06 98	.04 95		
Strawberries		.06 96	.04 94	.10 97	.07 84
Oranges	.09 91	.09 93	.05 90		
Alfalfa		.11 99	.06 87	.06 94	.13 79

Repeatability. The second measure of variability to be considered is repeatability. According to the AOAC (refer to Table I) repeatability is concerned with results of measurements on the same sample in the same laboratory but at different times. To determine the contribution of this measure of variability to robustness, we considered a variety of applications ranging from simple standards to some more complex and real samples (chlorpyrifos from grass and pesticides from fish tissue). Table VII contains the results for a simple standard (PES) having 20 ppm octadecane in iso-octane dispersed on filter paper. [The performance evaluation standard, PES, also contains malathion, which can be quantitated by GC using a range of detectors - AED, ECD, FPD, and NPD and azobenzene which can be quantitated with LC using UV/VIS detection]. In this work quantitation was done using a FID detector, thus the choice to quantitate on octadecane in Figure 4. The PES is used to track the chemical performance of the HP7680A, so that over the years of instrumentation development this sample has been run numerous times. Shown here are the repeatability results obtained on four different instruments, at four different times and four to seven replicate runs. The relative standard deviations range from 2.6% to 6.1% and are comparable to what could be obtained by a manual method such as "shake and shoot."

Shown in Figure 5 is the FID chromatogram of EPA surrogate spike and phenol spike mixtures extracted from sand. These are the same mixtures that the EPA used in a recent SFE round robin study. To determine repeatability, the peaks at 9.57, 14.86, 18.32, and 21.29 minutes were used. The precision results are listed in Table VIII. In this application four different instruments were used and relative standard deviations were calculated based on the average of replicate runs for each of the four peaks. For example, the relative standard deviation for the peak with a retention time at 9.57 minutes represents precision of a total of 14 extractions (each with subsequent GC analysis) done on four different instruments. The other entries are treated in the same manner. The precision for these samples is slightly lower (i.e., %RSD values are slightly higher) than in the previous applications but better than the certified values supplied by EPA or NIST which, in some instances, are as high as 20%.

The first "real-sample" application is given in Table IX, where chlorpyrifos was extracted from treated grass supplied by an outside collaborating laboratory. Two instruments were used for this application, extracting 11 samples on 4/26/89 and 8 samples on 10/13/90. The relative standard deviations were 3.9% and 1.7% respectively. The latter value is extraordinarily good since, in the work reported, precision due to the chromatographic analysis variability was between 1% and 3%. Further, the values of precision reported by the supplier for manual extraction of chlorpyrifos from similar matrices, leaves and roots, were 8% and 4% respectively.

A second "real-sample" application was the extraction of pesticides from freeze-dried fish tissue standards supplied by EPA Cincinnati to analytical

Table VII.   Repeatability: HP 7680A Performance Evaluation Standard [a]

Instrument ID	1	2	3	4
Time	09/5/89	08/30/89	11/15/89	08/11/89
n	4	5	4	7
Av Area Ct	281662	308471	517000	2.5848E6
Std Dev	17425	12996.2	13441	71149.7
Rel Std Dev	.061	.042	.026	.028
% Recovery	94.0	95.9	99.2	99.3

[a] 20 ppm Octadecane

30 August 1989   1 September 1990

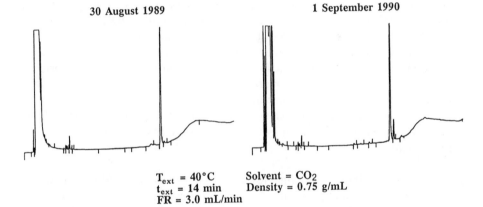

$$T_{ext} = 40°C \quad Solvent = CO_2$$
$$t_{ext} = 14 \ min \quad Density = 0.75 \ g/mL$$
$$FR = 3.0 \ mL/min$$

Figure 4.   Chromatographic analysis of extracted fractions produced by extracting a "check-out" sample, the Performance Evaluation Standard, for the HP7680A SFE on two occasions one year apart.

Table VIII.   Repeatability: EPA-Spiked Sand with Surrogate and Phenol Spike Mixture, (ppm)

Unit	n	t_R 9.57	14.86	18.32	21.29
		Area Counts			
1	5	570970	201690	109030	945290
2	3	483420	258480	103620	970990
3	3	541900	233370	103610	809580
4	3	569940	200170	90874	886980
Rel Std Dev		.076	.125	.076	.079
% Recovery		93.2	89.4	91.6	92.1

Based on ISTD

laboratories to test their performance in sample preparation. Our interest was to quantitate pesticides shown in Figure 6 and Table X and DDD and DDE (Table XI). This application was a comparison of the precision afforded by SFE to that given by an established manual procedure that has been run routinely over a long period of time, with quantitation provided by composite results from multiple participating laboratories. (Having accessibility to a well characterized standard which is also a "real sample" is a real help to evaluating alternative processes.) Note that DDD, DDE, and DDT are endogenous analytes to the matrix while lindane, aldrin, and endrin were spiked. The SFE precision decreases from 2.3% for lindane to 21% for DDT. This number appears to be unacceptable: however, when compared to the manual value of 50% listed in Table XI, it is clear that the precision is much better for the SFE method. Contributing to the value of 21% is a substantial chromatographic error due to poor resolution of the DDE and DDT peaks. Inspection of mean recoveries in Table XI reveals the same pattern among mean recoveries as that for the precision data. That is, both recovery and precision seem to be dependent on the ability to resolve the DDE and DDT peaks. This result is similar to the above results obtained for pesticide replicability on Celite, however the chromatographic conditions were quite different. The values listed in the first four columns in Table XI were reported by EPA for the manual extraction procedure. The values in the two remaining columns were calculated from SFE data.

In these last four applications we have demonstrated the variability of the SFE technique due to repeatability is very small - just as was the case with replicability.

Reproducibility. As a final test for the robustness of the SFE technique, the contribution of reproducibility to variability was determined. Again, the AOAC defines reproducibility as that measure of precision among results of measurements of same sample at different laboratories (Table I). In order to get an accurate picture of reproducibility the same application should be carried out by various laboratories. Although we did not design and carry out an inter-laboratory reproducibility study there were results from several intra-laboratory studies. Table XII contains results of different numbers of replicate extractions performed on three different instruments by three different users at three different times in three different laboratories. There is excellent agreement among precision values obtained by the three users shown in Table XII. What is even more striking about the reproducibility data is the fact that levels of familiarity with SFE technology differed from user to user. For example, User #2 had considerable knowledge of the hardware and little about sample preparation. User #3 was unfamiliar with the SFE technique and hardware, as well as this being his first experience with sample preparation and the subsequent chromatographic analysis. While the information in Table XII is limited experimental data, the relative standard deviations of 2.9% and 3% obtained under these conditions are promising with respect to the ease of use and the high

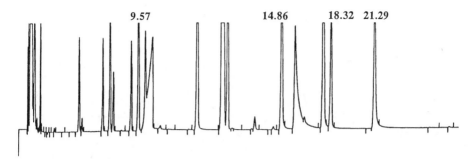

$$T_{ext} = 60\,°C \qquad Solvent = CO_2/500 \; \mu l \; CH_2Cl_2 \; added$$
$$t_{ext} = 20 \; min \qquad Density = 0.85 \; g/mL$$
$$FR = 2.5 \; mL/min$$

Figure 5.   An extracted fraction of a sample of sand spiked with two mixtures: EPA surrogate spike and phenol spike mixtures. Retention times designate individual components selected to evaluate the repeatability measure of the recovery data.

Table IX. Repeatability: **Chlorpyrifos from Grass**[a]

Instrument ID	2	3
Date	04/26/89	10/13/90
n	11	8
Av Area Count	1137	1142
Std Dev	19.71	19.35
Rel Std Dev	.039	.017
% Recovery	99.8	99.2

[a] *1 gram samples, 60ppm levels*

Table X. Repeatability: **Freeze-Dried Fish Tissue**[a]

Run No.	Lindane	Aldrin	Endrin	DDT
		Area Counts		
1	535170	917980	368840	278740
2	539950	931740	402760	280740
3	511760	656950	438090	172910
4	516770	675830	437160	213210
5	537580	691110	522600	261710
6	542850	764770	552070	154290
7	537210	799920	413460	212630
8	540470	823130	446690	132600
9	549690	675830	443140	499270
Av Area Ct	534606	770807	447201	245122
Rel Std Dev				
This work	.023	.136	.128	.21
EPA Cinn	-	-	-	.34
Mean Recovery				
This Work	94	90	100	91
EPA Cinn	-	-	-	68

[a] 1.7 g, EPA Cinn #1254, ppm

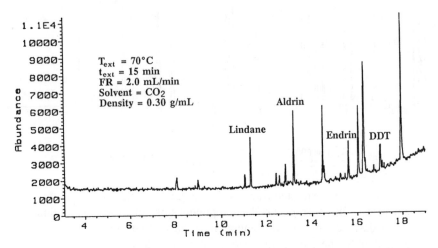

Figure 6.   Total ion current chromatogram of endogenous and spiked
pesticides extracted from an EPA standard fish tissue.

Table XI. Endogenous Pesticides in Fish

	EPA Composite Results for Manual Extraction[a]				HP7680A SFE	
	Mean Recovery (μg/g) (x)	Standard Deviation (s)	Relative Standard Deviation	95% CI (x ± 2s)	Mean Recovery (μg/g) (x)	Relative Standard Deviation
DDD	0.70	0.25	0.36	0.20–1.20	1.00	0.33
DDE	1.72	0.55	0.32	0.62–2.82	1.77	0.23
DDT	0.68	0.34	0.50	MDL–1.36	0.62	0.21

[a] Supplied with the sample material (EPA Quality Control Sample, "Toxaphene in Fish, Sample No. 4").

Table XII. Reproducibility: HP 7680A Performance Evaluation Standard[a]

Analyst	n	Day	Apparatus	Laboratory	Rel Std Dev	% Recovery
WM	5	08/06/90	1A	R&D	.021	99.3
MN	3	10/15/90	2A	Mfg	.030	95.1
EB	3	06/30/90	3A	QA	.029	95.9

[a] 20 ppm Octadecane

precision obtainable by SFE. Other applications focused on more complex samples (both in terms of matrix and analyte(s)) are underway to demonstrate reproducibility by working with other laboratories -- e.g., participating in round robins conducted by NIST and EPA. While octadecane (in the Performance Evaluation Standard) is not a particularly difficult compound to recover from a simple filter paper matrix using SFE, it is one of the compounds included in the NIST sample of diesels on soils and clays (28). The results from the PES checkout procedure function as a useful reference as chemists in collaborating laboratories migrate to more complex matrices and multiple analytes.

Summary

Supercritical fluid extraction (SFE) has become more commonplace in analytical laboratories as an alternative technique to more traditional, manual techniques. While SFE is a relatively new technology to the analytical laboratory, feasibility in terms of affording acceptable levels of recovery (accuracy) has been demonstrated for a real diversity of samples by many researchers. Other workers have evaluated many of the usual technical and practical considerations applied in developing a new technique and the resulting conclusions have not indicated any insurmountable issues associated with those considerations which would preclude the continued development of SFE as a powerful tool for the analytical chemist. One major consideration which has received somewhat minimal attention thus far is that of the robustness of the technique.

Utilizing existing measures of robustness which could be used routinely in reporting results derived from sample preparation methods employing supercritical fluid extraction has been a goal of this study. In particular, this discussion has proposed applying guidelines which are familiar to established analytical laboratories, being part of standard laboratory practices. With this in mind guidelines of the AOAC (Association of Official Analytical Chemists) were used as a vehicle to evaluate the robustness of SFE in the contributing areas of reliability and precision by examining precision and censored reliability data.

The following conclusions have been reached:
1. the AOAC guidelines provide a vehicle for clear, objective evaluation of results,
2. the robustness of supercritical fluid extraction as a sample preparation technique is typically comparable to (and often better than) that of established methods for a variety of samples ranging in complexity and analyte levels, and
3. current instrumentation not only validates the initial engineering premise that SFE would provide a powerful means to automate significant parts of sample preparation methods but also demonstrates that SFE is a technique which can be readily assimilated into the analytical laboratory, providing ease-of-use and a high degree of reliability.

Literature Cited

1. U.S. Patent 2,188,012 (Shell Development Co.; Pilat, S. and Godlewicz, M.), Filed (1936), Granted (1940).
2. U.S. Patent 2,188,013 (Shell Development Co.; Pilat, S. and Godlewicz, M.), Filed (1936), Granted (1940).
3. Giddings, J.C.; Myers, M.N.; McLaren, L. and Keller, R.A. *Science* **1968**, *162*, pp. 67-73.
4. Stahl, E. and Schilz, W. *Chem. Ing. Tech* **1976**,*48*, p. 773.
5. Bowman, L.M. Dense Gas Chromatographic Studies, Ph.D. Dissertation, University of Utah (1976).
6. Williams, D.F. *Chem. Eng. Sci.* **1981**, *36(11)*, p. 1769.
7. Randall, L.G. *Sep. Sci. Technol.* **1982**, *17(1)*, pp. 1-118.
8. Paulaitis, M.E.; Krukonis, V.J.; Kurnik, R.T., and Reid, R.G. *Rev. Chem. Eng.* **1983a**, *1 (2)*, p. 178.
9. Schantz, M.M. and Chesler, S.N. *J. Chromatogr.* **1986**, *363*, pp. 397-401.
10. Li, L. and Kiran, E. *Ind. Eng. Chem. Res.* **1988**, *27*, pp. 1301-1312.
11. Capriel, P.; Haisch A.; and Khan, S.U. *J. Agric. Food Chem.* **1986**, *34(1)*, pp. 70-73.
12. Hopfgartner, G.; Venthey, J.-L. Gulacar; F.O. and Buchs, A. *Org. Geochem.* **1990**, *15(4)*, pp. 397-402.
13. Barry, E.F. and Ferioli, P. *J. HRC and CC* **1983**, *6*, pp. 172-77.
14. Garfield, F.M. *Quality Assurance Principles for Analytical Laboratories;* Association of Official Analytical Chemists, Arlington, VA, "Methods of Analysis" 1984; pp. 61-66.
15. Brogle, H. *Chemistry and Industry* **1982**; pp. 385-390.
16. Lira, C.T. *Supercritical Fluid Extraction and Chromatography,* ACS Symposium Series No. 366, "Physical Chemistry of Supercritical Fluids" 1988, pp. 1-25.
17. McHugh, M. and Krukonis, V. *Supercritical Fluid Extraction: Principles and Practice,* Butterworths, 1986.
18. Tweeten, T.N., Wetzel, D.L., and Chung, O.K. *J. Am. Oil Chem. Soc.,* **1981**, *58(6)*, pp. 664-672.
19. Majors, R.E. *LC-GC,* **1991**, *9(1)*, pp. 16-20.
20. Official Methods of Analysis, 14th Edition, AOAC, Arlington, VA, Sec. 29.001 (1984).
21. Pipkin, W. *Am. Lab.* **1990**, November, pp. 40D-40S.
22. Randall, L.G. *J. Liq. Chromatogr.,* **1986**, *9(14)*, pp. 3177- 3183.
23. Schoeny, E.D. and Rollheiser, J.J. *Am. Lab.,* **1991**, pp. 42-47.
24. Berger, T.A. *Anal. Chem.* **1984**, *61(4)*, pp. 356-361.
25. Westendorf, R.G. "Automatic Analysis of Volatile Flavor Compounds", presented at the Symposium on Recent Developments in the

Characterization and Measurement of Flavor Compounds, 1984 Annual Meeting of the American Chemical Society, Phila., PA, August 26-31, 1984.
26. Lopez-Avila, V. *J. Chromatogr. Sci.,* **1990,** *28,* pp. 468-476.
27. Zahnow, E.W. J. Agric. Food Chem. 1985, 33(6), pp. 1206-1208.
28. Chesler, S. "The Round Robin Test Materials and a Summary of the Results," Consortium on Automated Analytical Laboratory Systems, 2nd CAALS Workshop on Supercritical Fluid Extraction of Solid Environmental Samples, November 8, 1991, NIST, Gaithersburg, MD.

RECEIVED December 2, 1991

Chapter 20

Analytical Applications of Supercritical Fluid Extraction—Chromatography in the Coatings Industry

William J. Simonsick, Jr., and Lance L. Litty

Marshall Research and Development Laboratory, E. I. du Pont de Nemours and Company, 3500 Grays Ferry Avenue, Philadelphia, PA 19146

Supercritical fluid extraction/chromatography can be used to characterize many of the components of today's high performance automobile coatings. For example, aliphatic isocyanates are highly reactive with protic solvents and are not chromophoric, therefore, their analysis poses a formidable task to the analytical chemist. Using carbon dioxide as the mobile phase in conjunction with flame ionization detection allows their quantification in coatings. Supercritical fluid chromatography is used to obtain molecular weight distribution information on nonchromophoric oligomeric materials which are not amenable to routine gel permeation chromatography. Furthermore, we use the accurate quantitative data afforded by SFC to corroborate and complement the qualitative data provided by spectroscopic techniques. We have used supercritical fluid extration to preferentially remove UV-stabilizers and/or polymer additives from cured films. Moreover, supercritical fluid extraction is ideal for removing undesirable matrices to facilitate other spectroscopic methods for identification.

Today's high performance automotive coatings contain a variety of chemical compounds. Coatings are typically composed of a binder system dissolved in an appropriate solvent formulation. In addition, stabilization packages (antioxidants, photostabilizers) are added to lengthen the usable lifetime of the coating. Pigment surfaces are coated with organic dispersing agents to minimize aggregation. The final products are the glamorous long-lasting finishes seen on today's automobiles. Each of the constituents listed above contain a variety of chemical functionalities encompassing a wide range of molecular weights. Hence, the complete characterization of such coatings represents quite a challenge to the analytical chemist.

0097–6156/92/0488–0288$06.00/0
© 1992 American Chemical Society

Solvents are usually mixtures of low-molecular-weight materials of high volatility which are removed from the paint formulation during the curing or drying process. Commercial stabilizers have molecular weights in the range of 300-2000 Daltons (Da) and are therefore, not volatilized during the curing process. Binders are traditionally high-molecular-weight polymers (>50,000 Da). These high-molecular-weight polymers can be synthesized from a host of monomers. Unfortunately, high-molecular-weight polymers require large amounts of organic solvents for dissolution which, during the curing process, are emitted into the atmosphere. Therefore, the use of low-molecular-weight oligomers (<5000 Da) reacted with highly functionalized crosslinkers, have gained popularity.

Higher solids and therefore reduced air emissions, are the benefits of employing reactive low-molecular-weight oligomers with crosslinkers (1). An alternative approach to reducing the amount of organic solvent emissions is by the use of architecturally designed waterborne or water dispersible resins that are synthesized from well-structured oligomeric building blocks. Many of the commercial oligomers, stabilizers, and crosslinkers are solubilized by supercritical carbon dioxide as their molecular weights rarely exceed 5000 Da.

We have several applications of supercritical carbon dioxide in the coatings industry both as an extracting solvent and as mobile phase for chromatography. The molecular weights of many of the coating constituents listed above are well within the solubilizing regime of supercritical carbon dioxide. Furthermore, routine gel permeation chromatography (GPC) is not suitable for low-molecular-weight materials and desorption mass spectrometric methods suffer in their quantitative aspects. Universal flame ionization detection (FID) was used in all chromatography experiments. Although the FID is a carbon counter we define it as a universal detector for organic coatings.

The attributes of supercritical carbon dioxide are well-known (2-4) and several reviews of applications have been published (5,6). Multiple detectors are possible although, FID has been our most successful method because many of the components of organic coatings do not possess a strong chromophore. However, employing FID we find that molecular weight data on nonchromophoric oligomers are possible. No modifiers were used in any of the following experiments. Using this simple system we addressed several problems faced by the coatings chemists.

Several researchers have combined the separating power of supercritical fluid chromatography (SFC) with more informative spectroscopic detectors. For example, Pinkston et. al. combined SFC with a quadrupole mass spectrometer operated in the chemical ionization mode to analyze poly(dimethylsiloxanes) and derivatized oligosaccharides (7). Fourier Transform infrared spectroscopy (FTIR) provides a nondestructive universal detector and can be interfaced to SFC. Taylor has successfully employed supercritical fluid extraction (SFE)/SFC with FTIR dectection to examine propellants (8). SFC was shown to be superior over conventional gas or liquid chromatographic methods. Furthermore, SFE was reported to have several advantages over conventional liquid solvent extraction (8). Griffiths has published several applications utilizing SFC with FTIR detection. Griffiths has compared sampling techniques for combined SFC and FTIR with mobile phase elimination (9). Solvent elimination allows the analyst to use more polar mobile phases, increase sensitivity through unrestricted spectral acquistion time, and eradicates spectral interferences from the mobile phase, especially in the mid-infrared region. Although we have not interfaced our SFC to any of the more elaborate spectroscopic detectors, we find the data afforded by the SFC/FID in conjunction with information provided by other off-line spectroscopic methods adequate for solving many of the problems encountered by today's paint chemist.

SFC provides complementary quantitative data to the structural information afforded by mass spectrometry. Thermally label materials such as isocyanates can be easily analyzed with minimal sample preparation. Supercritical carbon dioxide is nontoxic and can be obtained in high purity as measured by FID. The easy coupling of SFE with SFC makes the selective isolation and quantification of targeted analytes possible. Furthermore, we are in an age of increased environmental awareness. Solvent disposal is discouraged and has become very expensive. The waste disposal costs associated with supercritical carbon dioxide are negligible when compared to the solvent disposal costs generated by traditional Soxhlet methods.

Experimental

All experiments were performed on a Lee Scientific (Dionex Corporation - Lee Scientific Division - Salt Lake City, UT) - 602D Supercritical Fluid/Gas Chromatograph equipped with a 0.5 ml extraction cell. SFC grade carbon dioxide (Scott Specialty Gases, Plumsteadville, PA) was used as the extracting solvent and mobile phase in all experiments. All SFE and SFC investigations were performed under isothermal conditions. Flame ionization detection operating at $325^{\circ}C$ was used in all studies. The specific column, conditions, and parameters are listed in the Applications section.

The 3-(Diethoxymethylsilyl) propyl methacrylate was obtained from Union Carbide Corporation (Tarrytown, N.Y.). Hexamethylene diisocyanate was purchased from Eastman Kodak (Rochester, NY). The Tinuvin stabilizers were obtained from Ciba-Geigy Corporation (Ardsley, N.Y.). Standard solutions containing approximately 1% by weight were prepared in toluene or methylene chloride for direct injection.

Applications

Acrylic Monomers. The physical properties of a polymer are dependent upon the monomers used in the polymerization. Impurities contained in the monomers also affect the physical properties. Acrylosilane resins and crosslinkable emulsions can be prepared using 3-(Diethoxymethylsilyl) propyl methacrylate (I) (10). Free-radical polymerization using (I) will yield a resin with a methacrylate backbone possessing ethoxysilane sites for subsequent crosslinking reactions. Impurities contained in the monomer, specifically higher molecular weight condensation products (II, III), contain multiple vinyl functionality. During a free-radical polymerization such materials cause rapid molecular weight growth, increased viscosity, and possible gelation of the reaction. Hence, the identification and quantification of these condensates are important. Furthermore, structure-property studies are better conducted when the contents of starting materials are known.

Unfortunately, the condensation products exceed the volatility range amenable to gas chromatography (GC). Moreover, (I) does not possess a strong chromophore for easy high pressure liquid chromatography (HPLC) characterization. SFC with flame ionization detection obviates these difficulties.

Without authentic standards or a spectroscopic detection scheme, compound identification is difficult. Our choice of a desorption/ionization method is potassium ionization of desorbed species (K^+IDS) with mass spectrometric detection ($\underline{11},\underline{12}$) which provides a rapid qualitative tool for compound identification. Using K^+IDS, molecular weight data is available and fragmentation is minimal. Ions appear as $M[K]^+$, the mass of the analyte plus 39 Da, the mass of potassium. Hence, structure identification is possible based on a knowledge of starting materials and the molecular weight data afforded by K^+IDS.

Our approach to identifying eluting peaks in the absence of authentic standards has been to match the chromatographic profiles with the K^+IDS mass spectral results since both techniques address similar molecular weight regimes. We also use standards for quantitative comparison when possible. In most cases this approach has been successful. When this approach does not suffice, we resort to time-consuming fraction collection followed by spectroscopic identification.

The profile matching has worked both qualitatively and quantitatively for chemicals such as Tritons and ethylene oxide standards. However, we found quantification of the more volatile low-molecular-weight compounds by K^+IDS was not satisfactory. Therefore, K^+IDS is used to identify the constituents. SFC is subsequently used for quantification.

K^+IDS analysis of commercial (I) showed four peaks with molecular weights greater than (I) (260 Da). The normalized peaks from K^+IDS were 299 Da (100%), 457 Da (0.8%), 485 Da (1.9%), 597 Da (0.2%), and 671 Da (1%), respectively. We suspected that the relative abundance of the condensate products was exaggerated due to the evaporation of (I) under vacuum. Moreover, we suspect artifacts when using K^+IDS due to the high temperatures associated with the technique ($\underline{12}$). Figure 1 is a portion of the SFC chromotogram of (I). We also observe five peaks, four which elute after (I). We expect the condensation products (II, III) to elute after (I). SFC analysis showed that the K^+IDS results did bias the condensation products as we found I (100%), II (0.42%) and III (0.12%), respectively. Although we did not unambiguously identify all the peaks, such data is suitable for quality control/quality assurance (QC/QA) studies.

Structure-property investigations, however, require all compounds be identified and quantified.

The above example shows how we use SFC to furnish accurate quantitative data. In addition, SFC is used to insure that desorption/ionization methods do not yield artifacts particularly when analyzing reactive monomers. Under K^+IDS, the sample under study may be exposed to high temperatures giving rise to decomposition products or the re-association of thermal fragments. Using SFC, the sample integrity is preserved. Therefore, by simply comparing the number of peaks, a rapid evaluation of the desorption mass spectrometric results is provided.

Acrylic Crosslinkers. Butanediol diacrylate (IV) (BDDA) is a popular crosslinker used in the preparation of many polymers used for inks, paints, and plastics. Low-levels of impurities can adversely affect product properties. As previously discussed, K^+IDS provides a powerful qualitative technique, but yields poor quantitative data when analyzing volatile chemicals. BDDA is amenable to analysis by GC, unfortunately any higher-molecular-weight adducts exceed the volatility range amenable to GC. Moreover, BDDA is not chromophoric thereby HPLC characterization is also difficult.

Figure 2 is the K^+IDS mass spectrum of the BDDA (IV). The peak at 237 Da is due to the potassiated BDDA. The peak at 255 is 18 Da higher in molecular weight than (IV). We attribute this ion to structure (V) which is only monofunctional (n=2). Furthermore, this material has an available hydroxyl group for reaction in alkoxysilane formulations. The ion at 309 Da is 72 Da higher in molecular weight than the potassiated BDDA seen at 237 Da. Acrylic Acid (AA) weighs 72, thus we propose the 309 Da ion is due to structure (VI) where m+n=3. In a similar fashion, the ions at 381 Da, 453 Da, and 525 Da are adducts possessing 4, 5, and 6 AA units, respectively. The envelope shifted 18 Da higher in molecular weight (327 Da, 399 Da) is due to monofunctional material (V) possessing 3 and 4 AA units, respectively.

Unfortunately, K^+IDS does not afford accurate quantitative data owing to the volatility of BDDA. Notice the 237 Da ion, [BDDA]K^+, accounts for less than 20% of the total ion current. Unfortunately, most of the volatile BDDA is vaporized prior to K^+IDS analysis. Fortunately, SFC does not suffer from this difficulty as only the analyte's solubility in supercritical carbon dioxide is necessary for analysis. Figure 3 is the chromatogram of the BDDA obtained using supercritical carbon dioxide as the mobile phase with flame ionization detection. Notice the correspondence in peaks between the SFC and K^+IDS results. The purity calculated by SFC is 86.8%. Incidently, GC analysis showed a purity of 92.4%, owing to the fact that nonvolatile higher-molecular-weight adducts were not detected.

To summarize, SFC using FID extends the molecular weight regime currently accessible by GC. Using K^+IDS to identify the constituents of crosslinkers followed by quantitative analysis by SFC provides an accurate measure of the components in commercial

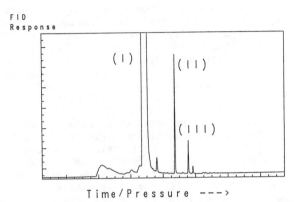

Figure 1.   Chromatogram of 3-(Diethoxymethylsilyl) propyl methacrylate (I). Conditions - 10M SB-biphenyl 30 column, 0.25um film thickness. Program - 200 atm for 10 min ramped to 415 atm at a rate of 10 atm/min. Column temperature - 100°C; Detector temperature - 325°C.

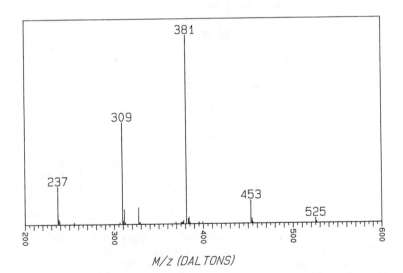

Figure 2.   K+IDS mass spectrum of butanediol diacrylate crosslinker (IV). Scan rate 100 - 1000 Da/sec, average of five scans. Consult reference 12 for specific details.

crosslinkers.   Undesirable monofunctional   materials are   easily
identified and quantified.

Acrylic Macromers. Thus far we have shown applications of SFC to
the characterizations of monomers and crosslinkers.   The next
couple applications will focus upon the analysis of oligomeric
methacrylates, specifically methacrylate macromers.   Methacrylate
macromers are frequently used as building blocks for larger
architecturally designed polymers.   Unfortunately, macromers far
exceed the capability of GC and do not possess a chromophore for
HPLC analysis.   Hatada et. al. has used packed column SFC to
analyzed the stereoisomers of oligomeric methylmethacrylate (MMA)
prepared by anionic polymerization (13).

     MMA macromers were prepared by using a cobalt chain-transfer
catalyst   (CoCTC)   and   azobisisobutyronitrile (AIBN)   as the
initiator   (14).   The   macromers   were   suspected to be low in
molecular weight (<1000 Da), therefore not suitable for accurate
GPC analysis.   K$^+$IDS is useful for structural characterization.
Unfortunately, K$^+$IDS does not yield satisfactory quantitative
data owing to the volatility of the MMA monomer. Fortunately, SFC
does not suffer from this difficulty as only the analyte's
solubility in supercritial carbon dioxide is necessary for
analysis.   Figure 4 shows the K$^+$IDS mass spectra of a MMA
macromer.   The   primary   ion   envelope   is   comprised   of a
distribution following the formula;

$$[M]K^+ = (39 + 100n) \text{ Da} \qquad (1)$$

where n refers to the number of MMA units.

     This envelope is vinyl terminated MMA macromer (VII), the
vinyl group is necessary for the synthesis of larger polymers.
We also observe a small distribution (~3%) following the formula;

$$[M]K^+ = (106 + 100n) \text{ Da} \qquad (2)$$

where n refers to the number of MMA units.

     We attribute this distribution to oligomeric MMA which
possesses an initiator fragment, $(CH_3)_2(CN)C$. This moiety weighs
68 Da, hence we attribute this distribution to a AIBN initiated
MMA oligomer which is vinyl-terminated (VIII).   The relative
ratio (~30/1) of the primary distribution to the secondary
distribution gives a measure of the CoCTC turnover number.
     As previously discussed, K$^+$IDS does not always provide
accurate quantitative data.   Therefore, SFC was performed on the
macromer.   The monomer volatility is not a problem in these
studies.   Figure 5 is chromatogram of the MMA macromer.   The
first large broad peak is due to monomer. Identification was
based upon retention time comparison to a standard.   Each of the
corresponding peaks are due to the oligomers up to the 15-mer.
We attribute the minor distribution to (VIII).

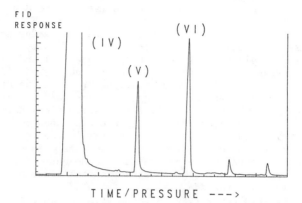

Figure 3.   Chromatogram   of   butanediol   diacrylate   (IV).
Conditions   -   10M   SB-biphenyl   30   column,   0.25um   film
thickness.   Program - 100 atm for 5 minutes ramped to 415 atm
at   10   atm/min.   Column   temperature   -   100°C;   Detector
temperature - 325°C.

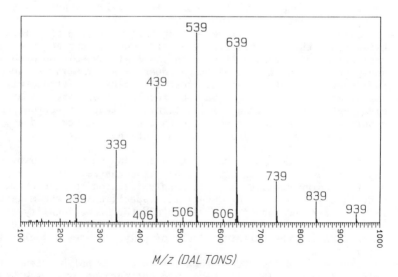

Figure 4.   K⁺IDS   mass   spectra   of   a   methymethacrylate
macromer.   Scan rate 100 -   1000 Da/sec, average of   at least
three scans.

Acrylic Copolymers. A widely used method to modify the physical and chemical properties of polymers is to prepare copolymers that contain monomer units chosen to give the desired properties. For example, copolymers of MMA are used in thermoplastic coatings where improved flexibility or resistance to degradation are needed (15).

Figure 6 is the supercritical fluid chromatogram of a 90:10 methylmethacrylate butylacrylate copolymer. Butylacrylate (BA) reduces the glass transition temperature (16) and improves the thermal stability of MMA. Previously published K^+IDS results showed three oligomer distributions (12). The distributions result from oligomers possessing zero, one, and two BA units in a MMA chain. We observe similar distributions using SFC (See Figure 6). The major distribution corresponds to the oligomer distribution possessing zero BA units in the MMA chain. The second most abundant distribution results from the oligomers which have one BA in the MMA chain. We do not observe a distinct third distribution which we expected from oligomers which have two BA groups in the MMA chain.

We explain these results based upon the specific arrangement of monomers in the oligomer chain. The exact sequence of monomers will affect their solubility in supercritical CO_2, therefore, one expects to see more than one envelope for the MMA oligomers which possess two BA units. In contrast, soft ionization mass spectrometry will not distinguish such isomers. Soft ionization mass spectrometric analysis preceded by SFC separation should yield molecular weight and sequence distribution data on copolymers.

Using SFC, we obtain good separation of the oligomers in excess of the dodecamer. However, we do not achieve good resolution of n-mers differing in chemical composition much past the pentamer. We attempted linear density programming to improve the separation of the higher order oligomers. Unfortunately, no improvement was obtained, but our results illustrate the ability of SFC to analyze low-molecular-weight nonchromophoric copolymers. Molecular weight data can also be computed and the SFC data can be used as a QC/QA tool for these oligomers.

Urethanes Precursors. Polyurethanes can be produced by the reaction of oligomeric diols with diisocyanates. The properties of the polyurethanes are intimately related to the chemicals contained in the starting materials. Specifically, the molecular weight distribution of the diols and the functionality of the isocyanates affect the properties. We have found SFC useful for characterizing the building blocks of polyurethanes, namely diols and isocyanates.

Figure 7 is the SFC chromatograms of two polybutylene (IX) oxide oligomers. The oligomers are not chromophoric and beyond the volatility range amenable to GC. Butylene oxide oligomers (top) were synthesized to achieve a molecular weight average of approximately 600. A molecular weight average of 665 was determined by SFC. The second synthesis was targeted at a slightly higher molecular weight. Experimentally, we observed a molecular weight average of 820. Routine gel permeation

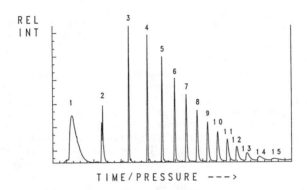

Figure 5.   Chromatogram   of   methylmethacrylate   macromer.
Conditions   -   10M   SB-biphenyl   30   column,   0.25um   film
thickness.   Program - 100 atm for 5 min   ramped to 415 atm at
10   atm/min.   Column   temperature   -   100°C;   Detector
temperature - 325°C.

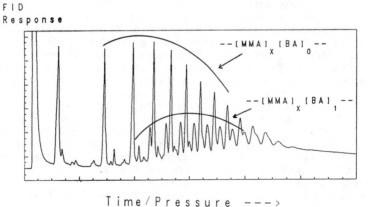

Figure 6.   Chromatogram   of   a   90:10   methylmethacrylate
butylacrylate   copolymer.   Conditions   -   2M   SB-biphenyl 30
development column, 0.25um film thickness.   Program - 100 atm
for   3   min   ramped   to   415   atm   at   10   atm/min.   Column
temperature   -   80°C;   Detector   temperature   -   325°C.

chromatography, the popular approach to obtain molecular weight data, is not suitable for such low-molecular-weight materials. Furthermore, during the synthesis of the polybutylene oxides oligomers, conditions can exist which favor the formation of undesirable macrocycles via a dehydration reaction. Fortunately, SFC resolves the linear from the cyclic oligomers.

Aliphatic isocyanates are popular crosslinkers for hydroxyl containing polyesters, polyethers, etc. The enduse properties of thin films and coatings are determined, in part, by the specific molecular structure of the isocyanate. SFC is a technique which provides these data in a reproducible fashion and is useful for QA/QC (2). Aliphatic isocyanates are reactive with protic solvents and are not chromophoric. Derivatization schemes circumvent this bottleneck for analytical detection (2), however, no time-consuming chemical pretreatment is necessary when using SFC. Figure 8 is the chromatogram of hexamethylene diisocyanate (HDI) (X) which was exposed to moisture. A solution of the HDI in toluene was prepared and injected. No derivatization step was employed prior to SFC analysis. K^+IDS analysis showed constituents having molecular weights of 168, 310, and 486 Da, however, we could not obtain reproducible results due to the reactivity of the HDI. We attribute the peaks to the monomer (X), urea dimer (XI), and biuret (XII). The relative abundancies of (X), (XI), and (XII) by SFC are 44%, 32%, and 24%, respectively. The precision using SFC was under 10% relative standard deviation in all cases.

We have found similar results with other isocyanates, that is, they react with moisture to form the amines which react with isocyanates and build molecular weight. SFC provides a quick check on this process. Furthermore, films crosslinked with a virgin HDI and one crosslinked with an aged material will exhibit very different physical properties. This example shows the utility of SFC for aliphatic isocyanates, but aromatics isocyanates can also be characterized using SFC (2).

Polymer Stabilizers. The useful lifetime of organic coatings are lengthened by the addition of stabilizers. Benzotriazole-type UV-stabilizers (UVA) are added to protect against degradation induced by exposure to sunlight. Hindered amine light stabilizers (HALS) slow the degradation caused by free-radicals. Both UVA and HALS are contained in many of today's high performance finishes. Identification of the stabilizers in coatings are important for durability studies, QA/QC, and in competitive intelligence studies.

Most stabilizers are relatively nonvolatile so they do not vaporize during a thermal curing process. Unfortunately, their low volatility make GC analysis impossible for many stabilizers. HPLC works well for the UVA, but HALS are not easily detected by conventional UV or fluorescent detectors. High resolution capillary SFC was shown to be an ideal separation method for twenty-one polymer additives (17). We chose SFC to characterize stabilizers contained in automotive coatings.

F I D
RESPONSE

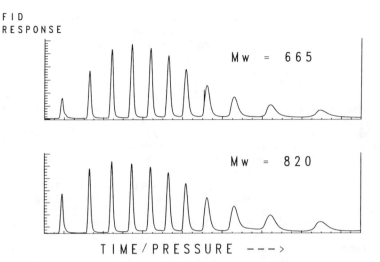

Mw = 665

Mw = 820

T I M E / P R E S S U R E — — ->

Figure 7. (Top) Chromatogram of polybutylene oxide oligomers (IX) targeted at a molecular weight of 600. (Bottom) Chromotogram of polybutylene oxide oligomers (IX) targeted at a higher molecular weight. Conditions - 10M SB-biphenyl 30 column, 0.25um film thickness. Program - 200 atm for 10 min then ramped to 415 atm at 10 atm/min and held at 415 atm for 10 min. Column temperature - 100°C; Detector temperature - 325°C.

F I D
Response

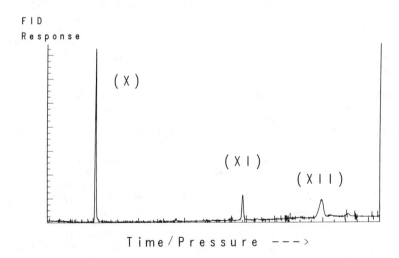

(X)

(X I)

(X I I)

Time / Pressure — — ->

Figure 8. Chromatogram of hexamethylene diisocyanate (X). Conditions - 10M SB-biphenyl-30 column, 0.25um film thickness. Program - 100 atm for 5 min ramped to 400 atm at 10 atm/min, and held at 400 atm for 15 min. Column temperature - 100°C; Detector temperature - 325°C.

Soxhlet methods are traditionally employed to remove stabilizers from cured finishes. Hawthorne has demonstrated supercritical fluid extraction to be ideal for removing materials from complex matrices (4). Because cured organic coatings are crosslinked networks of infinite molecular weight, SFE using CO_2 will not solubilize any of the polymer matrix. However, stabilizers can be quantitatively removed. The extracted compounds can be easily loaded onto a column, separated using CO_2, and detected by FID. Identification is based upon retention time matching with authentic standards. Less than 1 mg of cured paint is necessary to perform the analysis.

Figure 9 is a portion of the chromatogram of the extract from a cured paint film. The coating was extracted at 400 atmospheres (atm) at $125°C$ for twenty-minutes. The chromatographic conditions are described in Figure 9. The peak eluting at about twenty-five minutes corresponds to Tinuvin 440 (XIII) which was identified based upon retention time comparison to an authentic standard. Tinuvin 440 is a HALS and does not have a strong chromophore, but is easily detected using FID. The last peak appearing at thirty-four minutes is due to Tinuvin 900 (XIV), also identified by retention time comparison. We did not identify the peak at twenty-one minutes.

This analysis shows the power of SFE for isolation and concentration of low-molecular-weight additives from complex matrices. The benefits of SFE over conventional Soxhlet extraction methods are fast extraction rate, one step extraction and collection, low temperature operation, less costly, nontoxic solvent, and minimal waste disposal costs. We have found this approach applicable to many coatings, however, high-molecular-weight stabilizers can be difficult to quantitatively extract from highly crosslinked films. Fortunately, using static SFE with a co-solvent to swell the fill or raising the extraction temperature above the glass transition temperature (16) of the film facilitates the extraction process. This work is currently underway in our laboratory.

Isolation of Pigment Additives. Many pigments are coated with a long chain aliphatic organic compound to deactivate the surface and prevent aggregation. The isolation and spectroscopic analysis of such compounds can be challenging because of their relatively low concentration and interfering compounds. SFE can be used to selectively remove interfering materials from the analyte of interest. The analyte can be successfully isolated, collected, and identified by other spectroscopic methods.

Using SFE/SFC we were able to extract from a pigment, a material which gave a single chromatographic peak as seen in Figure 10. We surmise that this peak is due to the compound used to treat the pigment surface. The large earlier eluting peak corresponds to a high boiling solvent, 2,2,4-trimethyl-1,3-pen-tanediol-monoisobutyrate (Texanol). Owing to the large abundance of Texanol we were not able to identify the additives by routine gas chromatography/mass spectrometery.

F I D
Response

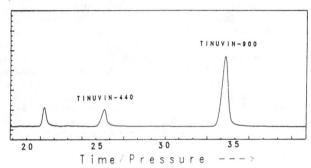

Figure 9. Chromatogram of an extracted paint film.
Conditions - 10M SB-biphenyl 30 column, 0.25um film
thickness. Program - 80 atm for 5 min ramped to 400 atm at 7
atm/min and held at 400 atm for 5 min. Column temperature -
125°C; Detector temperature - 325°C.

F I D
Response

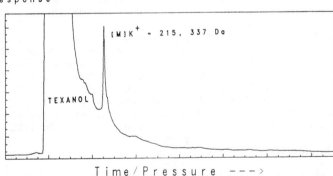

Figure 10. Chromatogram of the extracted pigment. Conditions
- 2.5 M SB-biphenyl - 30 column, 0.25um film thickness.
Program - 100 atm for 5 min programmed at 10 atm/min to 400
atm and held for 5 min. Column temperature - 125°C; Detector
temperature - 325°C.

In order to remove the undesirable Texanol, we extracted the pigment with supercritical carbon dioxide which was held at a low pressure (100 atm) for five minutes. The extract was found to contain Texanol. We increased the carbon dioxide pressure to 400 atm for five minutes to extract the pigment additive. The pressure control in SFE allows us to control the solubilizing power of the carbon dioxide, therefore, perform selective extractions.

Isolation of the pigment additive by SFE followed by K^+IDS analysis showed the additive had components of molecular weights 186 Da and 298 Da, respectively. We performed a literature search, using the Formula Weight Index of the Registry File of Chemical Abstracts Service (18,19) on these molecular weights and found two references each of which is a mixture of fatty acids and their corresponding methyl esters. These mixtures are used to stabilize pigments.

This analysis illustrates the utility of supercritical fluids for rapid selective extraction with no expensive solvent disposal problems, K^+IDS for molecular weight information, and the usage of on-line databases to assist in data interpretation.

Conclusions

We have demonstrated the utility of SFC and SFE using carbon dioxide with FID detection in the coatings area. SFC extends the molecular weight regime which is currently available using GC. Nonchromophoric compounds can be characterized with the universal FID detector. Accurate molecular weight data is provided on nonchromophoric oligomers. MMA macromers qualitatively characterized by K^+IDS are quantified by SFC. Preliminary data on copolymers is also promising. SFC provides quantitative data complementary to the qualitative information afforded by spectroscopic techniques.

SFE is ideal for removing low-molecular-weight additives from complex polymer matrices. Owing to the control of density, and therefore, solubilizing power one can selectively isolate targeted chemicals or remove complex matrices for ancillary methods of characterization. Any method which can be used to separate or simplify the constituents of today's coatings is welcomed by the coatings chemists. Supercritical carbon dioxide has proven to be a valuable tool to the coatings chemist.

Acknowledgements

The authors thank the Automotive Products sector of Du Pont for allowing publication of this work. Dr. M. J. Darmon synthesized the MMA macromers. We thank Dr. M. J. Mahon for his scientific and editorial assistance. Maryann Silva performed all of the mass spectrometric experiments. Audrey Lockton, Christie Connolly, Mary Clavin and Eileen Brennan are acknowledged for their clerical assistance in the preparation of this manuscript.

Literature Cited

1. Armour, A. G.; Wu, D$_{th}$ T.; Antonelli, J. A.; Lowell, J. H. Presented at the 186th American Chemical Society Meeting, Symposium of the History of Coatings, Sept 14, 1989, Paper 38.
2. Lee, M. L. and Markides, K. E. "Analytical Supercritical Fluid Chromatography and Extraction", Chromatography Conferences, Inc., Provo, Utah, 1990.
3. McNally, M.E. and Wheeler, J.R. J. Chromatogr. Sci., 27, 534 (1989)
4. Hawthorne, S.B. Anal. Chem., 62, 633A (1990).
5. Campbell, R.M., In: "SFC Applications," Compiled by K. Markides and M.L. Lee, Brigham Young University Press, Provo, Utah, (1988).
6. Campbell, R.M., In: "SFC Applications", Compiled by K. Markides and M.L. Lee, Brigham Young University Press, Provo, Utah, (1989).
7. Pinkston, J.D.; Owens, G.D.; Burkes, L. J.; Delaney, T.E.; Millington, D.S.; Maltby, D. A. Anal. Chem., 60, 962 (1988).
8. Ashraf-Khorassani, M. and Taylor, L. T. Anal. Chem., 61, 145 (1989).
9. Fuoco, R.; Pentoney, S. L.; Griffiths, P. R. Anal. Chem., 61, 2212 (1989).
10. Bourne, T. A.; Bufkin, B. G.; Wildman, G. C.; Grawe, J. R. J. Coat. Technol., 54, 69 (1982).
11. Bombick, D.; Pinkston, J. D.; Allison, J. Anal. Chem., 56, 396 (1984)
12. Simonsick, W. J. Jr.; Appl. Poly. Sci.: Appl. Polym. Symp., 43, 257 (1989).
13. Hatada, K.; Ute, K.; Nishimura, T.; Kashiyama, M.; Saito, T.; Takeuchi, M. Poly. Bull., 23, 157 (1990)
14. Janowicz, A. H., US Patent 4,694,054.
15. Solomon, D. H. The Chemistry of Organic Film Formers, John Wiley & Sons, Inc., New York, NY, 1967.
16. Beck, K. R.; Korsmeyer R.; Kurz, R. J. J. Chem. Ed., 61, 668 (1984).
17. Raynor, M. W.; Bartle, K. D.; Davies, I. L.; Williams, A.; Clifford, A. A.; Chalmers, J. M.; Cook, B. W. Anal. Chem, 60, 427 (1988).
18. Leiter, D.P.; Morgan, H. L.; Stobaugh, R. E. J. Chem. Document, 5, 238 (1965).
19. Dittmar, P. G.; Stobaugh, R. E.; Watson, C. E. J. of Chem. Inform. and Comp. Sci., 16, 111 (1976).

RECEIVED January 21, 1992

Chapter 21

Systematic Multiparameter Strategies for Optimizing Supercritical Fluid Chromatographic Separations

J. P. Foley[1] and J. A. Crow

Department of Chemistry, Louisiana State University, Baton Rouge, LA 70803–1804

Although supercritical fluid chromatography is now well established, little attention has been given to the method development and optimization of SFC separations of non-homologous, non-oligomeric analytes, in contrast to numerous reports of similar optimizations in GC and HPLC. This is unfortunate since the retention behavior of analytes is generally acknowledged to be more complex in SFC, due to the greater number of experimental parameters and the synergy among them. An organized approach to method development in SFC is discussed, and the importance of multiparameter approaches to optimization is illustrated. Research on the sequential simplex algorithm, recently introduced in SFC, is extended to more variables and more complex samples with analytes that exhibit both LC and GC-like behavior. Simultaneous methods are also discussed, with emphasis on interpretive methods that utilize response surface mapping ("window diagrams"); results obtained from a 3-level, 2-variable (density and temperature) factorial design and a mixture lattice design are reported. Response functions for the evaluation of SFC separations are also described.

The development of chromatographic methods can be a difficult task, especially when supercritical fluids are used as the mobile phase. In the supercritical region, there is a high degree of synergism between temperature and pressure (density), and small changes in one of these parameters can result in large changes in retention. Peak reversals are commonplace as the operating conditions are altered. There are many factors which must be considered when developing an SFC separation, including the proper choice of column, stationary phase, mobile phase, detector and operating conditions. It is essential that the development stage be governed by an organized approach that includes the following steps:

---

[1]Current address: Department of Chemistry, Villanova University, Villanova, PA 19085–1699

0097–6156/92/0488–0304$09.00/0
© 1992 American Chemical Society

1. sample characterization
2. selection of sample introduction method
3. column and stationary phase selection
4. mobile phase selection
5. initial chromatographic run
6. optimization of the separation

Only recently has SFC had the flexibility of instrumentation needed to optimize many parameters such as injection, detection, and column choice. As SFC matures, chromatographers will become better equipped to obtain quality separations through the optimization of many instrumental and operational parameters.

In this chapter we attempt to present an organized approach to method development in SFC, with most of our emphasis on step 6, the optimization of the separation. Although a detailed discussion of sample characterization (step 1) and system selection (steps 2-4) is beyond our present scope, we have included for the novice a brief summary of the choices available for columns, stationary phases, mobile phases, sample introduction, and detection at the beginning of this chapter. We assume the reader is familiar with the basics of SFC; if not, numerous reviews and monographs are available (1-5).

Choices in an SFC System

In order to optimize a separation, one should carefully consider each component of an SFC system, since any one of them may potentially limit the ability to achieve the separation of the analytes in a specific sample. In this section we summarize the choices available for each of the various SFC system components.

Supercritical Fluid. To be useful as a mobile phase in chromatography, a supercritical fluid must have a relatively low critical temperature and pressure, and a relatively high density/solvating power at experimentally accessible pressures and temperatures. The former criterion excludes water and most common organic solvents, whereas the latter excludes such low-boiling substances as helium, hydrogen, and methane. Commonly used fluids are listed in Table I.

Table I. Critical Parameters of Representative Supercritical Fluids

Substance	Symbol	T_c (°C)	P_c (atm)	ρ_c (g/mL)	$\rho_{400\ atm}$
Carbon Dioxide	CO_2	31.2	72.9	0.46	0.96
Pentane	$n\text{-}C_5H_{12}$	196.6	48.3	0.20	0.51
Nitrous Oxide	N_2O	36.5	71.7	0.45	0.94
Sulfur Hexafluoride	SF_6	45.5	37.1	0.74	1.61
Ammonia	NH_3	132.5	111.5	0.24	0.40
Xenon	Xe	16.6	58.4	1.10	2.30
Trifluoromethane	CHF_3	25.9	46.9	0.52	NA[a]
Chlorodifluoromethane	$CHClF_2$	96.0	48.4	0.525	1.12
Dichlorodifluoromethane	CCl_2F_2	111.8	40.7	0.56	1.12

SOURCE: Adapted from ref. (6).
[a] Not available.

Pure fluids. Carbon dioxide is often the mobile phase of choice for SFC, since it has relatively mild critical parameters, is nontoxic and inexpensive, chemically inert, and is compatible with a wide variety of detectors including the flame ionization detector (FID) used widely in GC and the UV absorbance detector employed frequently in HPLC (7). The usefulness of carbon dioxide as a mobile phase in many instances is somewhat limited, however, because of its nonpolarity (8), and many polar compounds appear to be insoluble in it. For a sample containing polar compounds, pure carbon dioxide may not be the proper mobile phase. The elution of polar compounds is often difficult and the peak shapes for these polar compounds are sometimes poor. This latter difficulty is commonly observed with nonpolar supercritical fluids and may be due to active sites on the stationary phase rather than any inherent deficiency in the fluid itself.

As shown in Table I, other fluids besides carbon dioxide are also available for SFC and should be considered during method development. Supercritical pentane has been used as a nonpolar mobile phase, almost exclusively with capillary columns in order to minimize the mobile phase volume and thus the risks associated with flammable gases. Nitrous oxide has recently been shown to be of considerable value for amines (9), and sulfur hexafluoride has been shown to provide excellent selectivity for group-type separations of hydrocarbons (7,10), although it requires the use of a gold-plated FID. Ammonia has been suggested as an alternative to carbon dioxide for polar compounds, but it tends to be fairly reactive with various high-pressure seals on commercial equipment, as well as a potential environmental hazard. The problems of seal degradation (leaks) are compounded by the fact that ammonia can then react with carbon dioxide to form the insoluble salt, ammonium carbonate, which could potentially plug the entire system. Xenon has the advantage of being very inert and completely transparent to infrared radiation, making it ideal for use with SFC/FT-IR. Unfortunately it is very expensive and is not a good solvent for polar compounds. The various freons listed near the bottom have been employed less frequently, although they have shown promise with polar compounds. Chlorodifluoromethane has been reported to be somewhat corrosive with respect to the flame ionization detector, but was much more effective than carbon dioxide in eluting phenolic compounds (11).

Modified Mobile Phases. In addition to pure supercritical fluids, much research has been performed on the use of modifiers with supercritical fluids. That is, rather than switching to a completely different supercritical fluid for the mobile phase, a small percentage of a secondary solvent can be added to modify the mobile phase while (hopefully) maintaining the mild critical parameters of the primary fluid. Through the use of modifiers, one can increase the fluid's dielectric constant, introduce hydrogen bonding, or alter mass transfer characteristics and the solvent viscosity (12). Modifiers allow the chemical tailoring of the mobile phase to meet a specific chromatographic need. Modifiers have been observed to increase solvent strength, enhance selectivity, and improve peak shape and column efficiency (13); improvements in peak shape and efficiency are often due to the deactivation (covering up) of active sites present on some types of stationary phases. The following modifiers have proven to be useful in one way or another with carbon dioxide (1,2,14): various alcohols including methanol, ethanol, isopropanol and hexanol; acetonitrile, tetrahydrofuran, dimethylsulfoxide, methylene chloride, dimethylacetamide, 1,4-dioxane, and even very small amounts of water, sometimes with ion-pairing agents.

A disadvantage of most organic modifiers is their incompatibility with the popular flame ionization detector (FID). We are presently investigating the modification of carbon dioxide with small amounts of highly oxidized, polar compounds that are invisible to the FID, such as formic acid. Our results with formic acid, at concentrations of about 0.3, 0.5, and 0.7% (w/w) in CO_2, may be summarized as follows (15), (Crow and Foley, manuscript in preparation): The addition of formic acid to carbon dioxide drastically reduced the retention and significantly altered the selectivity for several polar compounds, without disturbing the separation of nonpolar analytes (e.g., n-alkanes) to any significant degree. Thermodynamic studies from 55°C to 110°C revealed greater solute-fluid interactions for the polar analytes as well as a more ordered solvation environment, presumably as a result of clustering of formic acid molecules around the polar solutes. For a given amount of formic acid, larger reductions in retention were observed at higher densities and lower temperatures. Just 0.3% (w/w) of formic acid in CO_2 resulted in an average decrease in retention of about 50% for the polar compounds. Retention continued to be reduced upon increasing the formic acid concentration from 0.3% to 0.7% (w/w), but the changes were not significant at lower temperatures. At higher temperatures, however, retention of polar compounds continued to decrease; additional increases in the formic acid concentration at higher temperatures are likely to result in further reductions in retention.

More recently, Engel and Olesik have also concluded that formic acid is a good modifier for carbon dioxide in SFC. The results they obtained on porous glassy carbon stationary phases with 1.5% (w/w) formic acid in CO_2 (16) showed that formic acid was effective because of its strong H-bond donor and very weak H-bond acceptor characteristics; higher concentrations of formic acid (3%), however, were found to polymerize on the porous glassy carbon surface (17).

Polarity of Supercritical Fluids. In order to successfully perform any chromatographic separation, the analytes must be sufficiently soluble in the mobile phase. Efforts to ensure solubility have often been based on matching the "polarity" of the sample components and the mobile phase. For pure fluids, Giddings has reported a polarity classification based on the solubility parameter (18). In contrast to essentially constant values for incompressible fluids (liquids), the solubility parameter (mobile phase strength) of a supercritical fluid varies with its density.

With modified mobile phases, it is somewhat more difficult to estimate their polarity. The polarity can be measured with solvatochromic probes (19), but the results may sometimes be misleading due to specific probe-fluid interactions.

Sample Preparation and Introduction. There are many factors affecting sample introduction which should be considered when developing a separation. First, the sample solvent should be considered, making sure that it is soluble in the mobile phase, and that it solubilizes the sample. Second, the injection pressure and temperature can affect the separation if backflushing either by bulk flow or diffusion occurs. It is important to inject at conditions of lower diffusivity, higher viscosity, and/or higher pressure to prevent this effect. And finally, the injection volume is an important consideration. For packed columns, direct injection of up to about 1 μm onto the column can be used. For capillary columns, the injection volume must be much smaller (< 100 nL for a 10 m x 50 μm i.d. column for a 1% loss in resolution)(20). This can be achieved through split injection, or time-split injection discussed below. Although not always possible for every type of sample, a variety of solute focussing techniques allow much larger volumes to be injected (21).

Two types of injectors are frequently employed. For packed column SFC, a standard six port rotary valve with an external sample loop of 1-10 μL has proven to be quite reliable. For capillary column SFC, a similar rotary valve with an internal "loop" of 0.2 to 0.5 μL is typically employed. Frequently the rotor is pneumatically actuated in a very rapid fashion to allow only a small fraction of sample to be introduced ("time-split"); this is done to avoid column overload. Alternatively, the flow from the injector is split off in the same fashion as in GC. A disadvantage of the latter mode is the potential for sample discrimination.

Another powerful means of sample introduction is direct, on-line supercritical fluid extraction. A small amount of solid (or liquid) sample can be placed in an extraction vessel, followed by the introduction of the supercritical fluid. The material can be extracted at a specific temperature and pressure for a certain amount of time, and then transferred to the analytical column for analysis. As attested by the large number of chapters devoted to it in this monograph, supercritical fluid extraction is a field in its own right, and is playing an increasingly important role in sample preparation (as an alternative to solvent and Soxhlet extractions) as well as means of sample introduction in SFC.

Columns. Both capillary and packed columns have been used successfully in SFC, although the debate over which was better in terms of separating power, quantitative reproducibility, and compatibility with polar analytes was formerly a source of great controversy. Whereas packed columns at present nearly always provide greater efficiency and resolution per unit time (22), the much greater permeability (lower pressure drop per unit length) of capillary columns allows much greater lengths (1-10 m vs 3-20 cm for packed columns) to be employed and correspondingly greater total efficiency (peak capacity) to be achieved; this is sometimes desirable for complex samples. The lower pressure drop also facilitates the use of density or pressure programming. Modifiers are frequently useful for enhancing efficiency (particularly for packed columns), providing coverage of active sites and thereby eliminating the broadening due to secondary retention sites. Recently, better methods of deactivation and coating of porous silica particles for packed columns have renewed hope in the successful use of these columns for polar and basic compounds (23-25). Much larger injection volumes are possible with packed columns since the amount of stationary phase per unit length is much greater, and thus column overload is not as much of a concern as it is with capillary columns.

Although new types of columns will undoubtedly continue to be introduced, at present much research is being performed for the purpose of improving the existing ones. For packed columns the primary goal is the reduction in the number of active sites, while for capillaries it is the reproducible and uniform coating of capillaries with inner diameters closer to the theoretical optimum for mass transfer; the latter will unfortunately require further improvements in sample introduction and detection before the predicted improvements in resolution can be fully realized.

Detectors. One of the nice features of SFC (depending on the mobile phase selected) is the potential ability to use on-line virtually any of the detectors employed in HPLC (before decompression) or GC (after decompression). Although the flame ionization (FID) and absorption detectors (UV-vis) have seen the most widespread use until now, they provide little if any structural information, and applications utilizing mass spectrometers (MS) or FT-IR spectrometers continue to increase, along with those employing element specific detectors (1,26). Detectors that have been employed to date are as follows: flame ionization (FID), absorption (UV-vis), Fourier-Transform infrared spectrometer (FT-IR), mass spectrometer (MS), atomic emission (AE), inductively coupled plasma (ICP), nitrogen and phosphorous thermionic detector (NPD), chemiluminescence, flame photometric (FPD), electron capture (ECD), photoionization, light scattering and nuclear magnetic resonance (NMR).

Chromatographic Principles of Optimization

The efficient utilization of most, if not all, systematic optimization strategies requires an understanding of basic chromatographic principles. Such an understanding also greatly facilitates the interpretation of the results.

Like its two sister column chromatographic techniques (GC, HPLC), resolution in SFC is determined by the product of three terms—efficiency, selectivity, and retention—as shown in equation 1,

$$R_s = \frac{\sqrt{N}}{4} \ \frac{\alpha-1}{\alpha} \ \frac{k'}{1+k'} \tag{1}$$

where N is the number of theoretical plates, α is the selectivity, and k' is the retention (capacity) factor.

Each of the three terms in the fundamental resolution equation can be optimized to improve the separation (27). Retention (k') and selectivity (α) should first be optimized via changes in the density or composition of the mobile phase, the temperature, and the gradients, if any, associated with these variables. Optimization of retention and selectivity is clearly the first step, since the results obtained will indicate if the current mobile phase/stationary phase combination is adequate for the separation being considered. The efficiency of a column, N, is determined by its length and the nature of the stationary phase, including column diameter or particle size. Given the square root dependency of resolution on N, large changes in parameters controlling N (e.g., column length or linear velocity) will result in only moderate changes in resolution, and thus should only be considered if changes in selectivity and retention do not suffice.

Operational Variables. It is interesting to compare the variables available to the analyst for control of retention (and to a lesser degree, selectivity) in GC, HPLC, and SFC once the mobile and stationary phases have been selected. Whereas temperature is the primary variable used to modulate retention in GC, and mobile phase composition is likewise the primary variable in HPLC, *in SFC it is feasible to employ several variables to control retention*—density (ρ) or pressure (P), temperature (T), and composition (χ).

The typical effect of each of these variables on retention is shown in equations 2-4, where the β's are solute-dependent coefficients. The first order coefficients are almost invariably negative for density and composition; they can be positive or negative for reciprocal temperature, depending on whether the solutes exhibit GC-like or LC-like behavior (28), i.e., volatile/poorly solvated or nonvolatile/ highly solvated. The second order coefficients are usually negative and smaller; they are often neglected over the relatively narrow, but practical range of densities or mobile phase composition typically employed for most samples. Assuming all other variables are constant, equations 2-4 show that retention may be reduced by increasing density (or pressure), temperature (T), or the amount of stronger component in the supercritical fluid mixture (modifier). *Since the retention of different solutes does not depend on these variables in precisely the same way* (i.e., the first and/or second order coefficients in equations 2-4 are generally <u>not</u> the same for different solutes), *a change in selectivity frequently accompanies a change in these variables, particularly temperature and composition* (concentration of modifier).

$$\ln k' = \beta_0 + \beta_1 \rho + \beta_2 \rho^2 \tag{2}$$

$$\ln k' = \beta_0' + \frac{\beta_1'}{T} \tag{3}$$

$$\ln k' = \beta_0'' + \beta_1'' \chi + \beta_2'' \chi^2 \tag{4}$$

From thermodynamic and other considerations, density has been shown to be a more fundamental variable than pressure. For those supercritical fluids for which density can be predicted from pressure via an accurate equation of state (e.g., Peng-Robinson), it is easy to attain an accurate and precise density by simply controlling the pressure and temperature. The variation of density with pressure at constant temperature is frequently nonlinear near the critical temperature of the fluid and gradually approaches linearity as the temperature is increased. Although sufficient variation in retention is usually achieved with changes in density, greater variation is usually possible via changes in the mobile phase composition.

Multiparameter Control of Retention and Resolution. Although the key SFC variables—density (pressure), temperature, composition, and their respective gradients—can be utilized individually to vary retention, selectivity, and hence resolution, they can be employed collectively to exert even greater control. The simultaneous use of two or three variables to vary retention and selectivity instead of one is, however, an obviously more complex situation. Under ideal circumstances (in the absence of interaction), the relationships expressed in equations 2-4 might be written as

$$\ln k' = \beta_0{}^* + \beta_1{}^* \rho + \beta_2{}^* \rho^2 + \frac{\beta_3{}^*}{T} + \beta_4{}^* \chi + \beta_5{}^* \chi^2 \tag{5}$$

In reality, however, the synergistic effects of density, temperature, and composition on retention ($\ln k'$) will necessitate the inclusion of one or more cross terms in equation 5, at least for some sets of experimental conditions. Since the precise relationship depends on the parent fluid, the modifier, and the location of the experimental conditions relative to the critical point, it is often difficult to make intuitive predictions about which cross-terms, if any, should be included. Such predictions may become easier as the knowledge base of SFC continues to grow.

Despite the theoretical complications associated with multiparameter control of retention, the experimental rewards of improved resolution and/or shorter analysis times via simultaneous, multiparameter-induced changes in selectivity and/or retention can often be great. The feasibility of multi-variable control of retention in SFC and the interdependence of these variables are among the reasons that a systematic approach to optimization is essential in SFC.

Samples of Homologous or Oligomeric Compounds. As in GC and HPLC, the separation of homologous or oligomeric compounds in SFC is relatively straightforward. Peak reversals generally do not occur and the optimum conditions can generally be predicted from pertinent chromatographic principles. Some differences and other subtleties are worth mentioning, however.

In contrast to GC or HPLC, where a linear gradient in a single parameter (temperature or solvent strength) *is optimal for separating homologues and oligomers, in SFC the best approach for unmodified mobile phases is a combined asymptotic density/linear temperature gradient* (29). This can be explained as follows: First, in earlier work it was shown from chromatographic principles of retention and selectivity of homologues (and oligomers, by analogy) that an asymptotic density gradient is superior to a linear density or pressure gradient because it provides a more uniform band spacing (30). Second, by including a temperature gradient, some compensation is achieved for the reduction in column efficiency resulting from the continual increase in density, i.e., the decrease in solute diffusion coefficients and hence, mass transfer with increasing density. (Compared to GC and HPLC, for which losses in efficiency due to temperature or mobile phase programming, respectively, are rare, the decreased efficiency that occurs with density gradients is a unique disadvantage of SFC, albeit a minor one.)

Although the combined asymptotic density/linear temperature program is usually the best approach, such a gradient is difficult to achieve without computer control of the pump and oven. Even an asymptotic (or linear) density gradient alone is beyond the capability of some SFC pumps. When faced with such limitations, it is reasonable to use a linear pressure program instead. The curvature of the density pressure-isotherms for some supercritical fluids (e.g., pure CO_2) is such that a linear increase in pressure often approximates an asymptotic density gradient, particularly in regions near the critical point where the curvature is pronounced. Except when needed because of insufficient analyte solubility in the pure supercritical fluid, modifiers are rarely employed for the separation of homologues or oligomers, probably for a variety of reasons: First, polar modifiers have been shown to lower the separation (methylene) selectivity of homologues on both moderately polar to polar stationary phases (31). {Similar effects of modifiers have been observed in reversed phase liquid chromatography, and both can be explained on the basis of solubility parameter theory (32).} Second, homologous and oligomeric analytes differ only in the number of methylene and monomer units, respectively. The polarity range of these analytes is thus expected to be fairly small, and sufficient variation in retention can usually be achieved with the aforementioned density/temperature program. Third, accurate equations of state are frequently unavailable for modified supercritical fluids, and the presence of modifiers often precludes the generation of the preferred asymptotic density gradients. Fourth, as noted earlier, most modifiers complicate the detection process and are completely incompatible with the FID, a detector frequently employed for the numerous types of homologous and oligomeric analytes that lack a UV chromophore.

Despite these disadvantages, when modifiers become necessary for solubility or other reasons, the best approach for the separation of homologues or oligomers appears to be a combined pressure/composition gradient (31).

Samples with Non-homologous, Non-oligomeric Compounds.

Optimization of the separation of these samples is much more challenging than samples of homologues or oligomers. Basic chromatographic theory (equations 1-4 and related text) provides little direction for these separations, due to peak reversals that occur almost universally when conditions are changed, particularly temperature or composition. Historically, researchers have generally focused on only one or, at most, two experimental variables at a time, and have chiefly used trial-and-error as their optimization "strategy".

As noted earlier, however, numerous variables have been identified as significant in SFC, including temperature; the type of stationary phase; the polarity, density (or pressure), and modifier content of the mobile phase; and the corresponding gradients of temperature, density (pressure), and composition. Moreover, from chemometric principles it is clear that *any procedure which does not consider all the significant variables simultaneously will seldom, if ever, locate the true set of optimum conditions.* {This point is illustrated in the section below.}

Failure of Univariate Optimization.
In the univariate approach to optimization, all variables but one are held constant at arbitrary values while the remaining variable is changed until an "optimum" response is found. The process is then repeated for each successive variable, using the "optimum" value for any variables that have now been "optimized" and arbitrary values for all the remaining variables except the one that is presently being investigated.

Figure 1 illustrates the failure of the conventional univariate approach in finding the optimum conditions for a hypothetical SFC separation in which the temperature and density are to be optimized. The temperature is held constant at some arbitrary value while the density is varied (points A thru F). When the optimum density at that arbitrary temperature is located (point E), the temperature is then varied while holding the density constant at its "optimum" value (points G thru I) until an optimum temperature is found (point H).

Point H corresponds to the best result obtained by the univariate approach, but it is obviously not the true optimum (point 8). While the univariate approach could be repeated at other initial temperatures in the hope of finding the true optimum, the probability of success is low. Moreover, it takes 9 experiments to find a false optimum with the univariate procedure compared to 8 experiments to find the true optimum using the simplex approach (see later text for a more detailed discussion of the simplex component of this figure).

Clearly what is needed in SFC for samples with diverse, wide-ranging compounds is a systematic approach to optimization, applicable when two or more variables are changed simultaneously. *The remainder of this chapter is therefore devoted to the description of some systematic, multi-parameter approaches, with a natural emphasis on those strategies for which experimental results are available.*

Systematic Approaches to Optimization in SFC

Optimization methods can be classified in several ways, and the choice is largely subjective. For our purposes, it is convenient to categorize them as sequential or simultaneous. A sequential method is one in which the experimental and evaluation stages alternate throughout the procedure, with the results of previous experiments being used to predict further experiments in search of the optimum. In contrast, with a simultaneous optimization strategy, most if not all experiments are completed prior to evaluation. (Note that simultaneous has a different meaning here than in the previous section.)

Numerous approaches for the methodical optimization of SFC separations are potentially available. Virtually any systematic method that has been successfully employed in GC and HPLC is potentially amenable to SFC, given the common analytical features that these chromatographic techniques share. These include factorial design, simplex algorithm, mixture design, window diagrams, mixed lattice, adaptive searches, and step-search designs (33,34). Nontraditional approaches such as the optiplex algorithm, simulated annealing (35-37), and the venerable method of steepest ascent that have so far received little, if any, attention in chromatographic circles may also have promise.

In spite of the obvious problems associated with the trial-and-error method of optimization discussed momentarily (see Figure 1 and related text), the development and/or implementation of systematic optimization strategies in SFC is at a very early stage, with the majority of the above approaches as yet untested. The reason that the systematic optimization of SFC separations has not yet caught on in SFC is unclear, in light of the relatively large number of SFC applications that have been published. In the remainder of this chapter, we limit our discussion to methods whose potential in SFC has been demonstrated, i.e., to those methods that have actually been applied in SFC. We see little merit in discussing the other systematic approaches in general, hypothetical terms; we encourage any interested reader, however, to try some of these untested strategies and report the results at the earliest possible date.

Including the original results reported later in this chapter, only two systematic optimization procedures have been reported: simplex (38), (this chapter) and window diagrams (39), (this chapter). Due to space limitations, a somewhat greater emphasis will be placed here on original results rather than those published elsewhere. Experimental details common to the simplex and window diagram results obtained from our laboratory are summarized here for the sake of convenience and continuity of discussion.

Experimental (simplex and window diagram). The chromatographic system consisted of a Model 501 supercritical fluid chromatograph (Lee Scientific, Salt Lake City, Utah) with the flame ionization detector (FID) set at 375°C. The instrument was controlled with a Zenith AT computer. A pneumatically driven injector with a 200 nL or a 500 nL loop was used in conjunction with a splitter. Split ratios used were between 5:1 and 50:1, depending on sample concentration and the chosen linear velocity, while the timed injection duration ranged from 50 ms to 1 s. We found that the variation of both the split ratio and injection time allowed greater control over the

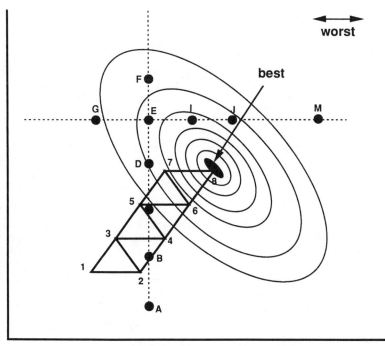

Figure 1. Comparison of univariate (one-variable-at-a-time) and simplex optimization of a simple response surface with two interdependent parameters (density and temperature). Ellipses represent contours of the response surface. See text for additional discussion.

amount of solute transferred onto the column. Data were collected with an IBM-AT computer using Omega-2 software (Perkin-Elmer, Norwalk, CT). The simplex program and the response function calculation programs were written in TrueBASIC (TrueBASIC Inc., Hanover, N.H.). Several capillary columns were used, ranging from a 0.55 meter to a 10 meter SB-Biphenyl-30 (30% biphenyl, 70% methyl polysiloxane), with a 50 μm internal diameter and a film thickness of 0.25 μm. The mobile phase was SFC grade carbon dioxide (Scott Specialty Gases, Baton Rouge, LA).

Simplex Optimization. The sequential simplex method is an example of a sequential multivariate optimization procedure that uses a geometrical figure called a simplex to move through a user-specified of experimental conditions in search of the optimum. Various forms of the simplex have been successfully used in different modes of chromatography, particularly HPLC (40-42) and GC (43-46).

In the simplex method, the number of initial experiments conducted is one more than the number of parameters (temperature, gradient rate, etc.) to be simultaneously optimized. The conditions of the initial experiments constitute the vertices of a geometric figure (simplex), which will subsequently move through the parameter space in search of the optimum. Once the initial simplex is established, the vertex with the lowest value is rejected, and is replaced by a new vertex found by reflecting the simplex in the direction away from the rejected vertex. The vertices of the new simplex are then evaluated as before, and in this way the simplex proceeds toward the optimum set of conditions.

The simplex optimization of a hypothetical SFC separation in which the density and temperature are to be optimized is illustrated in Figure 1, along with the univariate approach described earlier in this chapter. In this example, since there are two parameters to be optimized (density and temperature), the initial simplex consists of three vertices, i.e., the simplex is a triangle. Thus it is necessary to perform three initial experiments, corresponding to points 1-3 in Figure 1. Although all three vertices could be selected arbitrarily, it is arguably easier to specify the following— (i) one set of conditions; (ii) the step size; and (iii) the shape of the simplex—and let the algorithm itself select the remaining initial vertices. (This is particularly true for higher-order simplexes.)

Following the establishment of the initial simplex, the simplex begins moving away from the conditions that give the worst result (as described above) and systematically reaches the optimum set of conditions that yield the best separation (vertex 8). In contrast to the false optimum (point H) found by the univariate approach described earlier, the simplex method has located the true optimum density and temperature for this separation.

Details on the simplex algorithm are available elsewhere (33,34,47-50). The advantages and disadvantages of the simplex method pertinent to chromatographic applications are summarized in Table II.

With respect to the last deficiency in Table II, two options are available that will enhance the chances of finding the global optimum: (i) using a modified simplex which allows other movements besides reflections, such as expansions and contractions; and (ii) restarting the simplex in a different region of the parameter space. If the same optimum is found after restarting the simplex in a different area, it is more probable that the global optimum has been found. Both options were employed for the results reported here as well as those reported previously (38).

General aspects of the simplex method. Although the simplex algorithm can in principle be employed for the optimization of any kind or number of parameters of a particular process, for chromatographic applications it appears to be better suited for certain types of a limited number of variables.

Table II. Summary Evaluation of the Simplex Method for Optimization of Chromatographic Separations

Advantages

- little chromatographic insight is required
- computational requirements (relative to other statistical strategies) are minimal
- any number of parameters may be considered
- location of a true optimum (vs false optimum via univariate approach)

Disadvantages

- a large number of experiments may be required to find an optimum
- little insight into the response surface is provided
- a local rather than a global optimum may be found (33)

With regard to the *kind* of variable, because of its interactive, sequential nature, the simplex approach works best with continuous operational variables that can be changed rapidly (and ideally, automatically) between experiments. Thus, initial pressure is an ideal variable since it is continuously variable and easy to change between chromatographic runs. On the other hand, column/stationary phase parameters (phase type, polarity, length, particle diameter, etc.) are less suitable as simplex variables since they can only be changed to a small number of discrete values that may not always correspond to the conditions required for the next experiment; equally as important, the changing of these parameters (i.e., changing columns) would be time-consuming and difficult if not impossible to automate. Table III groups the important SFC variables into the two respective categories.

Table III. Suitability of Experimental Variables in the Simplex Optimization of SFC Separations.[a]

Ideal Variables	Less Suitable Variables
initial pressure (or density)	identity of the supercritical fluid
initial temperature	modifier identity
amount of modifier (mobile phase composition)	stationary phase type/polarity
gradients of each of the above	phase/column dimensions

[a] Criterion for evaluation of suitability: ability to be changed continuously, rapidly, and in an automated fashion.

One way to include the less desirable variables of Table III in the simplex optimization process would be to perform a simplex optimization for each of a limited number of user-selected combinations of the less suitable variables. The optimum results obtained for these selected combinations could then be compared, with the best overall result indicating which combination of fluid, modifier, stationary phase, etc. is best for that particular separation. A drawback of this approach, particularly if the individual optimization procedure is not automated, is its time-consuming, labor intensive nature.

With regard to the *number* of variables, three effects are observed almost universally as more parameters are considered (33): First, the optimum response improves asymptotically or logarithmically. Second, the number of experiments required to locate an optimum increases rapidly, usually exponentially. Third, the response surface becomes more complex, i.e., with potentially more local optima, and thus the global optimum becomes more difficult to locate and is also less robust.

Although the first effect is a positive one, it is usually overshadowed by the second and third (negative) effects which are generally much larger in magnitude. For this reason *it is advisable to limit the variables to only the ones believed to be important for the particular separation of interest.* To reiterate, when rapid global optimization and/or robustness of the optimum separation is desirable, the key is to consider only the most important variables. If certain variables can logically be excluded based on the retention/selectivity characteristics of the components of a particular sample, then by all means they should be.

The desire to restrict the number of variables when using the simplex algorithm introduces an interesting problem: Of the variables listed in Table III, how many should be included in the simplex procedure, and which ones? Clearly, from our earlier discussion the variables in the second column of Table III can be excluded, but that still leaves 6 "ideal" parameters: pressure (or density), temperature, modifier composition, and their respective gradients. How should one select from among these six parameters, since any of them may be important for a given sample?

In order to consider this question further, it is helpful to recognize all of the choices that are available, so as to put the question in perspective. Shown in Table IV is a complete listing of the possible combinations of the six ideal variables of Table III, from two-at-a-time to five-at-a-time. Before attempting to interpret the results, the reader should be aware of three important points: First, there is no reason to expect that for a given number of variables, the optimum result for all combinations will be equally as good. Second, it is not always safe to assume that the best result from the combination of n+1 variables will always be better than the best result from the combination of n variables, although as discussed above there is generally a slight trend along these lines. Finally, although it is convenient and reasonable to equate density with pressure (i.e., at constant temperature one cannot be varied without the other), it is not strictly correct to ascribe a one-to-one relationship to pressure and density gradients since pressure/density isotherms are nonlinear as noted earlier. Expressed another way, an optimization performed with linear density gradient as one of the variables will very likely give a different result than the same optimization with a linear pressure gradient.

Obviously the number of choices in Table IV is so large as to be almost bewildering. Nevertheless, the information is useful in a couple of ways. First, assuming (i) the viability of every combination; and (ii) the need to compare the results of different combinations with the same number of variables, Table IV suggests that it is better to use either a small (2) or large number (4, 5, or 6) of the total possible variables in the optimization, since an intermediate number (3) provides the greatest number of combinations and require comparisons. Second, Table IV illustrates the wealth of opportunities in the control and optimization of the retention/separation process in SFC. That is, the same long list of combinations in Table IV that potentially makes the selection of variable combinations in SFC difficult may also be the chromatographer's salvation in achieving the desired separation.

In some instances, the number of variables, and hence, the number of subsequent choices are reduced by equipment or user-specified constraints. For example, much of the commercial and home-made equipment is designed to accommodate only pure supercritical fluids or those with a fixed amount of modifier. That is, with some instrumentation the variables of modifier composition and modifier gradient are not convenient to consider. Under such instrument-limited circumstances, those combinations in Table IV that contain either of these variables (X or ΔX) could simply be ignored, at least on a practical level. Such a constraint would reduce the fifteen 4-variable combinations of Table IV to only one that we will discuss shortly: P, T, ΔP, and ΔT.

Table IV. Combinations of Experimental Variables for Simplex or Other Multiparameter Optimization in SFC*

	2-variable	3-variable	4-variable	5-variable
1.	P, T	P, T, X	P, T, X, ΔP	P, T, X, ΔP, ΔT
2.	P, X	**P, T, ΔP**	P, T, X, ΔT	P, T, X, ΔP, ΔX
3.	P, ΔT	P, T, ΔT	P, T, X, ΔX	P, T, X, ΔT, ΔX
4.	P, ΔP	P, T, ΔX	**P, T, ΔP, ΔT**	P, T, ΔP, ΔT, ΔX
5.	P, ΔX	P, X, ΔP	P, T, ΔP, ΔX	P, X, ΔP, ΔT, ΔX
6.	T, X	P, X, ΔT	P, T, ΔT, ΔX	T, X, ΔP, ΔT, ΔX
7.	**T, ΔP**	P, X, ΔX	P, X, ΔP, ΔT	
8.	T, ΔT	P, ΔP, ΔT	P, X, ΔP, ΔX	
9.	T, ΔX	P, ΔP, ΔX	P, X, ΔT, ΔX	
10.	X, ΔP	P, ΔT, ΔX	P, ΔP, ΔT, ΔX	
11.	X, ΔT	P, X, ΔP	T, X, ΔP, ΔT	
12.	X, ΔX	P, X, ΔT	T, X, ΔP, ΔX	
13.	**ΔP, ΔT**	P, X, ΔX	T, X, ΔT, ΔX	
14.	ΔP, ΔX	P, ΔP, ΔT	T, ΔP, ΔT, ΔX	
15.	ΔT, ΔX	P, ΔP, ΔX	X, ΔP, ΔT, ΔX	
16.		P, ΔT, ΔX		
17.		X, ΔP, ΔT		
18.		X, ΔP, ΔX		
19.		X, ΔT, ΔX		
20.		ΔP, ΔT, ΔX		

*Variables in Table III from which combinations were made: P = initial pressure, T = initial temperature, X = initial modifier, ΔP = pressure gradient, ΔT= temperature gradient, and ΔX = modifier gradient.

An example of a user-specified constraint might be one that excludes gradients of any type (pressure, temperature, or modifier) from the SFC method to be developed. Such a constraint is plausible in a QC or other high-throughput laboratory where it is important to minimize the time between analyses by eliminating the delays required to return to initial conditions.

It is clearly beyond the scope of this chapter to consider further the selection of which variables to use in the simplex optimization. To summarize our own relatively limited experience, however (boxes in Table IV represent combinations examined to date), we recommend the following: For a relatively simple separation, begin with a two-parameter simplex that includes either initial pressure (or density), using as many characteristics of the analytes and/or sample matrix to logically deduce which remaining variable to optimize. For a more complex separation, or one in which little is known about the sample, try a 4 or 5-variable simplex that includes the initial pressure and pressure gradient (or initial density and density gradient) as optimization variables.

Simplex Optimization Criteria. For chromatographic optimization, it is necessary to assign each chromatogram a numerical value, based on its quality, which can be used as a response for the simplex algorithm. Chromatographic response functions (CRFs), used for this purpose, have been the topics of many books and articles, and there are a wide variety of such CRFs available (33,34). The criteria employed by CRFs are typically functions of peak-valley ratio, fractional peak overlap, separation factor, or resolution. After an extensive (but not exhaustive) survey, we identified two CRFs that are straightforward and easy to use. We intentionally avoided the more complicated CRFs that include factors of maximum analysis time, minimum retention time, or other arbitrary weighting factors. As discussed by Schoenmakers (33), these complex CRFs are neither as versatile nor as desirable as previously believed. The "multiple" weighting factors of these CRFs can usually be reduced to a single weighting factor simply by rearrangement of the CRF.

One appropriate CRF uses a threshold criterion based on resolution between peaks (R_S), given by the equation:

$$R_{S,min} \geq x \qquad\qquad CRF\text{-}1 = \frac{1}{k_\omega} \qquad\qquad (6a)$$

$$R_{S,min} < x \qquad\qquad CRF\text{-}1 = 0 \qquad\qquad (6b)$$

In equation 6a, k_ω is the capacity factor for the last peak (retention time may be used instead when t_0 is not constant over the range of experimental conditions), and $R_{s,min}$ is the minimum acceptable resolution set arbitrarily by the user. CRF-1 favors chromatograms with a resolution greater than the arbitrary value, "x", for all peaks in the shortest amount of time possible. For chromatograms where $R_{s,min} < x$ for any pair of peaks, the response is set to zero. If $R_{s,min} \geq x$ for all pairs of peaks, the response is set equal to $1/k_\omega$. Thus as analysis time decreases, the response function value increases provided that the resolution does not fall below the threshold value. For our analyses, $R_{s,min}$ was chosen to be unity. A different value may be more appropriate under other circumstances.

An inherent problem with CRF-1 is its inability to distinguish between chromatograms with a resolution below the threshold. All such chromatograms would have a value of zero, among which the algorithm could not differentiate. A more continuous CRF is therefore frequently desirable.

A second CRF that we therefore considered is a continuous one based on the ratio of peak height to valley depth, i.e.,

$$CRF\text{-}2 = \frac{\prod_{i=1}^{n}\sqrt{P_{i,i-1}\cdot P_{i,i+1}}}{t_w} \qquad\qquad (7)$$

where, for the ith peak, $P_{i,i-1} = 1 - \frac{v_{i-1}}{h_i}$ and $P_{i,i+1} = 1 - \frac{v_{i+1}}{h_i}$ as shown in Figure 2.

Like CRF-1, CRF-2 also favors short analysis times and well resolved peaks. However, there is no threshold value for resolution, and the compromise between resolution and analysis time is not as well defined as in CRF-1. Inclusion of analysis time in the denominator of an objective function may result in the loss of some resolution, compensated for by a rapid analysis time (51). This is likely to occur to some extent when the resolution threshold is defined in excess of 1 to 1.25, since the peak-valley ratio utilized by CRF-2 does not diminish to an appreciable extent until the resolution falls below this range. Note that as the resolution drops below 1 to 1.25, however, CRF-2 decreases rapidly, and it is unlikely that a short analysis time will compensate for such poor resolution (33). Nevertheless, if a user-specified minimum resolution is an *absolute* requirement, it is probably better to use a threshold criterion such as CRF-1 in which the desired resolution is stipulated by the user.

While CRF-2 is a good criterion to use, an increase in the number of peaks during the optimization process could result in a lower response for CRF-2, if the peaks were collectively less well separated than the smaller group of peaks. There are two solutions to this problem: (i) restart the simplex from the vertex in which the number of peaks increased; or (ii) modify the CRF so that an increase in the number of peaks is allowed for. As shown in equation 8, the modification of CRF-2 is straightforward:

$$CRF\text{-}3 = n + \frac{\displaystyle\prod_{i=1}^{n}\sqrt{P_{i,i-1}\cdot P_{i,i+1}}}{t_w} \tag{8}$$

where n = the number of peaks observed in the chromatogram. Note that CRF-1 could be modified in the same way.

Minimizing the Time Required for Optimization. For a new sample, the time required for analysis must include the development of the method for separation. In the case of SFC, a great amount of time can be saved by performing the optimization (method development) with a very short column (52). This can be done efficiently in combination with the modified simplex algorithm described above or with other optimization procedures described later. Additional time will of course be saved if this short column proves to be sufficient, once the optimum conditions are located, for the final analysis. If, after optimization of retention and selectivity, the current column does not provide the required resolution due to insufficient efficiency, the resolution can be increased by a factor y via a y^2 increase in column length (equation 1, assuming N is proportional to length). If a gradient is being used, the gradient rate should be decreased appropriately. Assuming that the relationship between ln k and the variable of interest (density, pressure, or temperature) is linear, the gradient rate should be decreased by y^2 (53). Note that although an increase in column length results in a proportional increase in analysis time, hours if not days of analysis time have already been saved by first optimizing the separation on a short column.

If the separation is still unsuitable after optimizing the experimental conditions and column length, selectivity must be optimized further by changing the stationary phase, the type of column, or the mobile phase by changing it or adding a modifier. These choices were described earlier in this chapter; because the changes needed depend so heavily on the nature of the analytes and/or sample matrix, it is difficult if not impossible to provide concise, general recommendations.

Simplex Optimization Results. Of the 56 combinations of variables for use with the simplex in Table IV, only the 4 combinations highlighted with boxes have been utilized. Clearly the optimization of SFC separations via the simplex algorithm is still in its infancy, as are all other systematic methods of optimization for SFC. Nevertheless, as described below, the results provided by the simplex approach were quite good. The results of the 2 and 3-parameter simplexes are especially informative for the novice because their movement can be visualized in 3-dimensional space, in contrast to simplexes of four parameters and higher which cannot be depicted graphically.

Two and Three-Parameter Simplexes (38). Three combinations of variables were examined: (a) density gradient/temperature, (b) pressure gradient/temperature gradient, and (c) initial density/density gradient/temperature. Two samples were tested: (i) a synthetic mixture of 6 diverse, low-to-medium molecular weight solutes of varying polarity and functionality (acetophenone, propiophenone, bicyclohexyl, biphenyl, undecylbenzene and benzophenone); and (ii) a mixture of 3 difficult-to-separate sesquiterpene lactones—glaucolide A, burrodin, and psilostachyin A.

Importantly, good separations for both samples were obtained under conditions that could not be identified intuitively. Interestingly, the results for the different types of 2-variable simplexes were similar, although difficult to compare quantitatively because of differences in flow rate due to the use of different frit restrictors.

Four-Parameter Simplex. We have recently tested a four-parameter approach (density, temperature, their respective gradients) on samples that contain both volatile and nonvolatile analytes and therefore give complex changes in elution order as experimental conditions are changed. The success of the 4-parameter simplex in optimizing the separation of such samples is illustrated in Figure 3 and Table V for a 12-component sample containing alkanes, polyaromatic hydrocarbons, amines, ketones, esters, etc. Most significant, the optimization provided an increase in the number of peaks resolved from 10 to 12 (the 10-peak chromatogram is not shown). The provision for an increase in the number of peaks was an important part of CRF-3 (equation 7), the CRF utilized with these chromatograms. Note that to obtain baseline resolution for all components, either a longer column or a slower flow rate can be employed as described previously (38).

From Table V it can be seen that the last vertices in the simplex are not always the ones with the best response. This is due to the fact that once an optimum region is reached, the simplex may begin to step outside of the optimum and "circle" it, sporadically moving back to the optimum or a near-optimum every few steps. This is precisely what is illustrated in Table V, where the boxed-in vertices represent the six best results. Statistical treatment of these data indicate that a consensus has more or less been reached for the optimum conditions, particularly the initial density and initial temperature. One should not infer, however, that the mean values for all 4 parameters should be used instead of the (best) values corresponding to vertex 13. We have yet to perform sufficient comparisons to draw such a conclusion.

In the above 4-parameter simplex example, a linear density gradient was employed exclusively. For samples that contain some homologues or oligomers in addition to other more diverse compounds, it may be slightly advantageous to utilize an asymptotic density gradient instead. Although linear density gradients may result in a more predictable solvent strength program, as discussed earlier asymptotic density gradients have been shown to give better separations (band-spacing) of later eluting oligomeric peaks of higher molecular weight samples (30), particularly when a temperature gradient is performed simultaneously (29). Unfortunately, asymptotic gradients are more tedious to generate experimentally, and this disadvantage would be exacerbated during the course of a simplex run. As noted earlier, however, asymptotic density gradients can sometimes be approximated by a linear pressure program which is easier to achieve experimentally.

The simplex approach to SFC optimization has proven to be extremely helpful with real samples as well as synthetic mixtures. In recent work with Arochlor 1254 (Crow and Foley, manuscript in preparation), a mixture of polychlorinated biphenyls (PCBs), the number of resolved peaks increased from 19 to 28 in going from vertex 1 to vertex 21 (optimum). Resolution was improved sufficiently that baseline resolution of most if not all compounds could have been obtained by transferring the optimized method to a longer column as described earlier in the section on Minimizing the Time Required for Optimization.

Termination of the Simplex. The decision of when to stop the simplex is somewhat subjective. Upon reaching the vicinity of the optimum conditions, it is advisable to discontinue the simplex, as many experiments could be wasted as described by Schoenmakers (33). On the other hand, it is important to conduct a sufficient number of experiments to establish that the simplex is in this vicinity. Although statistical comparisons of the best results (e.g., best 20% as in Table II) may prove to be helpful, it should also be recognized that *it is not always necessary to fully optimize a given separation.* In our view, whenever a set of experimental conditions that provides the desired separation within the maximum specified analysis time has

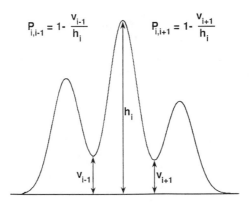

$$P_{i,i-1} = 1 - \frac{v_{i-1}}{h_i} \qquad P_{i,i+1} = 1 - \frac{v_{i+1}}{h_i}$$

Figure 2. Measurement of the peak valley ratio used in several chromatographic response functions (equations 6 and 7).

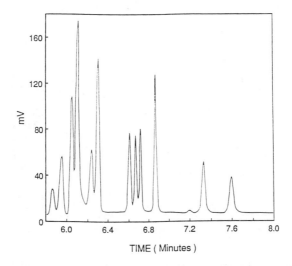

Figure 3. Four parameter, simplex-optimized SFC separation of a 12-component mixture. Chromatographic conditions as in Vertex 13 of Table II. Sample components: isoquinoline, n-octadecane (n-$C_{18}H_{38}$), naphthalene, quinoline, acetophenone, undecylbenzene, benzophenone, 2'-acetonaphthone, diphenylamine, o-dioctylphthalate, unidentified impurity, N-phenyl-1-naphthylamine, phenanthrene quinone. Other conditions as described in the experimental section.

Table V. Four Parameter Simplex Optimization Using CRF-3 (Equation 8) To Account For An Increase in the Number of Observed Peaks.

Vertex	Density (g/mL.	Temp. (°C.	Dens.Rate (g/mL/min.	Temp.Rate (°C/min.	Response (CRF-3)
1	0.15	75	0.20	5	11.155
2	0.34	86	0.22	9	11.398
3	0.19	86	0.29	9	12.006
4	0.19	121	0.22	9	11.055
5	0.19	86	0.22	24	12.007
6	0.24	45	0.25	14	9.011
7	0.21	102	0.23	11	12.039
8	0.31	105	0.28	21	11.056
9	0.19	83	0.22	9	12.029
10	0.06	92	0.26	17	-∞ a
11	0.27	88	0.23	11	12.016
12	0.23	93	0.16	18	12.068
13	0.25	97	0.09	22	12.072
14	0.26	99	0.16	3	11.103
15	0.21	89	0.21	18	12.046
16	0.17	98	0.14	19	11.041
17	0.24	90	0.21	13	12.035
18	0.27	107	0.15	23	11.051
19	0.21	89	0.20	13	12.060
20	0.20	98	0.16	19	12.033
21	0.23	92	0.20	15	12.018
22	0.25	81	0.12	23	12.011
23	0.22	97	0.20	14	12.048
24	0.22	94	0.15	19	12.024
25	0.22	93	0.17	18	12.061
26	0.24	99	0.12	15	12.041
27	0.22	91	0.19	17	12.008
28	0.23	88	0.12	21	11.188
29	0.22	95	0.18	16	12.049
30	0.23	95	0.13	17	12.047
$\bar{x}$ (6 best.	0.225	94.0	0.167	16.8	12.060
% RSD	6.1	3.2	24.5	19.3	0.08

a A boundary violation occurred, i.e., movement of the simplex to this vertex requires a pressure, density, and/or temperature that is not experimentally accessible. An infinitely negative response was given to this vertex to force the simplex to move in a different direction.

been found, the optimization procedure (method development) can be halted, even if the region located is not close to the optimum. What matters most is that the separation criteria are fulfilled and, to a lesser extent, that as robust a region (set of conditions) as possible is found. The latter guarantees that the separation will be *relatively insensitive* to small changes in the experimental variables. {Given the complexity of multi-component response surfaces, fairly precise control may still be necessary at the optimum.} Obviously more effort should be expended in optimizing what will become a routine analysis than on a one-time only separation.

Future work. As mentioned earlier, use of the simplex algorithm for the systematic optimization of SFC separations is still in its early stages. The success already achieved, however, merits continued research along these lines. Research opportunities include: (i) extension of the simplex method to less ideal variables and/or greater than 4 variables; (ii) investigation of the benefits of the simplex method to packed columns and modified mobile phases; and (iii) development of the capability to predict, *for a given type of sample, the best combination of variables* to optimize.

Optimization via Window Diagrams. The window diagram is an example of a simultaneous optimization method. Simultaneous methods are characterized by an initial, experimentally-designed collection of data, followed by an evaluation of the data. An important advantage of simultaneous approaches over sequential methods such as the simplex is the fact that simultaneous approaches, if properly utilized, will always locate the global optimum (within the user-specified parameter space). In contrast, although our implementation of the simplex (discussed earlier) was designed to be as global as possible (modified, with restarting), it was nevertheless potentially susceptible to local optima.

One type of simultaneous approach, termed an exhaustive grid search, requires the collection of a very large number of data points throughout the experimental set of conditions at regular intervals in order to map the response surface. Response surfaces are usually complex (complexity will increase with the number of solutes and the number of variables to be optimized), and thus require a large number of data points for accurate mapping.

A subset of simultaneous methods which overcomes the difficulty of mapping complex response surfaces by an exhaustive series of experiments are the interpretive methods, in which retention surfaces are modeled using a minimum number of experimental data points. Retention surfaces thus obtained for the individual solutes are then used to calculate (via computer) the total response surface according to some predetermined criterion. The total response surface is then searched for the optimum.

Alternatively, it is also possible to model other parameters such as resolution or the CRF's in equations 6-8 and 10-12 instead of retention. Although this may seem to be a more direct approach, it usually has a couple of important disadvantages. First, due to their greater complexity and irregularity, response surfaces are generally much more difficult to model with equivalent accuracy than the underlying retention surfaces of the individual solutes. Second, resolution and other CRF's are parameters that apply to *pairs* of peaks; since in a given chromatogram the total number of peak pairs (all combinations) will greatly exceed the number of individual peaks, the computational requirements of modeling responses surfaces for peak pairs instead of individual retention surfaces will be much larger.

When the interpretive approach to chromatographic optimization was originally developed, the criterion used was the minimum selectivity (relative retention) between any pair of peaks (54), and the resulting plot was called a "window diagram". The same or very similar methods are known by several different names, including minimum alpha plots (MAPs), overlapping resolution mapping (ORMs), minimum resolution mapping, selectivity surface mapping, and response surface mapping. All of these may be classified as window diagram methods since they provide windows where acceptable separations may be obtained.

Following the introduction of the window diagram approach in chromatography, it was shown that using relative retention (α) does not result in the true optimum with respect to the required number of theoretical plates (55). Depending on the specifics of the optimization process, a variety of alternative CRFs have been proposed. For our implementation of the window diagram approach in SFC, we recommend either of two CRFs from among those we have developed or modified for SFC, with the choice depending on whether or not the same length of column will be used during the optimization as in the final separation. Brevity permits only a limited discussion of these CRFs, and other references should be consulted for more details on these and related criteria (33,34,56). As with the simplex optimization method, we are the first laboratory to implement the interpretive (window diagram) approach in SFC {to the best of our knowledge}.

A disadvantage of simple interpretive methods is that the model to which the retention data (or other data) are fit must be fairly accurate. In other words, an interpretive approach may fail if one or more sample components exhibits anomalous retention. Although rare in SFC, such retention behavior is observed occasionally and is difficult to predict intuitively. Note, however, that by anomalous retention we do not mean behavior that is merely unusual, e.g., retention that decreases smoothly with increasing density (at constant temperature). Retention that varies in a regular (continuous) manner, even if unusual, can usually be modeled with a high degree of accuracy (vide infra).

Although it is beyond the scope of this chapter, more sophisticated interpretive methods can be employed to compensate for the anomalous retention behavior. That is, the complex retention surfaces are broken down into smaller parts in an iterative fashion; the smaller retention surfaces are more accurately modeled by simple functions. This iterative interpretive approach has recently been applied in micellar LC (MLC) for the optimization of organic modifier, surfactant concentration, and pH (57,58).

Retention Modeling of Individual Solutes. In order to calculate the total response surface, the retention of each individual solute in the sample must be accurately modeled. For the results that we report here, a four parameter equation based on the relationships in equations 2-5 was selected to describe the retention of each solute. The model employed for the fit was:

$$\ln k' = \beta_0 + \beta_1 \rho + \beta_2 \rho^2 + \beta_1' \frac{1}{T} + \beta_{11} \frac{\rho}{T} \tag{9}$$

where k' is the capacity factor, ρ is the density, T is the temperature and the coefficients (β's) are analogous to those in equations 2 and 3. The second order density term was included because of the broad density range over which the (global) optimization was conducted. An additional term (β_{11}) was included for the interaction of density and temperature. Although intended for the optimization of both density and temperature, equation 9 can also be used, for purposes of comparison, to individually "optimize" either variable while maintaining a constant value for the other one.

An important aspect of equation 9 is that, despite the incorporation of 5 independent coefficients (β's) into the retention model, only two variables (density and temperature) have been accounted for. This illustrates an important limitation of interpretive methods in which retention is being modeled directly: the number and type of variables that may be accommodated. With regard to the type of variable, equation 9 is also revealing in terms of important SFC variables (vide supra) that are missing, i.e., the gradients of density and temperature. Analogous to gradient elution in HPLC, solute retention under density and/or temperature gradients in SFC is difficult to model. In general the retention can be described with sufficient accuracy only if numerical methods are used to solve the complex equations, although simple analytical solutions have been proposed for various parameters for rapid linear density gradients at constant temperature (59). The accuracy of these analytical solutions has not been thoroughly evaluated, however.

Optimization Criteria for Interpretive Methods. As noted earlier in our discussion of the simplex methods, there are many chromatographic response functions (CRFs) for the evaluation and comparison of chromatograms during an optimization process. Here we discuss two CRFs that we employed successfully with this interpretive method of optimization. Since the retention behavior of every solute must be modeled prior to optimization, the number of sample components is known beforehand; it is thus unnecessary to include the number of peaks in these CRFs as was done in CRF-3 (equation 8) for the simplex.

If the optimization is carried out on the same column for which the final separation will be performed, a threshold CRF based on the separation factor (S) can be employed:

$$S_{min} \geq \varepsilon \qquad\qquad CRF\text{-}4 = \frac{1}{t_w}, \qquad\qquad (10a)$$

$$S_{min} \leq \varepsilon \qquad\qquad CRF\text{-}4 = 0 \qquad\qquad (10b)$$

where $S = \left(\dfrac{t_{R2} - t_{R1}}{t_{R1} + t_{R2}}\right)$ or $\left(\dfrac{k_2 - k_1}{k_1 + k_2 + 2}\right)$ is based on a Gaussian peak profile,

$\varepsilon = \dfrac{2R_{s,ne}}{N_c}$, $R_{s,ne}$ is the minimum acceptable resolution, and N_c is the efficiency of the column used.

CRF-4 is the same as CRF-1 introduced for the simplex, except that the parameter employed for the comparison is the separation parameter S instead of resolution R_s. An advantage of using S is that it can be calculated from retention data commonly supplied by electronic integrators and/or data systems, facilitating the window diagram search (vide infra).

The maximum value of CRF-4 will be obtained for a separation that provides the required resolution in the shortest amount of time, assuming that column parameters remain the same. Note therefore, that the retention factors (k's) of equation 9 must be converted to retention times (t_R's) via the simple relationship $t_R = t_0 (1 + k')$. Direct use of retention factors instead of times in equation 10a is not generally recommended unless it is known that t_0 is constant over the range of densities and temperatures employed.

On the other hand, if it is planned that the optimization column can be a different length (typically shorter) than the final separation column, another CRF can be used: the time-corrected normalized resolution-product, r^*_{tc}, given by

$$r^*_{tc} = \frac{\sqrt[n]{r^*}}{t_{ne}} \qquad\qquad CRF\text{-}5 \qquad\qquad (11)$$

$$\text{where } r^* = \prod_{i=1}^{n} \frac{S_{i,i\text{-}1}}{S_{avg}}, \qquad\qquad (12)$$

$$t_{ne} = \frac{S_{min}^2}{1 + k_w} , \tag{13}$$

n is the number of solutes, $S_{i,i-1}$ is the separation factor for the (i,i-1) pair, S_{avg} is the average separation obtained for all peaks, S_{min} is the lowest S for all pairs, and k_w is the capacity factor for the last peak.

Although CRF-5 appears to be quite complex, it is highly successful in predicting the optimum separation for equal spacing of all peaks in the shortest amount of time. The importance of the spacing, however, is secondary when S_{min} (equations 10 and 13) is used. Once the optimum is predicted, column length can be altered based on the value of S_{min} at the optimum to provide the desired resolution. The number of theoretical plates needed, N_{ne}, can then be calculated as

$$N_{ne} = 4 \left(\frac{R_s}{S_{min}} \right)^2 \tag{14}$$

Generation of Retention Surfaces. Two types of retention surfaces were generated: (i) an isothermal (two-dimensional) surface illustrating the dependence of retention on density; and (ii) a two variable (three-dimensional) surface illustrating the dependence of retention on temperature and density.

For the density-only retention surface, retention data were collected at 4 different densities (0.1, 0.2, 0.3 and 0.4 g/mL) at a temperature of 80°C. Data were fit to equation 2 via nonlinear regression, and the resulting retention equations for each solute could be used to calculate the response surface. Although the goodness of fit was difficult to estimate since there was only one degree of freedom (and it is easy for R^2 values to exceed 0.999 under these conditions), the good agreement of the predicted and measured retention values at the optimum (vide infra) provided additional support for the accuracy of equation 2.

For the second retention surface, data were collected according to a three-level, two-factor (density and temperature) experimental design. Each factor was assigned three different values (0.2, 0.3 and 0.4 g/mL; 75, 100 and 125°C), and experiments were conducted at the nine combinations. Data were fit to the model by multiple regression, and these retention surfaces were used to calculate the response surface.

Table VI gives the results for the multiple regression of the retention data. Values for each coefficient in equation 9 are given for each solute, as well as parameters describing the quality of the fit to the experimental data. As can be seen in Table VI, the fit was exceptional. Interestingly, when the regression was repeated without the interaction term, the fit was not as good, with R^2 values less than 0.988. Similar losses in correlation were also observed if the squared-density term was omitted from the model equation. Equation 9 thus appears to be the best model from among those that could be predicted from the relationships in equations 2 and 3.

Density Optimization. The *retention* surfaces as a function of mobile phase density at a temperature of 80 °C are shown in *Figure 4*. The quadratic dependence of ln k' on density is apparent over a wide range in density. More importantly, however, there are numerous peak reversals which demonstrate the need for a systematic optimization approach.

Shown in Figure 5 are the calculated *response* surfaces for both the threshold separation (CRF-4) and time-corrected normalized resolution product (CRF-5). As expected, the threshold response surface is discontinuous, with 3 local optima; in contrast, the response surface for CRF-5 is smooth, with no apparent local optima.

Because we chose not to vary column length in this study, the optimum separation predicted by CRF-4 in Figure 5 is more appropriate to examine than that by CRF-5. Shown in Figure 6 is the chromatogram corresponding to the optimum density of 0.204 g/mL (CRF-4). As noted earlier, a temperature of 80°C was arbitrarily selected

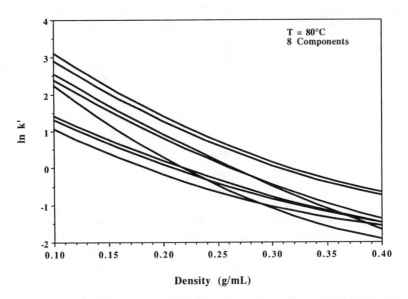

Figure 4. Isothermal *retention surfaces* of an eight component sample (Table III). The diversity of their chemical structure resulted in numerous peak reversals. Data used to generate these surfaces were collected at 80°C in density increments of 0.1 g/mL. Other conditions as described in the text.

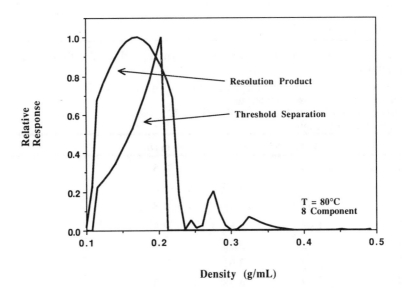

Figure 5. Comparison of the threshold separation and time-corrected, normalized, resolution-product *response surfaces* for the eight component sample (Table III). Response surfaces calculated via equations 9 and 10 using the isothermal retention surfaces of Figure 4.

Table VI. Regression Coefficients (Equation 9) of Individual Compounds in the Eight-Component Mixture to be Optimized by the Interpretive Approach

Solute[a]	β_0	β_1	β_2	β_1'	β_{11}	$R^2_{,adj.}$	$SS_{,res.}$
AcNap	-1.905	-14.88	2.676	22.17	-3.289	0.996	0.019
BzP	-2.819	-11.04	2.900	20.67	-4.281	0.998	0.010
n-C$_{18}$	-5.920	-6.876	3.983	28.17	-7.866	0.999	0.007
n-C$_{20}$	-3.955	-9.083	3.127	11.50	-4.008	0.997	0.021
IQ	-3.681	-4.800	2.456	12.83	-4.216	0.999	0.004
Nap	-2.985	-9.122	2.096	17.67	-3.152	1.000	0.001
Q	-3.819	-4.971	2.472	14.17	-4.265	0.999	0.004
UB	-2.707	12.00	2.615	16.17	-3.240	0.998	0.014

[a] Compound identification: AcNap = 2'-acetonaphthone, BzP = benzophenone, n-C$_{18}$ = n-octadecane, n-C$_{20}$ = n-eicosane, IQ = isoquinoline, Nap = naphthalene, Q = quinoline, and UB = undecylbenzene.

prior to the mapping of the retention and response surfaces. As seen in Figure 6, baseline resolution of all 8 components was achieved in about 37 minutes, except for components 2 and 3 which were almost baseline resolved. The accuracy of the retention model (equation 2) employed for this optimization is evident in Table VII, where differences in the predicted and actual retention factors were less than 8% for all 8 components.

Table VII. Density Optimization via an Interpretive (Window Diagram) Approach

Criterion: threshold separation factor (CRF-4, equation 9)
Preselected condition: temperature, 80 °C
Optimized condition(s): density, 0.204 g/mL
Chromatogram: Figure 6

Solute[a]	predicted k'	actual k'	% error
Nap	1.05	1.02	-2.5
Q	1.40	1.35	-3.8
IQ	1.55	1.48	-4.8
n-C$_{18}$	1.93	1.88	-2.8
UB	3.03	3.28	7.5
n-C$_{20}$	3.57	3.61	1.0
BzP	4.88	4.78	-2.0
AcNap	5.59	5.31	-5.3

[a] Compounds as in Table VI.

Simultaneous Optimization of Density and Temperature. Although near-baseline resolution was achieved for all eight sample components via the optimization of a single variable (density), as illustrated in Figure 1, a better (or in rare cases, equal) result will always be obtained if all variables of interest are optimized. The window diagram method is now considered for the *simultaneous* optimization of density *and* temperature for the separation of the eight component sample of Table VI, to provide a comparison with the SFC separation obtained with the density-only optimization (Figure 6).

Shown in Figure 7 is the three dimensional retention surface for quinoline, a representative sample component. Retention surfaces for the 7 other solutes, not shown for the sake of clarity, were generally as smooth and continuous, and resembled each other in a fashion analogous to the two dimensional ln k'/density surfaces of Figure 4. Similar peak reversals were also observed, and a systematic optimization scheme is again clearly warranted. Note that the individual effects of temperature and density, as predicted by equations 2 and 3, can be inferred from the appropriate cross sections of Figure 7.

A response surface was generated for the eight component mixture using CRF-4. Because we chose not to vary column length in this study, the optimum separation predicted by CRF-4 is more appropriate than CRF-5. The threshold separation factor response surface (CRF-4) is shown in Figure 8, where the tallest peak occurs at 0.19 g/mL and $2.65 \cdot 10^{-3}$ K^{-1} (104 °C). Although the precise locations of this global optimum and several other local optima are not readily discernible from Figure 8, they are readily apparent in the contour plot shown in Figure 9.

Shown in Figure 10 is the chromatogram acquired at the optimum predicted by CRF-4. Baseline resolution of all 8 components was achieved in about 27 minutes, except for components 2-4 which were almost baseline resolved. Additional evidence for the accuracy of the retention model (equation 9 and Table VI) employed for this window diagram optimization is evident in Table VIII, where predicted and measured retention factors differed by less than 15%. The slight positive bias observed for all solutes at the optimum conditions in Table VIII was coincidental; averaged over the entire parameter space the bias was almost completely random.

Table VIII.   Simultaneous Density and Temperature Optimization via an Interpretive (Window Diagram) Approach
Criterion: threshold separation factor (CRF-4, equation 9)
Optimum conditions: density, 0.19 g/mL; temperature, 104 °C
Chromatogram: Figure 10

Solute[a]	predicted k'	actual k'	% error
Nap	0.90	0.90	0.0
Q	1.17	1.21	3.3
IQ	1.29	1.34	3.7
$n\text{-}C_{18}$	1.47	1.69	13
UB	2.45	2.88	15
$n\text{-}C_{20}$	2.74	3.20	14
BzP	3.90	4.28	8.9
AcNap	4.51	4.74	4.8

[a] Compounds as in Table VI.

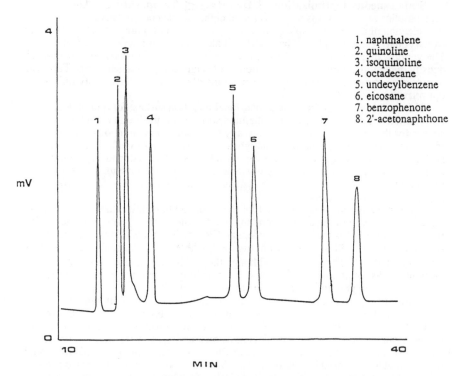

1. naphthalene
2. quinoline
3. isoquinoline
4. octadecane
5. undecylbenzene
6. eicosane
7. benzophenone
8. 2'-acetonaphthone

Figure 6. Density-optimized SFC separation of an eight component sample (Table III). Optimum density of 0.204 g/mL was determined according to CRF-4 (equation 9), as illustrated in Figure 5.

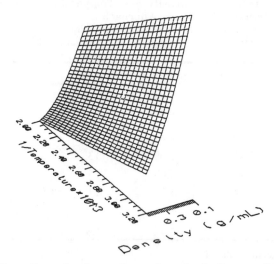

Figure 7. Three-dimensional *retention surface* of quinoline, as a function of reciprocal temperature and density (equation 8). Data used to generate these surfaces were collected as described in the text.

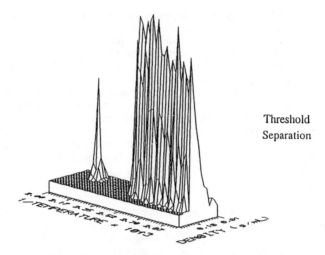

Threshold
Separation

Figure 8. Threshold separation factor *response surface* (equation 9) for the optimization of density and temperature in the SFC separation of the eight component sample.

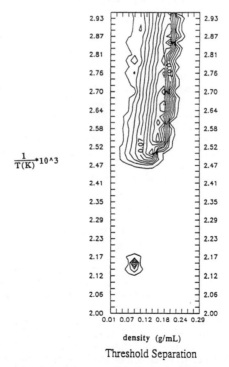

$$\frac{1}{T(K)}*10^3$$

density (g/mL)

Threshold Separation

Figure 9. Contour plot of the response surface of Figure 8. Of the two most readily apparent optima, $(2.15 \cdot 10^{-3} \text{ K}^{-1} \{191°C\}, 0.087 \text{ g/mL})$ and $(2.65 \cdot 10^{-3} \text{ K}^{-1} \{104°C\}, 0.19 \text{ g/mL})$, the latter is the global optimum over the range of experimental conditions we examined.

Comparison of the Optimized Density and Density/Temperature Separations. It is instructive to compare the results of the separations of the eight component mixture illustrated in Figures 6 and 10. Because a threshold CRF (CRF-4) was employed for both the density and density/temperature separations, the minimum resolution (directly related to S_{min}) was similar as expected, with slight differences attributable to the (minor) errors in predicting solute retention.

In contrast to the minor difference in the minimum resolution, however, the difference in the analysis time for the two optimized separations (Figures 6 and 10) was significant. The 30% shorter analysis time of 27 min for the optimized density/temperature method compared to 37 min for the density-only optimization clearly demonstrates the superiority of multiparameter optimization strategies. Had the arbitrarily selected temperature of 80°C ($2.83 \cdot 10^{-3}$ K^{-1}) not been relatively close to the true optimum temperature of 104 °C ($2.65 \cdot 10^{-3}$ K^{-1}), the difference would have been even more dramatic!

Mixture Design. The sole remaining example of optimization in SFC is the mixture design approach recently reported by Ong et al. (39) for the separation of eight nitroaromatic compounds. Like the previous window diagram method, the approach of Ong et al. is also considered a simultaneous, interpretive method. There are, however, some important differences between the two. First, the variables considered in the present mixture design strategy were pressure gradient and temperature, compared to initial pressure and temperature in the previous window-diagram approach. Second, the parameter space selected by Ong and co-workers was confined by design to a triangle rather than a rectangle, thus the use of a mixture design instead of the previous factorial approach. Finally, Ong et al. chose to model the resolution of all peak pairs (equation 15) rather than the retention of individual solutes (equation 9). They did so with no degrees of freedom compared to one degree of freedom in the previous optimization.

Figure 11 illustrates the parameter space defined by the equilateral triangle. The initial pressure and conditions for the 3 vertices of the pressure gradient/ temperature triangle were determined arbitrarily from the critical conditions of the supercritical fluid (carbon dioxide), the retention characteristics of nitroaromatic compounds, and the following criteria: (i) the first analyte should not co-elute with the sample solvent; and (ii) the retention factor of the last analyte should not exceed 30.

As implied by the vertices and closed circles in Figure 11, a minimum of seven experiments were required for the modeling and optimization process. The resolution between all pairs of peaks was measured and then modeled using equations of the form

$$R_s = a_1x_1 + a_2x_2 + a_3x_3 + a_{12}x_1x_2 + a_{13}x_1x_3 + a_{23}x_2x_3 + a_{123}x_1x_2x_3 \qquad (15)$$

where a_i are the coefficients and x_1, x_2, and x_3 correspond to the conditions at the respective vertices in Figure 11. The resulting 28 equations were then employed to generate an overlapping resolution map (response surface), from which a range of optimum conditions was then deduced (pressure gradient: 0.07-0.14 MPa/min; temperature: 77-88°C). Using these conditions, all 8 nitroaromatic compounds could be separated, although a few pairs of compounds were not baseline resolved.

Future work. As with the simplex algorithm, use of the window diagram approach for the systematic optimization of SFC separations is still in its early stages, although the success described here justifies research in this area. Research opportunities likewise include: (i) extension of this interpretive method to include variables of modifier content, as well as the gradients of density, temperature, and modifier; and (ii) investigation of the benefits of the window diagram strategy to packed column separations, particularly those that utilize modified mobile phases.

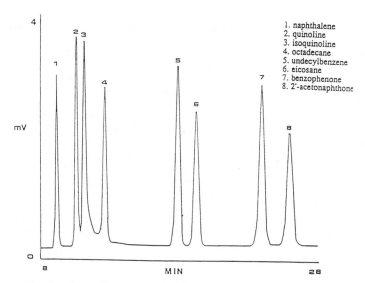

1. naphthalene
2. quinoline
3. isoquinoline
4. octadecane
5. undecylbenzene
6. eicosane
7. benzophenone
8. 2'-acetonaphthone

Figure 10. Density and temperature-optimized SFC separation of the eight component mixture of Table III. Result corresponds to the global optimum in Figures 8 and 9. Chromatographic conditions: 0.19 g/mL, 104°C.

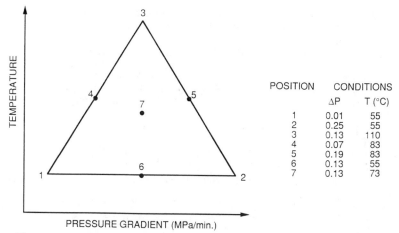

POSITION	CONDITIONS	
	ΔP	T (°C)
1	0.01	55
2	0.25	55
3	0.13	110
4	0.07	83
5	0.19	83
6	0.13	55
7	0.13	73

Figure 11. Mixture design strategy of Ong et al. for the optimization of pressure gradient and temperature. Adapted from ref. 39.

Summary

A systematic method development scheme is clearly desirable for SFC, and as shown in the present work, both the modified simplex algorithm and the window diagram method are promising approaches to the optimization of SFC separations. By using a short column and first optimizing the selectivity and retention, rapid separations are possible in the development stage, with the potential of optimizing efficiency later if needed. This saves hours of analysis time, especially if the short column proves to be sufficient for the final separation. For the modified simplex method, confidence that the global optimum has been found is provided by the convergence to the same optimum of two simplexes started at different points within the parameter space. With the window diagram method, the use of an interpretive approach helps to minimize the amount of experimental data required.

Given the mere handful of reports in the published literature (6,38,39,52), there are many avenues open in the development of systematic approaches to optimization in SFC. In addition to the opportunities mentioned in the sections on the simplex method and window diagram approach, others include the exploration of other sequential or simultaneous optimization strategies such as optiplex, simulated annealing, method of steepest ascent, etc. that are potentially useful in SFC.

Acknowledgments
Partial support for this research was provided by grants from the Exxon Education Foundation and the LSU Center for Energy Studies. J.A.C. is the recipient of an LSU Alumni Federation Fellowship.

Literature Cited

1. *Analytical Supercritical Fluid Chromatography and Extraction*; Lee, M. L.; Markides, K. E., Ed.; Chromatography Conferences: Provo, Utah, 1990.
2. *Supercritical Fluid Chromatography*; Smith, R. M., Ed.; Royal Society of Chemistry: London, 1988.
3. Chester, T. L.; Pinkston, J. D. *Anal. Chem.* **1990**, *62*, 394R-402R.
4. Schoenmakers, P. J.; Verhoeven, F. C. C. J. G. *TRAC—Trends Anal. Chem.* **1987**, *6*, 10-17.
5. Engelhardt, H.; Gross, A. *TRAC—Trends Anal. Chem.* **1991**, *10*, 64-70.
6. Foley, J. P.; Crow, J. A. In *Recent Advances in Phytochemistry*; N. H. Fischer, Ed.; Plenum Publishing Corporation: New York, 1991; Vol. 25.
7. Fields, S. M.; Grolimund, K. *J. Chromatogr.* **1989**, *472*, 197-208.
8. Lauer, H. H.; McManigill, D.; Board, R. D. *Anal. Chem.* **1983**, *55*, 1370-1375.
9. Ashraf-Khorassani, M.; Taylor, L. T.; Zimmerman, P. *Anal. Chem.* **1990**, *62*, 1177-1180.
10. Schwartz, H. E.; Brownlee, R. G. *J. Chromatogr.* **1986**, *353*, 77-93.
11. Ong, C. P.; Lee, H. K.; Li, S. F. Y. *Anal. Chem.* **1990**, *62*, 1389-1391.
12. Yonker, C. R.; McMinn, D. G.; Wright, B. W.; Smith, R. D. *J. Chromatogr.* **1987**, *388*, 19-29.
13. Lochmuller, C. H.; Mink, L. P. *J. Chromatogr.* **1989**, *471*, 357-366.
14. Geiser, F. O.; Yocklovich, S. G.; Lurcott, S. M.; Guthrie, J. W.; Levy, E. J. *J. Chromatogr.* **1988**, *459*, 173-181.
15. Crow, J. A.; Foley, J. P. *J. Microcol. Sep.* **1991**, *3*, 47-57.
16. Engel, T. M.; Olesik, S. V. *Anal Chem* **1991**, *63*, 1830-1838.
17. Engel, T. M.; Olesik, S. V. *J. Microcol. Sep.* **1992**, in press.
18. Giddings, J. C.; Meyers, M. N.; King, J. W. *Science* **1968**, *4*, 276-283.
19. Berger, T. A.; Deye, J. F.; Anderson, A. G. *Anal. Chem.* **1990**, *62*, 615-622.
20. Peaden, P. A.; Lee, M. L. *J. Chromatogr.* **1983**, *259*, 1.
21. Koski, I. J.; Lee, M. L. *J. Microcol. Sep.* **1992**, in press.
22. Schwartz, H. E.; Barthel, P. J.; Moring, S. E.; Lauer, H. H. *LC•GC: The Magazine of Separation Science* **1987**, *5*, 490-497.
23. Payne, K. M.; Tarbet, B. J.; Bradshaw, J. S.; Markides, K. E.; Lee, M. L. *Anal. Chem.* **1990**, *62*, 1379-1384.

24. Taylor, L. T.; Chang, H. K. *J. Chromatogr. Sci.* **1990**, *28*, 357-366.
25. Berger, T. A.; Deye, J. F. *J Chromatogr Sci* **1991**, *29*, 390-395.
26. Taylor, L. T.; Chang, H. K. *Chem. Rev.* **1989**, *89*, 321-330.
27. Snyder, L. R.; Kirkland, J. J. In *Introduction to Modern Liquid Chromatography*; 2nd ed.Wiley: New York, 1979; pp 663-694.
28. Chester, T. L.; Innis, D. P. *HRC&CC* **1985**, *8*, 561-566.
29. Later, D. W.; Campbell, E. R.; Richter, B. E. *HRC&CC, J. High Resolut. Chromatogr. Chromatogr. Commun.* **1988**, *11*, 65-69.
30. Fjeldsted, J. C.; Jackson, W. P.; Peaden, P. A.; Lee, M. L. *J. Chromatogr. Sci.* **1983**, *21*, 222-225.
31. Yonker, C. R.; Smith, R. D. *Anal. Chem.* **1987**, *59*, 727-731.
32. Tijssen, R.; Billiet, H. A. H.; Schoenmakers, P. J. *J. Chromatogr.* **1976**, *122*, 185-203.
33. Schoenmakers, P. J. In *Optimization of Chromatographic Selectivity: Guide to Method Development*Elsevier: Amsterdam, 1986; pp 119-169.
34. Berridge, J. C. In *Techniques for the Automated Optimization of HPLC Separations*; J. C. Berridge, Ed.; Wiley: New York, 1985; pp 19-27.
35. van Laarhoven, P. J. M.;   Aarts, E. H. L. *Simulated Annealing: Theory and Applications*; D. Reidel: Dordrecht, The Netherlands, 1988.
36. Güell, O. A.; Holcombe, J. A. *Anal. Chem.* **1990**, *62*, 529A.
37. Sutter, J. M.; Kalivas, J. H. *Anal. Chem.* **1991**, *63*, 2383-2386.
38. Crow, J. A.; Foley, J. P. *Anal. Chem.* **1990**, *62*, 378-387.
39. Ong, C. P.; Chin, K. P.; Lee, H. K.; Li, S. F. Y. *HRC-J. High Resolut. Chromatogr.* **1991**, *14*, 249-253.
40. Berridge, J. C. *J. Chromatogr.* **1982**, *244*, 1-14.
41. Berridge, J. C.; Morrissey, E. G. *J. Chromatogr.* **1984**, *316*, 69-79.
42. Balconi, M. L.; Sigon, F. *J. Chromatogr.* **1989**, *485*, 3-14.
43. Morgan, S. L.; Deming, S. N. *J. Chromatogr.* **1975**, *112*, 267-285.
44. Morgan, S. L.; Jacques, C. A. *J. Chromatogr. Sci.* **1978**, *16*, 500-504.
45. Bartu, V.; Wicar, S.; Scherpenzeel, G. J.; LeClercq, P. A. *J. Chromatogr.* **1986**, *370*, 219-234.
46. Dose, E. V. *Anal. Chem.* **1987**, *59*, 2420-2423.
47. Deming, S. N.; Morgan, S. L. *Anal. Chem.* **1973**, *45*, A278-A283.
48. Deming, S. N.; Parker, L. R., Jr. *CRC Crit. Rev. Anal. Chem.* **1978**, 187-202.
49. Shavers, C. L.; Parsons, M. L.; Deming, S. N. *J. Chem. Ed.* **1979**, *56*, 307-309.
50. Jurs, P. C. *Computer Software Applications in Chemistry*; Wiley: New York, 1986.
51. Smits, R.; Vanroelen, C.; Massart, D. L. *Z. Anal. Chem.* **1975**, *273*, 1-5.
52. Crow, J. A.; Foley, J. P. *HRC-J. High Resolution Chromatogr.* **1989**, *9*, 467-470.
53. Snyder, L. R. In *High Performance Liquid Chromatography, Advances and Perspectives*; C. Horváth, Ed.; Academic Press: New York, 1980; Vol. 1; pp 207-316.
54. Laub, R. J.; Purnell, J. H. *J. Chromatogr.* **1975**, *112*, 71-79.
55. Jones, P.; Wellington, C. A. *J. Chromatogr.* **1981**, *213*, 357.
56. Schoenmakers, P. J. *J. Liq. Chromatogr.* **1987**, *10*, 1865-1886.
57. Strasters, J. K.; Breyer, E. D.; Rodgers, A. H.; Khaledi, M. G. *J. Chromatogr.* **1990**, *511*, 17-33.
58. Strasters, J. K.; Kim, S. T.; Khaledi, M. G. *J. Chromatogr.* in press.
59. Smith, R. D.; Chapman, E. G.; Wright, B. W. *Anal. Chem.* **1985**, *57*, 2829-2836.

RECEIVED January 8, 1992

Chapter 22

Use of Modifiers in Supercritical Fluid Extraction

Joseph M. Levy, Eugene Storozynsky, and Mehdi Ashraf-Khorassani

Suprex Corporation, 125 William Pitt Way, Pittsburgh, PA 15238

The main limitation of the most often used supercritical fluid, carbon dioxide (CO_2) is its limited ability to dissolve polar analytes even at very high densities. Mobile phase selectivity and solubilizing power can be enhanced by the addition of polar organic compounds known as modifiers to the primary supercritical fluid phase. Depending on the situation, if the maximum solvent power of the primary supercritical fluid phase (i.e. CO_2) is not high enough to elute or separate the components of interest (i.e. polar or high molecular weight) then higher solubilizing power can be achieved by adding a modifier. Both the maximum solvent power and the solvent selectivity of the fluid mixture are determined by the chemical identity and concentration of the modifier. In the realm of supercritical fluid chromatography (SFC), the modifier effect manifests itself by altering solute retention, improving chromatographic efficiencies, and in some cases altering elution orders. For packed column SFC, adding as little as 0.5 percent of a modifier significantly altered capacity factors and separations factors by many orders of magnitude (1).

Several investigators have reported on the use of modifiers in SFC, Klesper and Hartmann used methanol as a modifier with normal n-pentane to separate polystyrene oligomers (2,3). Dichlorobenzene and pyridine were used as modifiers in supercritical dichlorofluoromethane ($CHCl_2F$) for the elution of metal chelates which were insoluble in the dichlorofluoromethane phase alone (4). Novotny et al. used isopropanol as a modifier with normal pentane to separate polynuclear aromatic hydrocarbons (PAHs) and monosubstituted aromatics (5). Conaway et al. used methanol, isopropanol and cyclohexane as modifiers with normal pentane to produce more nearly complete elution of polystyrene oligomers up to average molecular weights of 2200 amu (6). Other modifiers that have been used in SFC with carbon dioxide and normal n-alkane phases include straight chain and branched alcohols up to 1-decanol (7, 8, 9-18, 19), dioxane (13, 15, 16, 19), tetrahydrofuran (13, 15 16), diisopropyl ether (13, 15), acetonitrile (15-18), chloroform and methylene chloride (15-18), and dimethyl sulfoxide (16). Recent studies by Berger (20) have also been done comparing vapor-liquid equilibria measurements for methanol modified CO_2 with chromatographic results. Results using simultaneous pressure and composition (methanol and dioxane) programming have shown separations of additives in polystyrene and epoxy resins that were superior to programming only one or the other of these components (21). Recent advances in further enhancing the solubilizing powers of supercritical CO_2 have been made by the addition of secondary modifiers (i.e. ternary systems). Berger and Deye (22, 23) demonstrated

0097–6156/92/0488–0336$07.50/0

that polyhydroxybenzoic and polycarboxylic acids could be rapidly separated by SFC using ternary mobile phases consisting of very polar additives (citric acid, trifluoroacetic acid) dissolved in methanol and mixed with CO_2.

The use of modifiers in supercritical fluid extraction (SFE) has begun to be prevalent in the past few years (compared to SFC). Analogous to SFC, the use of modifiers in SFE is needed to enhance extraction recoveries by extracting the analyte out of the sample matrix (in SFC, modifiers aid in eluting the analyte out of the chromatographic column). Table I shows a listing of the various modifiers that have been used in supercritical fluid technology (24). The bulk of these modifiers have been utilized in SFC but can also be used in SFE. The modifiers that are listed in Table I have been used with CO_2 as the primary supercritical fluid. The advantages of CO_2 as a supercritical fluid is that it is chemically compatible and miscible with a large range of organic modifiers. Some of the modifiers in Table I have also been used in a limited fashion with such primary fluids as nitrous oxide and sulfur hexafluoride. The most common modifiers that have been used are alcohols (i.e. methanol).

A limitation to the use of modifiers in SFC has been the detector compatibility of the specific modifier. For example, most modifiers (i.e. methanol) provide a flame ionization detector response preventing use at a large concentration. In SFE, however, this limitation does not exist since modifiers can be used in a mixed mode with CO_2 and the sample. Figure 1 (taken from reference 25) shows an example of the use of modifiers in SFE. Here the comparison of extraction efficiencies obtained using CO_2 and CO_2 modified with methanol is shown. The matrices extracted in these SFE experiments were an XAD-2 sorbent resin and soils. The target analytes were dibenzo[a,i]-carbazole, diuron, 2,3,7,8-tetrachorodibenzo-p-dioxin (TCDD) and linear alkylbenzenesulfonate (LAS) detergent. For each of these respective analytes the extraction efficiency increased dramatically with the use of CO_2/methanol modifier compared to pure CO_2 only as the extracting fluid. This was even the case for the ionic compounds; namely, the linear alkylbenzenesulfonates (LAS), which were quantitatively recovered.

TABLE I: Modifiers That Have Been Used in Supercritical Fluid Technology With Carbon Dioxide as the Primary Supercritical Fluid

1,4-Dioxane	n-Heptanol
2-Butanol	n-Hexane
n-Butanol	n-Hexanol
t-Butanol	Isopropanol
Acetic acid	Methanol
Acetonitrile	2-Methoxyethanol
Carbon disulfide	Methyl-t-butyl ether
Carbon tetrachloride	Methylene chloride
Chloroform	N,N-dimethylacetamide
n-Decanol	n-Pentanol
Diethyl ether	2-Pentanol
Diisopropyl ether	n-Propanol
Dimethyl sulfoxide	Propylene carbonate
Ethanol	Sulfur hexafluoride
Ethyl acetate	Tetrahydrofuran
Fluoroform	Trichlorofluoromethane
Formic acid	Water

Studies have been done to characterize supercritical fluid solvents with solvatochromic dyes, such as nile red, and were used to observe solvatochromic shifts (26-29). In SFE, however, the selection of modifiers and the concentrations that would be utilized has been largely empirical because very little applicable solubility data exists for modified supercritical fluids. Moreover, the actual competitive mechanisms that exist during SFE are poorly understood. Specifically, in SFE, the mechanisms of solubility, diffusion and physical adsorption on the surface of the sample matrix all exist and distinctly depend upon the sample matrix and the target analyte of interest. Some of the work that has been done in SFC sheds light on how modifiers can be used in SFE. In SFC, the modifiers increase the solubility of the respective target solute in the mobile phase. In SFE, however, the modifiers serve to increase the solubility of the target solute in the extracting fluid. So the interactions that exist between the solute in the stationary phase can be paralleled between the interactions that exist between the solute and the surface of the sample matrix. Perhaps a suggested starting point for experimenting with modifiers to enhance efficiencies in SFE is to select a modifier that is a good solvent in its liquid state for the analyte of interest. The work presented in this chapter is intended to shed some light on the use of modifiers in SFE with respect to the following variables: actual identity of the modifier that could be used, the concentration of the modifier, the equilibration time of the modifier with the sample matrix before performing the actual extraction, and the type of sample matrix.

Practical Modifier Addition

Practically, modifiers have been introduced in a number of ways in both SFC and SFE. The first technique for modifier addition is the use of a pre-mixed cylinder that can be obtained from a commercial supplier. These cylinders have a discrete concentration level of a specific modifier in CO_2, for example. The cylinder is directly connected to the supply pump which directs the actual modified fluid to the SFE. The disadvantage of this technique is that the pump and the entire system can become contaminated with modifiers.

Another way modifiers can be added to the SFE system is by using two separate supply pumps. In this fashion, one supply pump is primarily used for CO_2 delivery and the second is used for the modifier delivery. Downstream of both of these pumps is a thermostated mixing tee. This tee mixes the modifier with CO_2, equilibrates the fluid in a thermostated zone and then delivers the mixed fluid to the extraction vessel.

Modifiers can also be added to the supply pump by using a high pressure modifier addition valve. This valve normally has a loop of known volume (i.e. 1-5 mL) that can be filled with liquid modifier and then channeled directly into the supply pump after valve actuation. The fluid is then mixed in the reservoir of the supply pump and delivered as a mixed fluid to the SFE oven. In most cases, this method of mixing should be viewed as being approximate in rapidly screening modifiers at different concentration levels.

The final and perhaps most effective way of delivering a modifier for SFE is by directly adding the modifier as a liquid to the sample matrix before filling the extraction vessel or while the sample matrix is already in the vessel but prior to the extraction. In this respect, the modifier is isolated from the actual primary supply pump thereby not contaminating the pump (and the entire system) with the modifier.

Moreover, you can screen a number of different of modifiers for the experiments in a relatively rapid fashion without flushing the entire system between different experimental runs. The work that is presented in this chapter has been accomplished by the addition of the modifier to the extraction vessel containing the sample matrix prior to the extraction.

Experimental Conditions

All of the experiments were performed using a Suprex stand-alone extractor model SFE/50 (with a syringe pump) and a Suprex Prepmaster^{TM} (with a Varipump^{TM}) stand-alone SFE. Details of these instruments can be found elsewhere (32). The modifiers were spiked into the extraction vessel with the appropriate weight of sample matrix before the extractions. The analyses were performed using on-line SFE/GC. Figure 2 shows a general schematic diagram of a typical on-line SFE interface to a GC. The test sample matrix was loaded into the extraction vessel with weights ranging from low milligram to gram quantities depending on the concentration levels of the analytes of interest. The extraction vessels utilized were stainless steel and 0.5 mL in volume.

Each of the experimental runs involved both static (pressurized vessel with no flow through vessel) and dynamic (flow through extraction vessel and into transfer line) extraction modes and controlled by electronic high pressure valves as shown in the flow schematic of Figure 2. All of the extractions were accomplished by flushing the extraction vessel with at least 3-5 times the void volumes to ensure efficient and quantitative extractions. The on-line transfer of the extraction effluent from the SFE to the capillary split/splitless injection port of the GC was accomplished by using a heated stainless steel transfer line. This line was specifically crimped to allow a flow of 150 mL per minute of decompressed gas to vent directly into the injection port. The stainless steel transfer line was inserted 40 cm directly through the unmodified gas chromatographic septum cap and septum of the injection port. The details of the on-line interfacing of SFE to GC can be found elsewhere (30-32). Both a syringe pump and a Varipump^{TM} (32) were used to deliver the CO_2 directly from the supply tank to the actual extraction vessels. Specifically the modifiers that were studied in the series of experiments that are presented in this chapter are the following: Propylyene carbonate, methanol, formic acid, acetonitrile, benzene, carbon disulfide, and hexane. The sample matrices that were used for the experiments were the following: shale rock, river sediment, and sludge/fly ash. All of these matrices were incurred samples and were not spiked with the respective target analytes.

The Effect of Changing Modifier Identities in SFE

One of the variables in terms of modifier usage in SFE for enhancement of extraction efficiencies is the identity of the modifier itself. Most of the applications to date in terms of modifier usage in SFC have been done using a primarily alcohol modifiers (i.e. methanol). In SFE very little has been conducted in terms of using different modifiers to show differences in selectivities or in enhancements of extractions efficiencies. Most of the applications in SFE have been demonstrated using modifiers such as methanol, ethanol, or methylene chloride, for example. To study the effect of varying modifier identities in SFE, experiments were done using the first test matrix; namely, sludge/fly ash. This matrix was difficult to extract using pure CO_2 as a supercritical fluid because of the nature of the matrix. Analytes

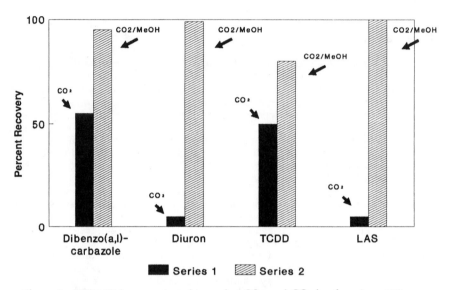

Figure 1. SFE efficiency comparison using CO_2 and CO_2/methanol modifier.

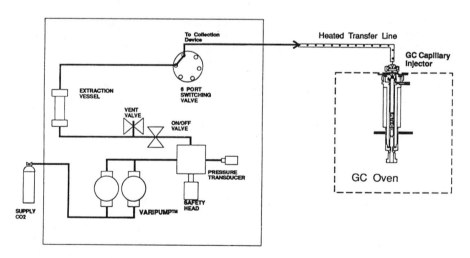

Figure 2. On-Line SFE/GC schematic diagram.

remained adsorbed on various active sites of this sample. In this series of experiments, a fixed amount of sludge/fly ash was used for the extraction (225 mg at a pressure of 375 atm, 50°C with 10 minute static and 7 minute dynamic extraction modes). These parameters were kept fixed, the only experimental variation was the actual identity of the modifier. A concentration level of approximately 10 percent was used for the modifiers (50 ul modifier added to a 500 ul vessel). On-line SFE/GC was used to separate the target analytes and determine their concentration levels. Table II summarizes the results in terms of listing the percent recoveries for the target analytes in the sludge/fly ash specifically; ethylbenzene (C_8H_{10}), cumene (C_9H_{12}), 2-chloronaphthalene ($C_{10}-H_2Cl$), and 1,2,4-trimethylbenzene (C_9H_{12}).

For reference, the percent recoveries obtained using CO_2 only for SFE are listed. These recoveries were calculated by comparison with values that were determined using a soxhlet extraction of the matrix followed by GC-MS analysis. As can be seen, the recoveries using supercritical pure CO_2 were low, approximately 70 percent. By the addition of different modifiers however, there was a distinct change in terms of percent recoveries obtained for the various target analytes. Propylene carbonate, for example, enhanced the percent recoveries to greater than 95 percent for all of the target analytes. This same type of enhancement was obtained using benzene. Methanol, on the other hand, did not achieve as high percent recoveries compared to either propylene carbonate or benzene. Of the other modifiers, acetronitrile achieved approximately the same percent recoveries as methanol. Formic acid, as a modifier, achieved only slightly higher recoveries than pure CO_2. Carbon disulfide and hexane achieved very little enhancement in terms of the extraction efficiencies when compared to pure CO_2.

Figure 3 shows a graphical representation of the comparison of all of the different modifiers that were used for the enhancement of extraction efficiencies of 2-chloronaphthalene from the sludge/fly ash matrix. The conclusions from this are that for this particular target analyte (2-chloronaphthalene) the highest extraction efficiency in SFE can be obtained with either propylene carbonate or benzene as the modifiers. To explain this, propylene carbonate is an organic compound that has three oxygens with lone pairs of electrons that can be available for bonding to active sites. In this respect, propylene carbonate effectively displaces analytes off of matrix surfaces, overcoming the actual binding forces that occur between the analyte and the surface of the actual sample matrix. Moreover, benzene is an organic compound that has a similar structure to the target analytes. Thus, the similarity in terms of the pi-electron clouds that were available on the benzene molecule and the pi-electron clouds of the target analytes could have achieved an interaction where the actual adsorptive forces were overcome for the target analyte and the surface of the sludge/fly ash particles. This result is consistent with some of the results that were obtained from other researchers that have used similar types of modifiers (i.e. benzene and toluene) for the extraction of dioxin from fly ash (33). It is interesting to note, for this particular sludge/fly ash matrix, that the mechanism predominating during the SFE step with modifier was the overcoming of physical adsorption. So the modifier molecule basically replaced the active sites on the sludge/fly ash matrix which were holding the target analyte.

Another matrix that was used to test the variation of modifier identity in SFE was shale rock. For these set of experiments a number of different compounds were chosen as target analytes, specifically pristane ($C_{19}H_{40}$), phytane ($C_{20}H_{42}$) and hexadecane ($C_{16}H_{34}$). Analogous to the study where the test matrix was sludge/fly ash, the modifier identity was varied and all other experimental parameters were kept constant (i.e. pressure, temperature, extraction duration, and modifier

TABLE II: Effect of Modifier Identity on Extraction Efficiencies for Sludge/Fly Ash Using On-Line SFE/GC

| | | | | Percent Recovery | | | | |
| | | | Propylene | | Formic | | Carbon | |
Compound	Benzene	Methanol	Carbonate	Hexane	Acid	Acetonitrile	Disulfide	CO_2
ethylbenzene	95	80	96	72	78	81	73	74
cumene	96	79	96	70	76	79	77	72
2-chloronaphthalene	92	82	93	66	70	83	67	66
1,2,4-trimethylbenzene	96	84	96	70	78	85	70	71

SFE: 225mg of sludge/ fly ash, 375 atm, 50°C, 10 minutes static, 7 minutes dynamic, 50 ul modifier added in 500 ul vessel

GC: 30 x 0.25 mm I.D. DB-WAX, 30°C (7 min) to 310°C at 7°C/minute, flame ionization detection

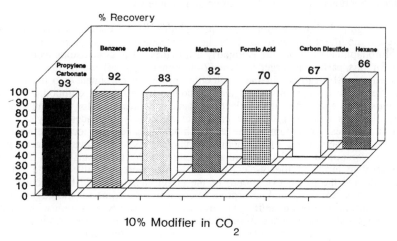

10% Modifier in CO_2

Figure 3. Comparison of different modifiers for the on-line SFE/GC characterization of sludge/fly ash.

concentration). Also on-line SFE transfer to a capillary GC was the means for the determination of the concentration levels. All calculated values were compared with recoveries that were determined by GC-MS analysis. Specifically, 300 mg of the finely ground shale rock was extracted at 350 atm, at a temperature of 60°C for a 30 minute static extraction and a 10 minute dynamic extraction duration with direct transfer of the effluent to the GC. All modifiers were spiked onto the shale rock as liquids prior to the extraction directly in the extraction vessel and all concentration levels were at 10 percent. Figure 4 shows the on-line SFE/GC (with flame ionization detection) characterization of the shale rock extraction effluent. In this case, the extraction was accomplished at 350 atm using CO_2 only. As can be seen in this particular shale rock, the approximate range of normal alkanes that were extracted was from C15 to C32 including the target analytes of pristane and phytane as well as b-carotene. Figure 5 is a chromatogram representing the characterization of the shale rock effluent that was extracted at an increased pressure, 450 atm. Compared to Figure 4, the extraction effluent characterization shown in Figure 5, demonstrates a clear increase in regards to the peak heights (i.e. concentration levels) of the respective analytes. This indicated that there were higher percent recoveries of the analytes from the shale rock at higher pressure (densities). At a pressure of 450 atm and a temperature of 60°C with the 40 minutes combination of static/dynamic extractions, all of the hydrocarbons were extracted from the shale rock. However, for the purpose of this study, the extraction pressure setpoint was lowered to 350 atm so the addition of the various modifiers could be examined and demonstrated.

The first modifier tested is shown in Figure 6 where carbon disulfide was added to the shale rock. Due to the impurities in the carbon disulfide (CS_2), a slight response was obtained from the flame ionization detector as can be seen early in the chromatogram. Comparing pure CO_2 extraction to the extraction where carbon disulfide has been added, however, shows increases across the entire range of hydrocarbons (the attenuation of all of the chromatograms were kept exactly the same for purposes of a graphical comparison). Figures 7 and 8 showed the results using hexane modified CO_2 and methanol modified CO_2, respectively. For both hexane and methanol, solvent peaks corresponding to the modifiers can be seen in the resultant GC-FID chromatograms after the on-line transfer of the extraction effluent. But, there was a gradual increase in terms of the overall response of the various hydrocarbon peaks with hexane and even a much more dramatic increase in the extraction efficiency when methanol was used. Table III shows a summarized comparison of the percent recoveries of the target analytes (C17, pristane, and phytane) as they were obtained using the different modifiers.

TABLE III: Effect of Modifier Identity on Extraction Efficiencies for Shale Rock Using On-Line SFE/GC

| Compound | Percent Recovery | | | |
	Carbon Disulfide	Hexane	Methanol	CO_2
C17	96	99	101	78
Pristane	90	91	102	81
Phytane	89	92	101	82

SFE: 350 atm, 60°C, 30 minutes static, 10 minutes dynamic, 10% modifier concentrations.

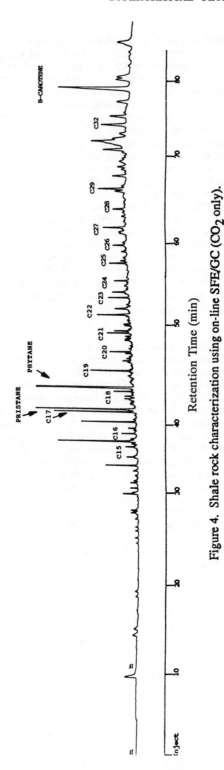

Figure 4. Shale rock characterization using on-line SFE/GC (CO_2 only).

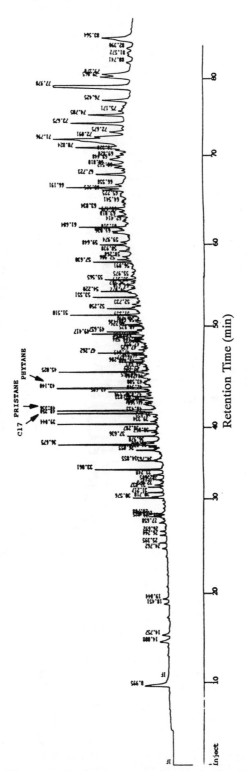

Figure 5. Shale rock characterization using on-line SFE/GC (CO_2 only).

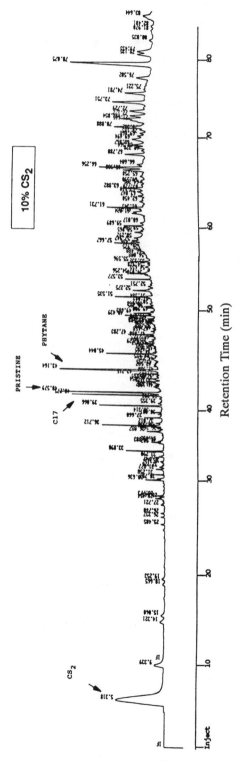

Figure 6. Shale rock characterization using on-line SFE/GC with modifier.

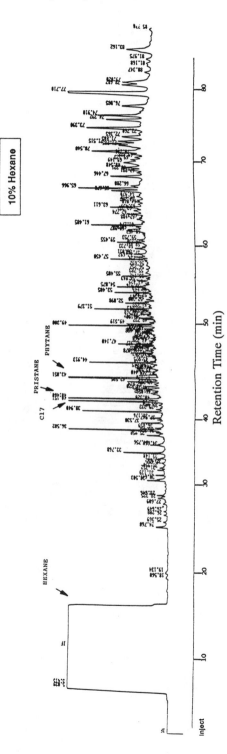

Figure 7. Shale rock characterization using on-line SFE/GC with modifier.

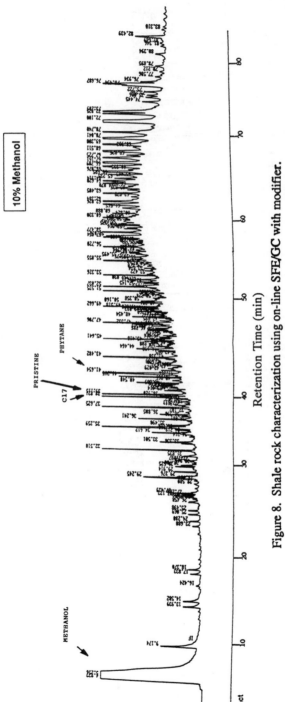

Figure 8. Shale rock characterization using on-line SFE/GC with modifier.

In comparison with the percent recoveries that were obtained using pure CO_2 (at 350 atm pressure), an increase was seen in the extraction efficiencies for all the analytes using the different modifiers. However, the greatest increase in percent recovery was seen using methanol as the modifier compared to hexane or carbon disulfide. For this particular shale rock sample matrix, methanol could be utilized as an effective modifier for enhancing extraction efficiencies. Moreover, the results have shown that an equivalent extraction efficiency could be obtained using lower density conditions (low pressures) with the aid of modifier addition. This could be important in many different application areas. For example, there are samples where higher density conditions are not desirable due to the extractability of additional respective organic analytes in a complex matrix that could be interferences. Therefore, extractions at lower pressures would reduce these interferences and increase the analytical reliability for the determination of target analytes.

The Variation of Modifier Concentrations in SFE

In addition to varying modifier identities to achieve different extraction efficiencies in SFE, another parameter that can be varied is the concentration of the modifier that is added to the primary supercritical fluid. In many cases as the concentration of the respective modifier in the primary supercritical fluid increases the critical temperature needs to be adjusted to accommodate the change. The adjustment of the corrected critical temperature is normally done on a mole fraction basis (moles of CO_2 versus moles modifier).

To illustrate the effect of changing the modifier concentration in SFE, a series of experiments were performed using a shale rock sample matrix and on-line SFE/GC transfer of the extraction effluent for determination of the analyte concentrations. The results of this study can be seen graphically in Figures 9 and 10 with the extraction of the shale rock using 5 percent carbon disulfide modified CO_2 and 12 percent carbon disulfide modified CO_2, respectively. For purposes of this illustration, the chromatograms in Figures 9 and 10 were both kept at the same attenuations. As can be expected, the higher concentration levels of modifiers yielded a higher extraction efficiency. Table IV shows the results that were obtained when 5,10, and 12 percent levels of carbon disulfide were spiked on to the shale rock matrices before SFE.

TABLE IV: Effect of Modifier Concentration on Extraction Efficiencies for Shale Rock Using On-Line SFE/GC

Compound	Percent Recovery Carbon Disulfide			
	5%	10%	12%	CO_2*
C17	90	96	101	78
Pristane	83	90	99	81
Phytane	86	89	102	82

*CO_2 extractions for comparison were performed at 350 atm, although higher recoveries are attainable at higher pressures.

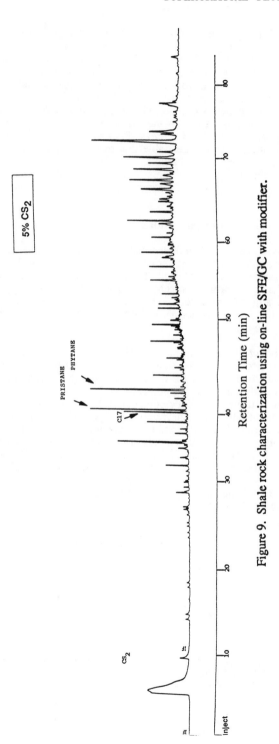

Figure 9. Shale rock characterization using on-line SFE/GC with modifier.

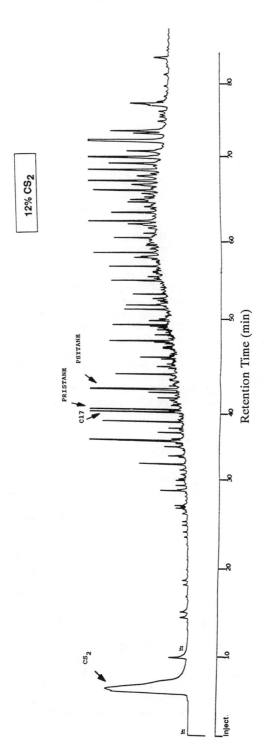

Figure 10. Shale rock characterization using on-line SFE/GC with modifier.

For all three of the target analytes (C17, pristane, and phytane) there was an increase in the percent recovery as the concentration level of the modifier increased. For purposes of this comparison, the extraction pressures were kept at 350 atm and the recoveries obtained with pure CO_2 are also listed. As discussed earlier, with increasing concentration levels of the modifiers, additional active sites were covered by the modifier molecule thereby releasing the target analytes for extraction.

Equilibration of Modifiers in SFE

Depending upon the modifier identity, the nature of the analytes of interest, as well as the type of sample matrix, there is a need to apply an equilibration period when using modifiers in SFE. If the modifier has not reached a level of equilibration with the primary supercritical fluid (i.e. existing as a one phase system) as well as the sample matrix, there could be situations where the enhancement of the extraction efficiencies using modifiers are not seen. In these cases, what happens is that the modifier is displaced out of the extraction vessel into the effluent and off-line into the collection vial or on-line into a chromatograph without achieving an interactive extraction. The actual rate determining mechanisms are not specifically known. Mechanisms of solubility, diffusion and/or physical adsorption, which will all be distinctly dependent upon the type of sample matrix, are mostly likely the controlling parameters. To illustrate the importance of the equilibration time in using modifiers in SFE, a shale rock sample matrix was spiked with a concentration of 12 percent of carbon disulfide and was extracted for a period of only 2 minutes statically before the dynamic transfer of the effluent to the GC. The result can be seen in Figure 11 where poor extraction efficiencies were obtained. This can be contrasted to Figure 10 which shows that higher recoveries were obtained when the 12 percent carbon disulfide CO_2 was allowed to equilibrate statically for 30 minutes compared to 2 minutes as was done in Figure 11. In the most situations, when a modifier is needed to be used to enhance SFE efficiencies a static equilibration stage is recommended and also times of static extraction need to be optimized.

The Effect of the Sample Matrix in SFE

One important aspect of method development in SFE is that the extractability of analytes from sample matrices is strongly dependent upon the nature or type of the sample matrix. The efficiency of the extraction of a discrete set of analytes needs to be optimized for that set of analytes from a standard matrix and also optimized for the extractability of those specific analytes from a particular sample matrix. However, the optimum set of parameters for both of these cases may be distinctly different. This is due to the analytes, depending on the nature of the sample matrix, having an affinity for the physical outside surface of the matrix or actually being present within the sample matrix particles. This affinity results in either the predominance of a solubility mechanism, a diffusion type mechanism or a physical adsorption type mechanism. Therefore, the optimum extraction conditions will vary if the sample of interest is a polymer, soil or vegetable matrix.

A study to explore the effect of the sample matrix in SFE was done with a set of polychlorinated biphenyl (PCBs) and a river sediment matrix. Figure 12 shows the on-line SFE/GC characterization of a PCB standard ranging from monochloro to decachlorolbiphenyl compounds present at concentration levels from 5 to 100 microgram per milliliter. An election capture GC detector (ECD) was used for the determination of these PCB's. Since this PCB standard was made up in liquid

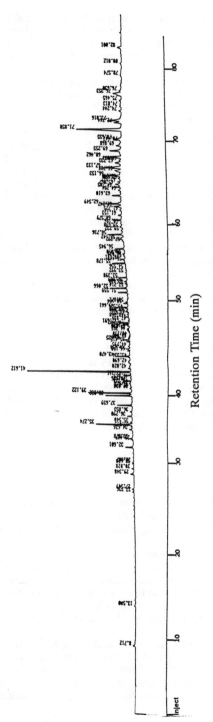

Figure 11. Shale rock characterization using on-line SFE/GC with modifier.

methylene chloride, the entire mixture was spiked in discrete aliquots onto alumina adsorbent beads inside an extraction vessel prior to the extraction. Using the extraction conditions of 450 atm pressure, 60°C and a 25 minutes static/dynamic extraction duration, the complete (98%) recovery of the PCB standards was obtained. In addition as can be seen in Figure 12, a large CO_2 contamination peak was detected early in the chromatogram as a result of the impurities in the CO_2 responding to the election capture detector. These PCB standards were present in a river sediment matrix and the result of the on-line SFE/GC-ECD characterization of this matrix can be seen in Figure 13. Since the chromatograms were obtained at the exact same attenuations, it is evident that the PCB's were not extracted efficiently using the identical SFE conditions as were utilized with the alumina spiked adsorbent. The only difference between these two characterizations was the type of sample matrix (i.e. alumina adsorbent vs. river sediment). In order to achieve a more efficient extraction of the PCB's from the river sediment, it was necessary to use modifiers. These results can be seen in Figures 14 and 15 with the use of 5 percent carbon disulfide modified CO_2 and 5 percent propylene carbonate modified CO_2, respectively. For comparison, the chromatograms were purposely kept at the same attenuation. Obviously the carbon disulfide increased the extraction efficiencies of the PCB's from the river sediment compared to pure CO_2 and propylene carbonate increased the SFE efficiency even more. In addition, the propylene carbonate modifier had an unknown impurity that yielded an ECD response that eluted after the PCB envelope. The data is summarized in Table V for each of the different PCB standards; including carbon disulfide, propylene carbonate and acetonitrile as modifiers for the extraction of the river sediment.

For the river sediment, the use of CO_2 only for the extraction did not achieve high percent recoveries for any of the individual PCB compounds. However, propylene carbonate achieved a nearly complete extraction of each of the PCB compounds from the river sediment with acetonitrile yielding slightly lower recoveries and carbon disulfide achieving the lowest recoveries of the three modifiers examined. It was also interesting to note that there was no preference in terms of the extractability of the monochlorobiphenyl as opposed to the decachlrorobiphenyl regardless of the modifier identity.

Conclusions

Since the early days of SFC, there always has been a desire to extend the useful range of the technique to more polar molecules. A similar type of desire exists in SFE. The hope for achieving efficient extractions of polar molecules from polar as well as non-polar substrates can only be realized with the use of more polar primary supercritical fluids or by the use of modifiers. Many of the more primary supercritical fluids that exists; namely, ammonia or water, are not effectively usable in the analytical laboratory due to instrumental as well as safety restrictions, therefore, the need to do more research on the use of modifiers in SFE is greatly necessitated. Based upon the limited study that was done within the scope of this chapter, a few conclusions can be drawn. These conclusions are summarized in Figure 16.

For the various matrices that were studied, specifically the sludge/fly ash, river sediment and shale rock, certain modifiers were more effective that others in terms of achieving high extraction efficiencies. For the target analytes (PCB's, aromatics, chlorinated aromatics and hydrocarbons) propylene carbonate and benzene achieved the highest extraction efficiencies compared to pure CO_2.

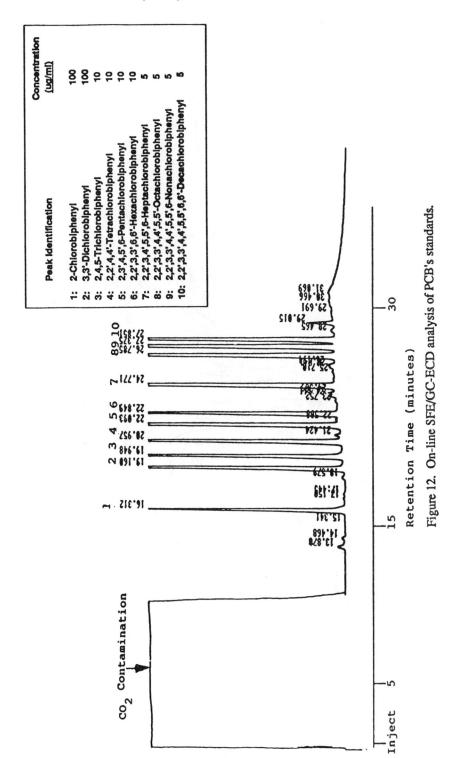

Figure 12. On-line SFE/GC-ECD analysis of PCB's standards.

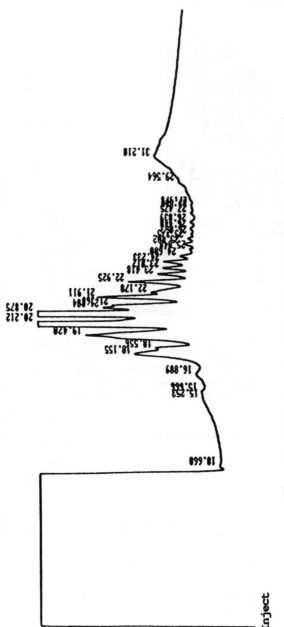

Figure 13. On-line SFE/GC-ECD analysis of PCB's in river sediment (CO_2 only).

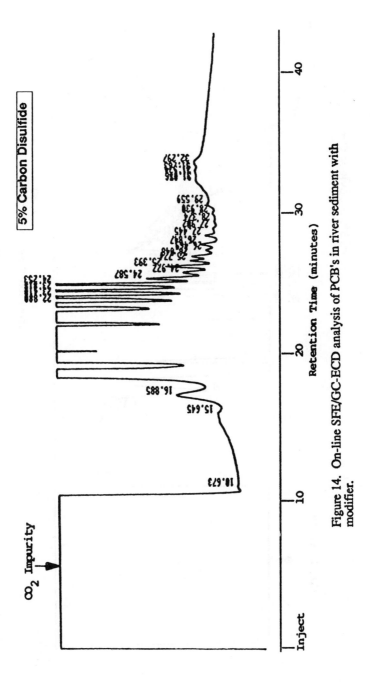

Figure 14. On-line SFE/GC-ECD analysis of PCB's in river sediment with modifier.

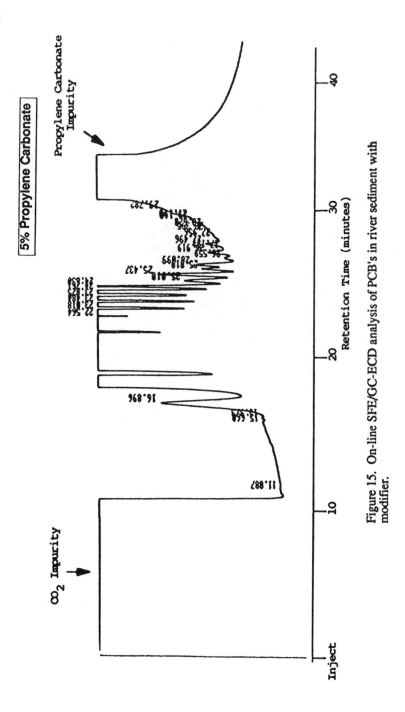

Figure 15. On-line SFE/GC-ECD analysis of PCB's in river sediment with modifier.

TABLE V: Effect of Modifier Identity on Extraction Efficiencies of River Sediment Using On-Line SFE/GC

Compound	Carbon Disulfide	Percent Recovery Propylene Carbonate	Acetonitrile	CO_2*
2-Chlorobiphenyl	72	98	80	36
3,3'-Dichlorobiphenyl	70	96	86	32
2,4,5-Trichlorobiphenyl	68	96	87	38
2,2',4,4'-Tetrachlorobiphenyl	65	95	85	40
2,3',4,5',6-Pentachlorobiphenyl	63	94	84	41
2,2',3,3',6,6'-Hexachlorobiphenyl	78	93	82	39
2,2',3,4',5,5',6-Heptachlorobiphenyl	76	93	81	36
2,2',3,3',4,4',5,5'-Octachlorobiphenyl	74	92	87	37
2,2',3,3',4,4',5,5',6-Nonachlorobiphenyl	73	90	86	43
2,2',3,3',4,4',5,5',6,6'-Decachlorobiphenyl	70	90	80	44

SFE: 60 mg of river sediment
        CO_2 at 450 atm, 60°C, 25 minutes static, 5 minutes dynamic
        Modifiers at 450 atm, 60°C, 25 minutes static, 5 minutes dynamic
        Modifier concentrations all at 5% levels

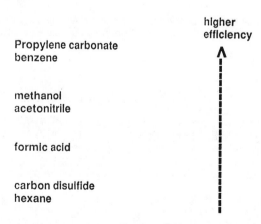

Figure 16.   Modifier effectiveness in SFE.

Methanol and acetonitrile were intermediate modifier, formic acid achieved lower efficiencies and carbon disulfide and hexane were the least effective modifiers for this set of analytes and for these particular matrices. Moreover, all of these results were obtained with a very necessary static equilibration step that needed to be performed prior to the dynamic extraction.

Within the scope of this study, on-line SFE/GC characterization was used very effectively to monitor the extraction procedures and to determine the concentration levels of the respective analytes. This presents another advantage in terms of the use of modifiers in SFE since the modifiers appeared as solvent peaks in the gas chromatographic separations and did not interfere with the target analyte determinations. An additional possible advantage in regards to the practical use of modifiers in SFE is the fact that they can be added as liquids to the sample matrices in the extraction vessels prior to the actual extraction, thereby, not contaminating the primary supply pump which delivers the unmodified CO_2. This aids in cutting the method development time as well as in parameter optimization. Another important aspect in terms of modifier usage is a realization in regards to the types of sample matrices that are being extracted. The predominance of the mechanisms (solubility, diffusion and adsorption) will be dictated depending on the type of sample matrix as well as the type of target analyte that is being extracted. Further work needs to be done to understand the fundamental mechanisms that occur during the extraction step. In some cases, and depending on the extraction conditions, a modifier could be used even as a derivatizing agent to distinctly enhance the selectivity or extractability of a specific set of analytes from a complex sample matrix. These in-situ/on-line SFE derivatizations need to be also further studied and coupled with the use of modifiers with the hopes of achieving class fractionations from complex matrices.

Acknowledgments

The following people need to be identified and acknowledged for their experimental work which was described in this chapter: Robert Ravey and Carol Hamilton.

References:

1. J.M. Levy and W.M. Ritchey. J. of High Resolution Chromatogr. and Chromatogr. Comm. 8 (1985) 503-509.
2. E. Klesper and W. Hartmann. Poly. Lett. Ed. 15 (1977) 707-712.
3. W. Hartmann and E. Klesper, Poly. Lett. Ed. 15 (1977) 713-719
4. N.M. Karayannis and A.H. Corwin. Anal. Biochem. 26 (1968) 34.
5. M. Novotny, W. Bertsch and A. Zlatkis. J. Chromatogr. 61 (1971) 17-28.
6. J.E. Conaway, J.A. Graham and L.B. Rogers. J.Chrom. Sci. 16 (1978) 102-110.
7. J.A. Nieman and L.B. Rogers. Sep. Sci. 10 (1975) 517-545.
8. J.B. Crowther and J.D. Henion. Anal. Chem. 57 (1985) 2711-2716.
9. J.A. Grahma and L.B. Rogers. J. Chrom. Sci. 18 (1980) 75-84.
10. F.P. Schmitz and E. Klesper. Polymer Bull. 5 (1981) 603-608.
11. Y. Hirata. J. Chromatogr. 315 (1984) 31-37.
12. C. Fujimoto, Y. Hirata and K. Jinno. J. Chromatogr. 332 (1985) 47-46.
13. Y. Hirata and F. Nakata. J. Chromtogr. 295 (1984) 315-322.
14. A.L. Blilie and T. Greibrokk. Anal. Chem. 57 (1985) 2239-2242.
15. P.A. Mourier, E. Eliot, M.H. Caude, R.H. Rosset, and A.G. Tambute. Anal. Chem. 57 (1985) 2819-2823.

16. L.G. Randall, Ultrahigh Resolution Chromatography. A.Ahuja (editor), ACS Symposium Series, 1984.
17. R. Board, D. Gere, D. McManigill and H.Weaver, presented at the Pittsburgh Conference on Analytical Chemistry, 1982.
18. T. Doran. Gas Chrom. Proc. of Int. Symp. of Gas Chrom. (Europe) 9 1972) 133-143.
19. F.P. Schmitz, H. Hilgers, B. Lorenchat and E. Klesper. J. Chromatogr. 346 (1985) 69-79.
20. T.A. Berger. J. of High Res. Chrom. 14 (1991) 312-316.
21. M.A. Morrissey, A. Giogetti, M. Polasek, N. Pericles and H.M. Widmer. J. of Chrom. Sci. 29 (1991) 237-242.
22. T.A. Berger and J.F. Deye. J. of Chrom. Sci. 29 (1991) 141-145.
23. T.A. Berger and J.F. Deye. J. of Chrom. Sci. 29 (1991) 26-30.
24. M.L. Lee and K.E. Markides. Analytical Supercritical Fluid Chromatography and Extraction. Chromatography Conferences, Inc. (1990) 377.
25. S.B. Hawthorne. Anal. Chem. 62 No. 11 (1990)
26. T.A. Berger and J.F. Deye, Anal. Chem. July, 1991.
27. J.F. Deye, T.A. Berger and A.G. Anderson. Anal. Chem. 62 (1990) 615-622.
28. C.R. Yonker, S.L. Frye, D.R. Kalwarf and R.D. Smith. J. Phys. Chem. 90 (1986) 3022-3026.
29. J.M. Levy and W.M. Ritchey. J. of High Res. Chroma. and Chrom. Comm. 10 (1987) 493-496.
30. J.M. Levy and A.C. Rosselli. Chromatographia 28 No. 11/12 (1989) 613-616.
31. J.M. Levy, A.C. Rosselli, D.S. Boyer and K. Cross. J. High Res. Chrom. 13 (1990) 418-421.
32. J.M. Levy. Amer. Lab. August, 1991.
33. N. Alexandrou and J. Pawliszyn. Anal. Chem. 61 (1989) 2770-2776.

RECEIVED January 8, 1992

Author Index

Affiliation Index

Subject Index

Production: Margaret J. Brown
Indexing: Deborah H. Steiner
Acquisition: Cheryl Shanks and Anne Wilson
Cover design: Amy Meyer Phifer

Printed and bound by Maple Press, York, PA